ESTATES GAZETTE LAW REPORTS

Estates Gazette Law Reports

2011

Volume 2

Edited by
HH Judge Hazel Marshall QC

Routledge
Taylor & Francis Group
LONDON AND NEW YORK

EG Books

First published 2011
by Routledge
2 Park Square, Milton Park, Abingdon, Oxon OX14 4RN

Simultaneously published in the USA and Canada by Routledge
711 Third Avenue, New York, NY 10017

EG Books is an imprint of the Taylor & Francis Group, an informa business

British Library Cataloguing in Publication Data
A catalogue record for this book is available from the British Library

Library of Congress Cataloging in Publication Data
A catalog record for this book has been requested

ISBN: 978-0-415-50666-3 (hbk)
ISSN: 0951-9289

Typeset in Times New Roman by Estates Gazette

MIX Paper from responsible sources FSC® C004839 www.fsc.org

Printed and bound in Great Britain by the MPG Books Group

CONTENTS

Cases reported in "Estates Gazette" [2011] issues 18-34

SUPPLEMENT

Cases not reported in "Estates Gazette"

TABLE OF CASES

Titles of cases shown in bold type
Names of parties reversed shown in ordinary type

INDEX OF SUBJECT MATTER

CIVIL PRACTICE AND PROCEDURE

Pritchard and others *v* Teitelbaum and others

CHANCERY DIVISION

[2011] EWHC 1063 (Ch)

20 April 2011

MORGAN J

Estates Gazette 30 July 2011

[2011] 30 EG 58

Civil practice and procedure — Trespass — Writ of possession — RSC Ord 113 r 7 — CPR 55 — Sale of property with completion deferred for three years — Grant of long lease to purchaser with vacant possession pending completion — Vendor resiling from arrangements and refusing to give up possession — Purchaser obtaining possession order and writ of possession — Vendor evicted — Whether vendor entitled to be reinstated in property pending determination of application to set aside possession order — Whether writ of possession validly issued — Whether vitiated by failure to give notice to vendor of intended eviction

The first claimant exercised her statutory right, under the Housing Act 1985, to buy the property of which she was a secure tenant at a discount and contracted to sell it on to the fifth defendant at a profit of £20,000. The sale to the fifth defendant was not to be completed for three years, so that the first claimant would not incur an obligation under the Act to repay the discount. In the meantime, the first claimant was to grant a lease of the property to the fifth defendant with vacant possession. The necessary documents were signed in August 2004, although the first claimant had until the end of that month to vacate the property.

The first claimant then sought to resile from the arrangement and refused to give up possession. The fifth defendant obtained a possession order against her in January 2011 (see *Fineland Investments Ltd* v *Pritchard* [2011] EWHC 113 (Ch); [2011] 06 EG 102 (CS)) and sent her a copy. The first claimant did not relinquish possession on the required date in March 2011. Instead, the fifth defendant was sent a copy of a claim form by which the first claimant and her son, the second claimant, sought damages and possession of the property and the third defendant claimed to be its owner pursuant to an alleged agreement of June 2004. The first claimant applied to have the possession order stayed or set aside; a hearing was fixed for May 2011.

Meanwhile, the fifth defendant obtained a writ of possession. It considered that it did not require the court's permission to issue the writ since the case fell within r 7 of RSC Ord 113 as a possession claim against trespassers. The writ of possession was enforced and the first and second claimants were evicted in April 2011.

The claimants applied to the court for an order requiring the defendants to allow them to enter or re-enter the property. The first claimant submitted that the writ of possession was unlawful since she should have been given notice of the application for the writ and of the intended eviction.

***Held*: The claim was dismissed. (1) In deciding whether to grant the relief sought by the claimants, it was relevant to consider whether the court would set aside the order for possession or the writ of possession. The first claimant's prospects of success in obtaining an order setting aside the order for possession were slight and insufficient to justify restoring her and her son to the premises pending the consideration of her application for that relief. Nor were there any grounds for setting aside the writ of possession. The fifth defendant had not required permission of the court to issue the writ of possession since the possession order had been made in "a possession claim against trespassers" within r 7 of RSC Ord 113. That phrase should be given the same meaning as in CPR 55; accordingly, the definition in CPR 55.1(b) applies and "a possession claim against trespassers" refers to a claim for the recovery of land alleged to be occupied by persons, other than existing or former tenants or subtenants, who have entered or remained on the land without the consent of a person entitled to possession of it. The first claimant had not entered as a trespasser but had remained in possession as a trespasser after August 2004. She was not a tenant or subtenant because her secure tenancy had ended when she took a transfer of the freehold from the local authority on her right-to-buy acquisition. Although she had thereafter been a licensee of the fifth defendant for a period, that did not mean that the possession claim had to be dealt with as a claim by a licensor or former licensor under CPR 55.2(1)(a)(iii), rather than a possession claim against trespassers within CPR 55.2(1)(b). The two provisions are not mutually exclusive; the former is concerned with the party bringing the claim and the latter with the party against which the claim is brought, and a single claim can fall within both provisions. The fifth defendant's claim qualified as a possession claim against trespassers notwithstanding that it had also claimed other relief in the same proceedings. Accordingly, the writ of possession had been properly issued. (2) The writ of possession was not otherwise invalidated merely by the failure to give notice to the first claimant of the intended eviction. Neither the High Court rules nor the forms for the writ of possession require notice to be given to the party against which the order for possession is to be executed. (3) Further, the fifth defendant had been entitled to enforce the possession order in its favour without issuing and enforcing a writ of possession. Proceeding in that way would not have infringed section 3 of the Protection from Eviction Act 1977 since the licence that the first claimant had previously held had been gratuitous and was therefore an excluded licence within section 3A(7).**

The following cases are referred to in this report.

Aglionby v *Cohen* [1955] 1 QB 558; [1955] 2 WLR 730; [1955] 1 All ER 785

Fineland Investments Ltd v *Pritchard* [2011] EWHC 113 (Ch); [2011] 06 EG 102 (CS)

Fleet Mortgage & Investment Co Ltd v *Lower Maisonette 46 Eaton Place Ltd* [1972] 1 WLR 765; [1972] 2 All ER 737; (1972) 222 EG 1755, Ch

Forcelux Ltd v *Binnie* [2009] EWCA Civ 854; [2010] HLR 20

Greater London Council v *Jenkins* [1975] 1 All ER 354; [1975] 1 WLR 155

Hackney London Borough Council v *Findlay* [2011] EWCA Civ 8; [2011] HLR 15

Jephson Homes Housing Association Ltd v *Moisejevs*; *sub nom Moisjevs* v *Jephson Homes Housing Association* [2001] 2 All ER 901; [2001] 3 EGLR 14; [2001] 41 EG 186; (2001) 33 HLR 54, CA

McPhail v *Persons, Names Unknown*; *Bristol Corporation* v *Ross* [1973] Ch 447; [1973] 3 WLR 71; [1973] 3 All ER 393; 72 LGR 93, CA

This was the hearing of an application by the claimants, Mrs Janice Pritchard, Mr Robert Carter and Mr John Carew-Reid, for an order that the defendants, Mr Nathan Teitelbaum, Dennis Matthews Solicitors, Ingram Winter Green Solicitors, Bude Nathan Iwanier, Fineland Investments Ltd, Fineland Properties Ltd and Northern Holdings, permit them to enter or re-enter residential premises in respect of which a possession order had previously been made and a writ of possession executed.

The first claimant appeared in person; Marie-Claire Bleasdale (instructed by Bude Nathan Iwanier) appeared for the first, fifth and sixth defendants; Francis Bacon (instructed by Barlow Lyde & Gilbert LLP) appeared for the third defendant; James Purchas (instructed by Fishburns LLP) appeared for the fourth defendant; the second and seventh defendants did not appear and were not represented.

Giving judgment, **Morgan J** said:

[1] The principal application that is before me is an application by the claimants for an order requiring some or all of the defendants to permit the claimants to enter, or re-enter, into possession of the dwelling-house at 1 Garratt Close, Beddington, Surrey, alternatively, for the claimants to be permitted to protect and/or collect their possessions that are in those premises. The immediate background to this application is that the first and second claimants, Mrs Janice Pritchard and her son Mr Robert Carter, were evicted from the premises by the fifth defendant, Fineland Investments Ltd (Fineland), on 1 April 2011.

[2] The events that led up to the eviction on 1 April 2011 essentially begin with a transaction entered into between Mrs Pritchard and Fineland on 9 August 2004. The nature of that transaction was described in detail in a judgment given by Ms Alison Foster QC, sitting as a deputy judge of the Chancery Division, on 31 January 2011, in proceedings brought by Fineland against Mrs Pritchard. Those proceedings were commenced on 29 June 2006 and I will refer to them as "the 2006 proceedings". The neutral citation of the judgment given on 31 January 2011 is [2011] EWHC 113 (Ch)*.

[3] In very brief summary, prior to 9 August 2004, Mrs Pritchard was a secure tenant of the premises, holding the same from Sutton London Borough Council. She was entitled to, and did, exercise the right to buy under the Housing Act 1985 (the 1985 Act). On 9 August 2004, Sutton London Borough Council transferred the freehold in the premises to Mrs Pritchard at a price discounted in accordance with the 1985 Act. On that date, her secure tenancy of the premises came to an end: see section 139(2) of the 1985 Act. Also on 9 August 2004, Mrs Pritchard contracted to sell the freehold in the premises to Fineland. The price agreed with Fineland was £20,000 more than the price paid by Mrs Pritchard. To avoid Sutton London Borough Council reclaiming all or part of the discount on the sale to Mrs Pritchard, the contract to sell to Fineland provided for completion at a date just over three years thereafter. Also on 9 August 2004, Mrs Pritchard granted to Fineland a lease of the premises for a term of 20 years. As the term of the lease was only 20 years, this grant did not oblige Mrs Pritchard to pay back to Sutton London Borough Council any part of the discount. Under the formal documents, Mrs Pritchard was obliged to give vacant possession of the premises to Fineland on 9 August 2004. However, Fineland informally agreed to allow Mrs Pritchard a short time to vacate the property. The deputy judge held that this meant that Mrs Pritchard was entitled to remain in the property until 30 August 2004.

[4] Shortly after 9 August 2004, Mrs Pritchard sought to resile from the arrangements she had made with Fineland. She declined to vacate the premises. Eventually, Fineland brought the 2006 proceedings to claim possession of the premises and various other heads of relief. Mrs Pritchard filed a defence to those proceedings. Initially, she was represented by solicitors but later in the course of those proceedings she acted in person.

[5] In late 2009, directions were given fixing the date of the trial of the 2006 proceedings so that it would take place in October 2010. On 22 September 2010, Mrs Pritchard applied for an adjournment of the trial. That application was heard by Floyd J on 5 October 2010 and was dismissed.

[6] The trial of the 2006 proceedings duly began on 11 October 2010. Mrs Pritchard had written to the court just before this date and the deputy judge considered her letter as a further application to adjourn the trial. The deputy judge considered that application and decided not to adjourn the trial. The trial then proceeded on 11 and 12 October 2010. Fineland was represented by counsel. Mrs Pritchard did not attend the trial. The deputy judge was addressed in detail on behalf of Fineland and Fineland called oral evidence in accordance with witness statements that had earlier been served.

[7] The deputy judge reserved her judgment, which was handed down on 31 January 2011. Mrs Pritchard attended on the handing down of the judgment. The approved judgment extends to some 30 pages. The deputy judge considered the documents that had been entered into by the parties, the allegations made by Mrs Pritchard in her defence and the oral evidence. The deputy judge found in favour of Fineland.

[8] A detailed order was drawn up to give effect to the judgment of 31 January 2011. Mrs Pritchard was ordered to give possession of the premises on 14 March 2011. The order also provided for the transactions entered into on 9 August 2004 to be perfected by appropriate registrations at the Land Registry. In particular, it was provided that Fineland should be registered as the proprietor under title number SGL 660273 in respect of the lease granted by Mrs Pritchard on 9 August 2004. The deputy judge refused permission to appeal. Mrs Pritchard has not, at any time, issued an appellant's notice seeking to appeal to the Court of Appeal against the order of 31 January 2011.

[9] On 4 and 24 February 2011, solicitors for Fineland wrote to Mrs Pritchard with a copy of the order of 31 January 2011. The solicitor specifically referred to 14 March 2011 as the date on which possession was to be given up. In the second letter, the solicitor reminded Mrs Pritchard that she would need to remove her possessions from the property by that date.

[10] On 9 March 2011, Mrs Pritchard appeared before David Richards J seeking a stay in respect of the possession order that was expressed to take effect on 14 March 2011. The judge directed that Mrs Pritchard should decide whether she wished to pursue that application and, if she did, he would hear an application to that effect on 11 March 2011. Mrs Pritchard did not later indicate to the judge or to Fineland that she wished to pursue the application and she did not appear before the judge on 11 March 2011, although Fineland was ready to appear through counsel on that date if an application were made.

[11] Although I think that Fineland were not previously aware of this fact, Mrs Pritchard told me at the hearing on 13 April 2011 that she appeared before Peter Smith J and sought a stay of the order of possession. She did not give me the date of that appearance but she did tell me that the judge said that he could not, or would not, extend the date for possession and informed Mrs Pritchard that if she wished to challenge the order for possession she would have to appeal to the Court of Appeal.

[12] Mrs Pritchard did not vacate the premises on 14 March 2011. On that date, the solicitor for Fineland was contacted by Mr Carew-Reid. On 15 March 2011, Mr Carew-Reid sent to that solicitor the claim form in proceedings that were issued by the claimants on 11 March 2011. In the details of claim set out in the claim form, Mr Carew-Reid claimed to be the owner of the premises pursuant to an alleged agreement of June 2004. Mrs Pritchard and her son also claimed damages and possession of the premises as a result of what they said was immoral behaviour, conspiracy to defraud and a fraudulent trick by the defendants. The defendants included Fineland and also an associated company of Fineland, a director of Fineland and three firms of solicitors.

[13] On 21 March 2011, Mrs Pritchard issued an application in the 2006 proceedings. The order that she sought was that there be a directions hearing in respect of a later full hearing of an application

* Editor's note: Reported at [2011] 06 EG 102 (CS)

to stay and/or to set aside the order made on 31 January 2011. That application was returnable on 7 April 2011 when directions were apparently given for a hearing on 5 May 2011.

[14] On or around 30 March 2011, Fineland applied for a writ of possession, and a writ in form no 66 was issued on 30 March 2011. Fineland did not apply for permission to issue this writ. It took the view that the case fell within RSC order 113 r 7 rather than RSC order 45 r 3. It should be noted, however, that the appropriate writ under order 113 r 7 is form no 66A and not form 66 but I would regard the difference as an irregularity that, on the facts of this case, the court should waive.

[15] On 1 April 2011, the writ of possession was enforced at the premises by two enforcement officers. Mrs Pritchard was evicted. Her son, Mr Carter, was not present at the beginning of the eviction but he returned in the course of it. The premises were secured. Various possessions belonging to Mrs Pritchard and her son remained on the premises. There is a considerable dispute about the way in which the eviction was carried out and about the conduct of the enforcement officers, the police (who were in attendance) and Mrs Pritchard herself. On the view I take as to the legal position, it is not necessary in this present judgment to investigate the serious allegations and counter-allegations that have been made about the course of the eviction and the behaviour of the parties.

[16] On 7 April 2011, Mr Carew-Reid appeared before me, without notice to Fineland. He wished to make an application in accordance with an application notice issued that day; that application notice is the one that I later heard on 13 April 2011 and that is the subject of this judgment. Mr Carew-Reid gave me a very abbreviated history of the matter that omitted many highly relevant matters. The associate in court on that day went in search of the court file that contained the judgment handed down on 31 January 2011. Having read that judgment to myself, I indicated to Mr Carew-Reid that it was not appropriate for me to make the orders that he sought when I had not heard from the other side, who indeed knew nothing of his application.

[17] At the hearing on 13 April 2011, Mrs Pritchard appeared in person. She spoke on behalf of herself and her son, who was not present. Mr Carew-Reid was not present although it may be that Mrs Pritchard was able to speak on his behalf.

[18] In considering whether to grant the relief sought by the application notice of 7 April 2011, I think that I should consider whether this is, or might be, a case in which the court will set aside the order for possession made on 31 January 2011 and also consider whether anything that happened in respect of the issuing of the writ of possession on 30 March 2011 or the circumstances in which the eviction took place on 1 April 2011 should result in the court setting aside the writ of possession of 30 March 2011 and/or allowing the claimants to enter, or re-enter, the premises. If a court were in due course to set aside the order for possession, it would seem to follow that the court should then set aside the writ of possession and it would be open to the court, if it thought fit in all the circumstances, to restore the claimants to the position they were in before the writ of possession was executed on 1 April 2011.

[19] At the hearing on 13 April 2011, Mrs Pritchard did not focus on her application to set aside the order for possession but instead focussed on the circumstances in which she was evicted on 1 April 2011. None the less, it seems to me that the first topic that I should consider is whether a court would or might set aside the order for possession in this case.

[20] In my judgment, the court's powers to set aside the order for possession of 31 January 2011 are governed by CPR 39.3. In this case, the hearing on 11 and 12 October 2010 amounted to a "trial" for the purposes of that rule: contrast *Forcelux Ltd* v *Binnie* [2009] EWCA Civ 854*. In any event, the court should normally apply the provisions of CPR 39.3(5) by analogy, even if they do not strictly apply in a case where judgment was given at a hearing that was not a "trial": see *Hackney London Borough Council* v *Findlay* [2011] EWCA Civ 8*, G
in [24].

[21] Although the application that Mrs Pritchard has made to set aside the order for possession was not for hearing on 13 April 2011, if the possibility of the order for possession being set aside is to be a ground for me restoring Mrs Pritchard to possession of the premises, I must form some assessment on the material before me of Mrs Pritchard's prospects of success in that application. As to the requirements of CPR 39.3(5), on the material before me, it is strongly arguable that Mrs Pritchard did not apply promptly to set aside the order of 31 January 2011. Further, in view of the refusal of Floyd J and of the deputy judge to adjourn the trial, I do not think that Mrs Pritchard has put forward a good reason for H
not attending the trial. In addition, it is strongly arguable that she has not shown a reasonable prospect of success at the trial. I would describe Mrs Pritchard's prospects of success in obtaining an order setting aside the order for possession as being only slight. In my judgment, those prospects are not sufficient to persuade me to restore her and her son to the premises in the meantime, while an application to set aside the order for possession is further considered.

[22] I next turn to consider the various arguments that I heard about the circumstances in which the writ of possession was issued and in which Mrs Pritchard and her son were evicted on 1 April 2011.

[23] The principal complaint made by Mrs Pritchard was that she J
was not given notice of the intended eviction on 1 April 2011. She was told, and reminded, in advance of 14 March 2011, that she was expected to deliver up possession by that date. However, I find that Fineland did not attempt to tell Mrs Pritchard that it had applied for, and obtained, a writ of possession and was proposing to execute that writ on 1 April 2011. I also find that when the enforcement officers attended the premises on 1 April 2011, this was a complete surprise, indeed a shock, to Mrs Pritchard. I think that Mrs Pritchard thought that she would have notice of any intended eviction and she may have thought that the proceedings issued on 11 March 2011 and the application made on 21 March 2011 were an effective means of keeping Fineland at bay K
while those proceedings took their course.

[24] Mrs Pritchard submitted to me that the writ of possession was unlawful and the eviction should not have happened without notice having been given to her. I therefore need to consider, among other things, whether Fineland should have informed Mrs Pritchard of its intention to seek a writ of possession and, separately, whether (following the issue of the writ of possession) Fineland should have notified Mrs Pritchard of the intended eviction on 1 April 2011.

[25] I have already mentioned that Fineland relied on RSC order 113 r 7 as justifying its decision not to seek the permission of the court to issue the writ of possession. Such permission is normally L
required under RSC order 45 r 3 but order 113 r 7 is an exception to that normal requirement.

[26] Whether this case is governed by order 113 r 7 or order 45 r 3 is not an altogether easy point. Order 113 r 7 refers to "an order for possession in a possession claim against trespassers under Part 55". In my judgment, the phrase "a possession claim against trespassers" just quoted from order 113 r 7 is to be given the same meaning as in CPR 55, where the phrase is defined by CPR 55.1(b) to mean: "a claim for the recovery of land which the claimant alleges is occupied only by a person or persons who entered or remained on the land without the consent of a person entitled to possession of that land but does not M
include a claim against a tenant or subtenant whether that tenancy has been terminated or not".

[27] If one applies the above definition in CPR 55.1(b) to this case, it would appear that the possession order in this case was made in "a possession claim against trespassers".

Mrs Pritchard had not entered the premises as a trespasser but after 30 August 2004 she remained in possession as a trespasser: see *Greater London Council* v *Jenkins* [1975] 1 WLR 155, as to the application of this definition to an ex-licensee. Further, I do not think

* Editor's note: Reported at [2010] HLR 20

* Editor's note: Reported at [2011] HLR 15

that the claim to possession against Mrs Pritchard was a claim against a tenant or subtenant; although she had been a tenant of Sutton London Borough Council up until 9 August 2004, she then became a transferee as freeholder from Sutton London Borough Council (her tenancy thereupon being terminated) and then for some 21 days she was the licensee of Fineland.

[28] There are other matters that I ought to consider. In CPR 55.2(1)(a)(iii), there is a reference to a possession claim brought by a "licensor (or former licensor)". A separate provision, CPR 55.2(1)(b), refers to "a possession claim against trespassers". The claim to possession in the present case would seem to fall within CPR 55.2(1)(a)(iii); does that mean that it must fall outside CPR 55.2(1)(b)? In my judgment, these two provisions need not be mutually exclusive. Paragraph (a)(iii) refers to the person bringing the claim: para (b) refers to the person against whom the claim is brought. A single claim can come within both provisions. As I have explained, this case would seem to fall also within the definition of "a possession claim against trespassers".

[29] I have also considered whether the claim to possession was "a possession claim against trespassers", given that the 2006 proceedings also included claims for other heads of relief. If one works through the many provisions in CPR 55, which refer to "a possession claim against trespassers", it is sometimes straightforward, and sometimes not, to apply those provisions to the 2006 proceedings. However, even if the individual provisions do not always fit with more complex proceedings such as the 2006 proceedings, that does not compel the conclusion that the claim to possession in the 2006 proceedings did not fall within the definition in CPR 55.1(b). I also note para 1.7 of PD 55A, which refers to CPR 55 applying to proceedings that include a claim to possession but also includes other claims. That would seem to cover the 2006 proceedings.

[30] I have also considered the policy behind the special rule in RSC order 113 r 7. It seems to have been considered that a claim against a trespasser raises different issues and is to be dealt with in a different way from other possession claims and, in particular, without the safeguard of the need for permission pursuant to RSC order 45 r 3. Had Fineland made a claim to possession against Mrs Pritchard on the same basis as in the 2006 proceedings but using a separate set of proceedings, where the only remedy claimed was an order for possession, I would without hesitation have held that such proceedings were a claim to possession against a trespasser. Had Fineland then obtained an order for possession in such proceedings, it would have been able to take advantage of RSC order 113 r 7. I do not see any good policy reason for treating the order for possession in the 2006 proceedings any differently just because it was in a set of proceedings that claimed relief in addition to an order for possession.

[31] My conclusion is that the writ of possession in the present case was properly issued even though Fineland did not seek the permission of the court for the issue of the writ.

[32] For the sake of completeness, I have considered what would probably have happened had Fineland sought permission under RSC order 45 r 3. In my judgment, Fineland would have been entitled to seek that permission by an *ex parte* application. RSC order 46.4(1) says so. The decision in *Fleet Mortgage & Investment Co Ltd* v *Lower Maisonette 46 Eaton Place Ltd* [1972] 1 WLR 765* is a little difficult to square with the wording of RSC order 46.4(1) and may have to be confined to a case within order 46 r 2(1)(d): see how the matter was described by Simon Brown LJ in *Jephson Homes Housing Association Ltd* v *Moisejevs* [2001] 2 All ER 901†, in [44] and [47]. The present case is not within order 46 r 2(1)(d). If permission had been sought by an *ex parte* application, I find that it would have been within the power of the court to give permission on such an application, without directing that the application be served on anyone. The application would have had to have satisfied RSC order 45 r 3(3). Fineland would have been able to satisfy the court that Mrs Pritchard's son, who was living with her, was likely to have received sufficient notice of the proceedings. There was no material put before me to show that her son had any rights binding on Fineland. There was no material before me to show that Mr Carew-Reid was in actual possession of the premises.

[33] I now need to consider Mrs Pritchard's separate point that even if the writ of possession was properly issued, she should have been given notice of the intended eviction. However, there is simply no requirement in any High Court rule or in the forms for the writ of possession that notice needs to be given to the persons against whom the order for possession will be executed.

[34] It follows from the above that the writ of possession was validly issued and there was no procedural defect in the manner of execution of the order for possession.

[35] The order for possession in the present case was enforceable against Mr Carter who had no rights as against Fineland, entitling him to retain possession. As to Mr Carew-Reid, I had no evidence that he had been in possession of the premises on or before the 1 April 2011. Further, in so far as Mr Carew-Reid relies on the contract that he says he made with Mrs Pritchard in June 2004, that contract is not binding on the leasehold interest held by Fineland as the contract was not at any relevant time protected by registration at the Land Registry. Fineland's entitlement to possession is pursuant to its title as lessee of the premises. Further, although I need not decide anything on this point, the genuineness of the alleged contract of June 2004 calls out for investigation.

[36] I have held that the writ of possession was valid and that possession has been recovered pursuant to it. In any case, Fineland was entitled to enforce the order for possession in its favour without issuing and enforcing a writ of possession: see *Aglionby* v *Cohen* [1955] 1 QB 558 and *McPhail* v *Persons, Names Unknown* [1973] Ch 447, at p457. Proceeding in that way would not infringe section 3 of the Protection from Eviction Act 1977 since the licence that Mrs Pritchard enjoyed from 9 August 2004 to 30 August 2004 would appear to have been a gratuitous licence and therefore an excluded licence pursuant to section 3A(7) of that Act.

[37] It is apparent from the entirety of the proceedings and the applications on the part of Mrs Pritchard that she contends that unlawful violence was used in connection with the eviction. Conversely, Fineland alleges that Mrs Pritchard acted unlawfully in resisting the eviction and it refers to the fact that she was arrested on a charge of aggravated trespass pursuant to section 68 of the Criminal Justice and Public Order Act 1994.

[38] If one or other of the parties involved in the eviction on 1 April 2011 behaved unlawfully, such behaviour is not to be condoned. However, any unlawful actions on the part of Fineland or the enforcement officers or the police on or after the eviction do not invalidate the recovery of possession by Fineland, with or without the benefit of a valid writ of possession. If Mrs Pritchard is able to establish such unlawful behaviour, I would not regard that as a justification for making an order that she be restored to the premises. In view of the judgment of the deputy judge on 31 January 2011 and my finding that Mrs Pritchard has not shown a real prospect of the order for possession being set aside, the effect of restoring her to possession of the premises would be that Fineland would then be required to evict her a second time, which, on my findings, it would be able to do. I would not regard the grant of an injunction leading to a second eviction as conferring any worthwhile benefit on Mrs Pritchard.

[39] The above conclusions deal with the claimants' application for an order allowing them to enter, or re-enter, the premise. There is also the question of the claimants having access to the premises to remove their possessions. Fineland did not oppose the claimants having a suitable opportunity under suitable supervision to remove the relevant items. At the conclusion of the hearing on 13 April 2011, I indicated the order that I would make in this respect and I need not repeat that matter in this judgment.

[40] The application that was made on 7 April 2011 was served on the third defendant and on the fourth defendant. The third defendant

* Editor's note: Also reported at (1972) 222 EG 1755

† Editor's note: Also reported at [2001] 3 EGLR 14

was the solicitor that had acted for Fineland at an earlier stage in the 2006 proceedings until it was replaced by the fourth defendant. In addition to joining these defendants to the application of 7 April 2011, the claimants also gave notice that they required a number of individual solicitors from the third and fourth defendants to attend the hearing on 13 April 2011 so that they could be cross-examined. The result was that the third and fourth defendants separately instructed counsel to represent them at the hearing.

[41] I held at the hearing on 13 April 2011 that the claimants' conduct in giving notice to the third and fourth defendants as respondents to the application of 7 April 2011, which really only concerned Fineland, and also the application to cross-examine individual solicitors at the two firms was open to criticism on the grounds that it was unnecessary to require the third and fourth defendants to attend the hearing and it was wrong to apply for an order that individual solicitors be cross-examined. C I therefore dismissed any application in these respects against the third and fourth defendants. I further ruled that I would certify that the applications against the third and fourth defendants were totally without merit and that the claimants should pay the costs of the third and fourth defendants. They applied for costs to be assessed on the indemnity basis. I directed that they should have permission to apply at a later point for costs to be on the indemnity basis. My reason for postponing that particular decision is that the claimants have made allegations against the third and fourth defendants that remain to be dealt with at a future hearing and so as not to prejudge what might be determined at a future hearing, I will not rule at this stage on whether the costs of the D hearing on 13 April 2011 should be on the indemnity basis.

Claim dismissed.

LANDLORD AND TENANT
GENERAL

Grand *v* Gill

COURT OF APPEAL

[2011] EWCA Civ 554

19 May 2011

THOMAS, LLOYD AND RIMER LJJ

Estates Gazette 9 July 2011

[2011] 27 EG 78

Landlord and tenant — Repairs — Breach of covenant — Landlord and Tenant Act 1985 — Damp and mould to plasterwork in flat caused partly by inadequate ventilation — Claim by tenant for breach of landlord's repairing obligations — Section 11 of 1985 Act — Whether damp resulting from inherent design of flat such that no actionable disrepair arising — Whether plasterwork forming part of structure of flat so that damage constituting structural damage for which liability arising

The appellant was the tenant of a top-floor flat let to her by the respondent landlord. The respondent held a long leasehold interest in the flat but no interest in the remainder of the building. The appellant brought a claim against the respondent for breaches of the landlord's repairing obligations in the lease and section 11 of the Landlord and Tenant Act 1985. Her main complaint was damp and mould throughout the flat, which had become apparent shortly after she moved in and had subsequently worsened, rendering one bedroom uninhabitable, until curative works were finally carried out in 2009. She also complained of the poor functioning of the boiler, resulting in a lack of proper heating.

In the county court, the judge found that the damp had been caused primarily by the design and structure of the flat, particularly its lack of ventilation, although it had been aggravated by water ingress from the roof and guttering, which were not the respondent's responsibility since they did not fall within his demise. He held, applying *Quick* v *Taff-Ely Borough Council* [1985] 2 EGLR 50; (1985) 276 EG 452, that there was no actionable disrepair where severe condensation was caused by the inherent design of the property, rather than by any structural damage calling for repair. Since he found that the damage to the internal plasterwork caused by damp was merely decorative and not structural, he concluded that the respondent's liability for damp should be confined to only 10% of the full liability figure of £6,000, to reflect the extent to which the lack of adequate heating and a broken window had contributed to the problem. Of a total award of £5,600 damages, therefore, most related to the boiler problem and only £600 reflected damage caused by the damp.

The appellant appealed. The central issue on the appeal was whether the plasterwork could be regarded as part of the structure of the flat so that damage to it should be treated as structural damage giving rise to the landlord's repairing obligation.

***Held*: The appeal was allowed. The plaster finish to an internal wall or ceiling forms part of the "structure" of a dwelling-house for the purposes of the covenant implied by section 11(1) of the 1985 Act, rather than being merely decorative. In the past, when lath-and-plaster ceiling and internal partition walls were more common, the plaster played an essential role in creating and shaping the ceiling or partition wall, serving to give a dwelling-house its essential appearance and shape. Moreover,** F **plasterwork generally, including that applied to external walls, is ordinarily in the nature of a smooth, constructional finish to which decoration can be applied, rather than a decorative finish in itself. The wall and ceiling plasterwork in the appellant's flat was therefore properly to be regarded as part of the structure of that flat. The damage to that plasterwork was therefore damage to the "structure" of the flat, for which the respondent was responsible under his repairing obligations by virtue of the tenancy agreement and section 11(1) of the 1985 Act. The respondent had breached his repairing obligations in that respect and was liable to compensate** G **the appellant appropriately. The full liability figure should not have been discounted by 90%. The correct award was: (i) full compensation for the plasterwork damage in the sum of £750; and (ii) discounted compensation of 10% for the remainder of the damage by damp to which the inadequate heating had contributed, which, assuming that this accounted for the remaining £5,250 of the judge's full liability figure, would be £525. The total award for the damp was therefore £1,275, resulting in overall damages of £6,275.**

The following cases are referred to in this report.

Irvine v *Moran* [1991] 1 EGLR 261; (1992) 24 HLR 1, QB H
Marlborough Park Services Ltd v *Rowe* [2006] EWCA Civ 436; [2006] 2 P&CR 8; [2006] 2 EGLR 27; [2006] 23 EG 166; [2006] HLR 30
Niazi Services Ltd v *van der Loo* [2004] EWCA Civ 53; [2004] 1 WLR 1254; [2004] 1 EGLR 62; [2004] 17 EG 130
Quick v *Taff-Ely Borough Council* [1986] QB 809; [1985] 3 WLR 981; [1985] 3 All ER 321; [1985] 2 EGLR 50; (1985) 276 EG 452; (1986) 18 HLR 66, CA
Staves v *Leeds City Council* [1992] 2 EGLR 37; [1992] 29 EG 119; (1991) 23 HLR 107, CA

This was an appeal by the appellant tenant, Tanya Grand, from a decision of HH Judge Karsten QC, sitting in Central London County Court, assessing damages on a claim against the respondent landlord, J Param Gill, for breach of the landlord's repairing covenants in section 11 of the Landlord and Tenant Act 1985.

John de Waal (acting pro bono) appeared for the appellant; the respondent appeared in person.

Giving the first judgment, **Rimer LJ** said:

Introduction

[1] This appeal by Ms Tanya Grand, the claimant, is against an order made by HH Judge Karsten QC on 7 May 2009 in Central London County Court. The defendant, respondent to the appeal, is K Mr Param Gill, Ms Grand's landlord. Before the judge, Ms Grand represented herself, and Mr Gill was represented by counsel. Before this court, the position was the reverse. Ms Grand had the benefit of pro bono representation by John de Waal of counsel, for whose assistance I express my gratitude; and Mr Gill represented himself.

[2] Ms Grand's claim against Mr Gill was for damages for his breaches of covenant as her landlord. The judge awarded her damages of £5,600. By her original grounds of appeal, Ms Grand complained about the trial and its outcome in a number of respects, but in the event she obtained permission from Patten LJ to appeal only on grounds 2 and

9. The terms of Patten LJ's permission show that he was in fact referring to ground 10, not ground 9.

[3] Ground 2 complains of the judge's failure to: (i) deal with Ms Grand's claim for special damages (pleaded as being for £392, but strictly for at most £343, £49 of the pleaded claim being in respect of costs for printing documents for the proceedings); and (ii) award her any interest on the damages he did award. I need say no more about the special damages point since Mr de Waal, on instructions, abandoned it at the hearing. He did, however, pursue the interest point. Ground 10 is somewhat discursive, but the only element advanced to us, in line with Patten LJ's permission, was based on what was said to be the judge's mistaken treatment of Ms Grand's claim for general damages, with the consequence that he is said to have under-compensated her.

Facts

[4] Ms Grand was formerly Mr Gill's tenant of the second-floor (top) flat in a three-storey building at 9B Marlborough Parade, Uxbridge Road, Hillingdon, Middlesex. The ground floor consists of commercial premises and there is another flat on the first floor. She originally had a 12-month assured shorthold tenancy under a tenancy agreement dated 21 November 2004; and following the expiration of the original term, the tenancy continued as an assured periodic tenancy. Ms Grand occupied the flat with her daughter Alison, who was 17 at the time of the trial in May 2009. The flat comprises two bedrooms and a living room, kitchen, bathroom and lavatory. Mr Gill has a long leasehold interest in the flat but no interest in the remainder of the building. Ms Grand made claims against him under several heads. They included claims for harassment under the Protection from Harassment Act 1997 and for personal injury, which both failed and with which this appeal is not concerned. She also advanced a claim for breach of Mr Gill's covenant for quiet enjoyment, which succeeded and for which she recovered damages of £350 that formed part of the total damages awarded of £5,600. There is no need to refer further to that head of claim. I turn to the issues before the judge with which the appeal is concerned.

A. Damp and mould

[5] Ms Grand's main complaint was of damp and mould throughout the flat. The damp was apparent shortly after she moved in and got worse over time. It was evident on the walls of the living accommodation and became so bad that the smaller bedroom (Alison's) became uninhabitable and Alison had to move into the living room. Curative works to the damp problem were only done in April 2009. The judge found the complaint about the damp to be established.

B. Water ingress

[6] Another complaint found by the judge to be associated with the damp and mould and causally connected with it was about the undisputed ingress of water into the flat through the ceiling, which was from two sources. The first was a leaking roof above the ceiling. The second was the guttering at the level of the roof. The roof was not, however, part of the premises let to Ms Grand and the judge found that the responsibility for the repair of the roof and guttering lay with the head landlord.

C. An inadequate boiler

[7] The complaint here was that the gas-fired boiler was old, extensively corroded, broke down frequently and worked inefficiently. The judge accepted Ms Grand's evidence that during a total of 207 days (give or take a day or so) between 22 November 2004 and 24 November 2007 the boiler did not function at all. It was replaced in November 2007, following which there was consistent proper heating in the flat. During the periods prior to that when the old boiler was working, it failed to produce sufficient heat to provide a warm or comfortable environment and to provide hot enough water for a hot shower. The ambient temperature that it achieved was just 15 degrees centigrade, which the judge found was inadequate. Ms Grand's evidence in relation to the boiler was not challenged.

Judge's further findings and conclusion

[8] The judge found, in [21], that Ms Grand had given sufficient notice to Mr Gill of the various defects and matters complained of to activate such obligations to which he was subject to effect necessary repairs. For the reasons he gave, he rejected, in [25], an assertion that any damages to which she was in principle entitled ought to be reduced on the ground that she had, so it was asserted, failed to afford access without good reason to workmen whom Mr Gill had sent to the flat.

[9] The judge dealt in [26] and following with the claim in respect of the damp and mould. He held that the experts' reports showed that the primary cause of it was the design and structure of the flat, essentially its lack of ventilation. The penetrating damp had been aggravated by the ingress of water from the roof and the guttering, but those areas were not within Mr Gill's demise. He held that to the extent that damp or mould had been caused by condensation because of a lack of ventilation, a landlord's repairing obligation under section 11 of the Landlord and Tenant Act 1985 (the 1985 Act) did not arise until there was some condition calling for repair. He relied on *Quick* v *Taff-Ely Borough Council* [1986] QB 809* for the proposition that there was no actionable repair where severe condensation was caused by the design of the property.

[10] The judge explained as follows his damages awards in respect of: (i) the damp and mould; (ii) Mr Gill's failure to resolve the boiler problem over a period of three years from November 2004 to November 2007, resulting in a lack of proper heating to the flat during this time, and (iii) his breach of the covenant for quiet enjoyment:

> 26. … It seems to me that, save to the extent that it can be shown that any structural damage is caused which could constitute actionable disrepair, the principles of *Quick* v *Taff Ely* apply to this case. The only possible structural damage arising on the facts of this case might be damage to the plasterwork, but there is no evidence that the condensation required more than the walls to be cleaned, so that the damage caused was decorative rather than anything else.
>
> 27. Essentially, it seems to me that the principal cause of the damp and mould is therefore the design defect in the property, for which [Mr Gill] is not to be held responsible, but I am satisfied that it was contributed to to a certain extent by the non-functioning and/or poor functioning of the boiler during the first three years of the tenancy, and that is disrepair of the installation for which [Mr Gill] is plainly responsible. It was also contributed to to a small extent by the fact that for a year or so one of the double glazed windows had been smashed and was to be repaired, so that one of the two layers of double glazing was absent. In addition, as I touched on towards the beginning of this judgment, the ingress of water contributed to the condensation, but the lack of adequate heating in my judgment did make a contribution to the condensation problem.
>
> 28. In his first report, the expert says, at paragraph 5.02.1, that "condensation occurs when moisture in the air condenses on cold surfaces, consequently the control of condensation is governed by the presence of moisture within the air and the presence of cold surfaces." He goes on to say that "the presence of cold surfaces is governed by thermal capacity and performance of the structure, together with the presence and use of heating system." It was the failure or lack of any adequate heating system for these three years [November 2004 to November 2007] that made a contribution, as I have found, to the fact that there were these cold surfaces upon which damp and mould was able to grow.
>
> 29. In addition, there is one area which relates to the exterior of the flat itself. It is referred to in the Scott Schedule to the first report, at paragraph 1.02, where "dampness with minor efflorescence and minor mould growth" is noted at the top of the front wall of the living room, requiring all defective plaster to be hacked off down the wall, approximately 600 millimetres from the ceiling. This was caused by the guttering, but it is a structural defect.
>
> 30. As to the lack of heating at all for periods of time and to an adequate standard for the remainder of the first three years of tenancy, plainly this created an extremely difficult situation for [Ms Grand] and her daughter. The daughter (or her mother) describes how at the beginning they shared a bed and had to cuddle up and cover themselves as best they could against the cold. The lack of the possibility to have a shower and to wash clothes properly must have made living conditions extremely difficult, and there is no doubt that [Mr Gill], having been given notice as I have found, failed to take reasonable steps to get the matter put in order over this three year period, until eventually the boiler was replaced. …

---

* Editor's note: Also reported at [1985] 2 EGLR 50

33. I now turn to assess the damages. … In assessing the damage and disrepair I have regard to and follow the approach approved by the Court of Appeal in *Wallace* v *Manchester City Council* [1998] 3 EGLR 38. I have regard to the rent that was payable. I have regard to the discomfort and inconvenience. I approach this case on the basis that it is right to take both aspects into account, so that I approach it as a mixture of the two. I take into account the rent in assessing damages for distress, discomfort and inconvenience. The most important damage, so to speak, was the consequence of the damp and mould, but it seems to me that on the evidence it is not possible to say that more than a ten per cent contribution to the damp and mould was made by the lack of adequate heating coupled with the small matters of the ingress of water into the living room from the guttering and the broken window which was repaired after a year or so. On the basis of full liability for that part of the claim I would have considered that damage in the region of £2,000 per annum would be appropriate, that is to say, £6,000 in all, but since I find the defendant is responsible only to the extent of ten per cent, the award under this head is only £600.

34. A much bigger award arises in relation to the boiler problem. The 207 days, amounting to approximately 30 weeks, during which there was no heating should, in my judgment, be dealt with by an award of damages of £1,750. The remainder of the first three years of the tenancy, during which there was inadequate heating coming from the boiler, should be compensated at the rate of £1,200 per annum. In setting that figure I have regard to the decision mentioned in Islington v. Spence in July 2001, Legal Action 26 in Clerkenwell County Court, referred to by [counsel for Mr Gill] in paragraph 43 of his skeleton argument, in which the court awarded £1,100 per annum for the period during which, after heating had been restored, the radiators failed to heat up properly, and I add a little to take account of [inaudible] since 2001, so that produces a figure of £3,600 for three years, less approximately £700 for 30 weeks covered by the award of damages for there being no heating at all, so producing a total under this head of £2,900. This means that damages for disrepair will be awarded at a total figure of £5,250.

35. As to the damages for breach of the covenant for quiet enjoyment, in the circumstances it seems to me it is a relatively modest sum. It will be appropriate to cover the distress and inconvenience suffered by [Ms Grand] and I fix that at £350. So the total damages will be £5,600.

Appeal

[11] The main ground of appeal is that it is said by Mr de Waal that the judge was wrong to apply the 90% discount to the damage caused by the damp. The judge's reasoning was that only 10% of the cause of the damp and mould could be attributed to fault on the part of Mr Gill, mainly because of the inadequacy of the boiler. Mr de Waal submitted that in that respect the judge was wrong and he asked us to increase the damages awarded under this head from £600 to £5,000. The judge's error, he submitted, was in his application of the decision in *Quick*.

[12] Mr Gill was subject to repairing obligations under both the tenancy agreement and section 11 of the 1985 Act. By clause 2.3 of the tenancy agreement, Mr Gill agreed, so far as material, to keep in repair:

2.3.1 The structure and exterior of the Property (including drains, gutters and external pipes)…

2.3.3 The installations at the Property for space heating and heating water.

[13] By section 11(1)(*a*) of the 1985 Act, Mr Gill was subject to an obligation in the like terms as clause 2.3.1, with section 11(1A) providing that in a case, as here, in which the lease is of a dwelling-house forming part only of a building, the covenant implied by section 11(1) shall have effect as though:

the reference in paragraph (a) of that subsection to the dwelling-house included a reference to any part of the building in which the lessor has an estate or interest;…

[14] It is accepted that Mr Gill was not responsible for the repair of the defective roof and guttering, which caused some of the water ingress. The judge, however, made the findings that he did as to the consequential damage to the internal plasterwork in [26] and [29] of his judgment; and it is said that Mr Gill was liable to repair that plasterwork.

[15] Mr de Waal referred us to the evidence before the judge as to such damage. It was contained in the two expert reports of Mr Ian Lovatt, a chartered surveyor. The first report followed Mr Lovatt's inspection of the flat in March 2006. Item 1.02 in the appended Scott Schedule, relating to the living room, referred to damaged plasterwork, caused **G** by penetrating dampness, in the upper part of the front wall extending downwards to the extent of about 600mm. There was also a damaged area to the left side of the main window. The repair required was stated as follows:

Hack off all defective plaster and properly re-plaster to a smooth and even finish. Prepare and decorate front wall complete.

The judge referred to this damage in [29] of his judgment. The only other reference to damaged plasterwork in the schedule was in para 5.01, relating to the kitchen ceiling, where significant dampness was noted. The repair required was in terms identical to that just quoted, save that **H** the second sentence read "Prepare and redecorate ceiling complete".

[16] Mr Lovatt's second report followed an inspection of the flat he made in February 2008. Item 1.01 of the Scott Schedule referred again to the living room and noted damage to the ceiling in the nature of a "small area of fallen lath plasterwork", for which the required repair work was "Cut back defective plaster, tack board and skim". Item 5.01 referred to kitchen ceiling, said to be dry at the time of the inspection, but requiring preparation and redecoration. This item made no reference to damaged plasterwork.

[17] The judge found, in [28], that the inadequacy of the heating **J** system, caused by the defective boiler for which Mr Gill was responsible, contributed to the damp problems from which the flat suffered. In [33], the judge assessed that, if Mr Gill were liable for all the damage, an award of £6,000 would be appropriate. As, however, he was only 10% responsible for the damp damage, an award of £600 was appropriate. Where, Mr de Waal submitted, the judge went wrong was to overlook that Mr Gill was not just in breach of his obligations in respect of the heating problems but also in relation to the omission to repair the damaged plaster. The judge ought not, therefore, to have discounted his total damages figure of £6,000 by as much as 90% because that ignored that Mr Gill was 100% responsible for the damage flowing from the disrepair of the plasterwork. *Quick* shows, as the judge **K** recognised, that liability under repairing covenants such as Mr Gill's for conditions caused by condensation resulting from a design defect does not arise unless they result in physical damage covered by the landlord's covenant, when the repairing obligation will apply. In this case, however, the judge found (or ought to have found) that there was physical damage, resulting in disrepair, to what is said to have been part of the structure of the property, namely the plasterwork in the living room (front wall and ceiling) and in the kitchen (ceiling).

[18] In my judgment, assuming that the plasterwork damage identified by Mr Lovatt is correctly characterised as damage to the "structure" of the flat, it was damage for which Mr Gill was responsible **L** under his repairing obligations under both the tenancy agreement and the provisions in section 11(1) of the 1985 Act (see *Quick*, *per* Dillon LJ, at p818E). A key question, however, which Mr de Waal very properly raised with us, is whether such plasterwork did form part of the "structure" of the flat. Only if it did was Mr Gill liable to repair it.

[19] The judge may have assumed in [26] that the plaster formed part of the structure of the flat, but I find it difficult to read his last sentence of that paragraph as making a finding to that effect. In *Quick*, the landlord council conceded for the purposes of the appeal that the plaster in the house was part of its structure (see at pp820B and 822G) and so that decision provides no authority on the point. In *Staves* v *Leeds* **M** *City Council* (1991) 23 HLR 107*, another decision of this court, a like concession was also made by the landlord council (see *per* Ewbank J, at p110, and *per* Lloyd LJ, at p112). Given the concession, that case also cannot be regarded as authority on the point. In *Niazi Services Ltd* v *van der Loo* [2004] EWCA Civ 53; [2004] 1 WLR 1254†, an issue came before this court as to whether plasterwork forms part of "the structure" of a dwelling-house within the meaning of section 11 of the 1985 Act.

* Editor's note: Also reported at [1992] 2 EGLR 37

† Editor's note: Also reported at [2004] 1 EGLR 62

A Having recognised it as a difficult question, the court explained in [7] why it did not propose to answer it.

[20] The point was, however, the subject of a decision by Mr Recorder Thayne Forbes QC in *Irvine* v *Moran* (1992) 24 HLR 1; [1991] 1 EGLR 261. The lease in question there was one to which section 32 of the Housing Act 1961 applied, but section 11(1) of the 1985 Act is, so far as material, in identical terms. Before the judge were preliminary issues as to which of several items including "internal wall plaster" were part of "the structure and exterior of the dwellinghouse" and so within the landlord's repairing covenants imposed by section 32. Dealing first with the meaning of the word "structure… of
B the dwellinghouse" in that section, Mr Recorder Thayne Forbes said, at p5:

> I have come to the view that the structure of the dwellinghouse consists of those elements of the overall dwellinghouse which give it its essential appearance, stability and shape. The expression does not extend to the many and various ways in which the dwellinghouse will be fitted out, equipped, decorated and generally made to be habitable. I am not persuaded by Mr Brock [counsel for the landlord] that one should limit the expression "the structure of the dwellinghouse" to those aspects of the dwellinghouse which are load bearing in the sense that that sort of expression is used by professional consulting engineers and the like; but what I do feel is, as regards the words "structure of
> C the dwellinghouse", that in order to be part of the structure of the dwellinghouse a particular element must be a material or significant element in the overall construction. To some extent, in every case there will be a degree of fact to be gone into to decide whether or not something is or is not part of the structure of the dwellinghouse. It is not easy to think of an overall explanation of the meaning of those words which will be applicable in every case and I deliberately decline to attempt such a definition. I am content for the purposes of this case to say that I accept Mr Brock's submission that "structure of the dwellinghouse" has a more limited meaning than the overall building itself and that it is addressed to those essential elements of the dwellinghouse which are material to its overall construction.
>
> That I think is as far as I am able to go. In coming to that conclusion I have been guided to some extent and have sought to follow the approach suggested
> D by Lord Justice Megaw in *Campden Hill Towers Limited* v *Gardner* [1977] 1 QB 823, in particular the passage that appears at page 834F.

[21] Having so directed himself as to the "structure… of the dwellinghouse", Mr Recorder Thayne Forbes considered which of the items in issue fell within it. Dealing with "internal wall plaster", he said, at p6:

> As I have said, section 32(1)(a) and the words "structure of the dwellinghouse" mean something less than the dwellinghouse overall and a limited to the essential material elements that go to make up the structure of the dwellinghouse. It seems to me that internal wall plaster is more in the nature of a decorative
> E finish and is not part of the essential material elements which go to make up the structure of the dwellinghouse. I therefore hold that internal wall plaster and, for the same reasons, the door furniture do not form part of the structure of the dwellinghouse, bearing in mind I have held that those words mean something less than the overall construction.'

[22] In this court's decision in *Marlborough Park Services Ltd* v *Rowe* [2006] EWCA Civ 436; [2006] HLR 30*, Neuberger LJ (as he then was), with whose judgment Sedley and Tuckey LJJ agreed, cited from the first passage that I have quoted from *Irvine* (relating to the phrase "the structure… of the dwellinghouse"), saying of it:

> F 17. While I accept, as I have emphasised, that words such as "structure" or "main structures" must take their meaning from the particular document, lease or statute in which they are found, and from the surrounding circumstances, and while it can be said that any attempt to define them will, to an extent, raise as many questions as it answers, it seems to me that that is a good working definition to bear in mind, albeit not one to apply slavishly.

The issue in *Marlborough* was whether wooden joists in the intermediate floor of a two-storey maisonette formed part of "the main structures of the Property" and the manner of its resolution is of no direct assistance for present purposes.

* Editor's note: Also reported at [2006] 2 EGLR 27

[23] Mr de Waal, for Ms Grand, submitted that the decision in *Irvine* as to plaster on internal walls not forming part of the structure of a dwelling-house was in principle wrong. He accepted that, applying Mr Recorder Thayne Forbes's guidance as to the meaning of "the structure… of the dwellinghouse", internal plasterwork does not give the house stability. But it does, he said, contribute to its appearance and shape. It is, he said, artificial to regard internal plasterwork, whether on walls or ceilings, as purely decorative. The relevant distinction is, he said, between decoration and fittings on the one hand and everything else making up the dwelling-house on the other.

[24] I have not found this question a straightforward one, the answering of which was not helped by the fact that there was no expert evidence as to the plasterwork in the flat or plasterwork generally, we have no findings about the matter by the judge and had no argument on it from Mr Gill. Despite these disadvantages, I consider that the court must, however, answer the question.

[25] For myself, although I would accept and adopt Mr Recorder Thayne Forbes's observations as to the meaning of "the structure… of the dwellinghouse" as providing for present purposes, as Neuberger LJ put it, a good working definition, I am respectfully unconvinced by his holding that the plaster finish to an internal wall or ceiling is to be regarded as in the nature of a decorative finish rather than as forming part of the "structure". In the days when lath-and-plaster ceiling and internal partition walls were more common than now, the plaster was, I should have thought, an essential part of the creation and shaping of the ceiling or partition wall, which serve to give a dwelling-house its essential appearance and shape. I would also regard plasterwork generally, including that applied to external walls, as being ordinarily in the nature of a smooth constructional finish to walls and ceilings, to which the decoration can then be applied, rather than a decorative finish in itself. I would therefore hold that it is part of the "structure". I would accordingly accept that the wall and ceiling plaster in Ms Grand's flat formed part of the "structure" of the flat for the repair of which Mr Gill was responsible.

[26] It follows that I consider that the judge should also have found that the disrepair of the damaged plasterwork meant that Mr Gill was in that respect in breach of his repairing obligations and should have compensated Ms Grand appropriately. I admit to at least some uncertainty as to whether, in arriving at his "full liability" figure of £6,000 in [33] the judge was including the plasterwork damage, but I infer that he was; and, consistently with that, Mr de Waal's complaint was that he was wrong to discount the £6,000 figure by as much as 90%. I propose to proceed on that basis. He should, in my view, therefore have awarded Ms Grand: (i) full compensation for the plasterwork damage; and (ii) discounted compensation for the remainder of the damage to which the inadequate heating contributed (there was no challenge to the principle of the latter approach that the judge adopted in [33]).

[27] Mr de Waal submitted that, because of the judge's error in relation to the plasterwork, an appropriate figure to substitute for his £600 one was the figure of £5,000. I respectfully disagree. Although I agree that Mr Gill should be 100% liable for the plasterwork disrepair, I consider that to regard that element of his liability as representing a figure approaching £5,000 of the judge's "full liability" £6,000 figure as unrealistic. This court is not in as good a position as was the judge to make the required assessment but the sums involved are, in my view, of too modest a nature to justify a remission for a reassessment by the judge. I consider, therefore, that we should make the assessment ourselves, although I recognise that that will require the wielding of a fairly broad brush. I would assess the plaster damage as representing £750 of the judge's £6,000; and so would assess the balance of the damage at £5,250. On that basis, the figure I would substitute for the judge's £600 figure is £750 + (£5,250 x 10%), or £1,275. The result is that the overall damages award that the judge made ought, I consider, to be increased from £5,600 to £6,275 and paragraph 1 of his order of 7 May 2009 adjusted accordingly. I would so order.

Interest

[28] For unexplained reasons, the judge did not award any interest on the damages, although I understood from Mr de Waal that it was applied for (it was claimed in the original and amended particulars of claim). I consider that an award of interest ought to have been made. Mr de Waal suggested a rate of 2%, with which I agree, and that it should run on the damages from the date of the issue of the claim on 6 November 2006, with which I do not agree. I disagree because the damages awarded include a material element of compensation in respect of a period subsequent to November 2006, when the boiler was still malfunctioning, and it cannot be just to require Mr Gill to pay interest from that date on sums that relate to damage suffered by Ms Grand subsequently. During the argument, the court suggested that a fairer date to take would be 1 December 2007 which followed the replacement of the boiler, after which the flat enjoyed proper heating with which Mr de Waal agreed. I would order Mr Gill to pay interest at 2% on the adjusted damages of £6,275 from that date to the date of the judge's order (7 May 2009); and [1] of the judge's order should be varied to include such an interest award. I consider that it also follows that the references to "damages" in [5] and[7] of the order should be varied to read "damages and interest".

Costs

[29] As I have said, Mr de Waal appeared pro bono but, were the appeal to succeed, he invited us at the hearing to make an award of costs against Mr Gill in favour of a prescribed charity as permitted by section 194 of the Legal Services Act 2007. He provided a breakdown of his time spent in preparation for the appeal, which totalled 12.5 hours and, at a standard hourly rate of £250, produced a total figure of £3,125. Save that the breakdown included: (a) work done for the abandoned appeal in respect of special damages; and (b) a slight over-estimate for the length of the hearing of the appeal, I consider that it was a moderate and fair one. I would, however, adjust it to take account of both these matters and reduce it to £2,500. As required by section 194(3) of the 2007 Act, I consider that, had Mr de Waal's representation of Ms Grand been provided on a fee-paying basis, it would have been appropriate to order a proportionate part of his fees to be paid by Mr Gill, a respondent who resisted the appeal unsuccessfully. I would therefore summarily assess at £2,500 the costs that the respondent, Mr Gill, must pay. I would direct that, within 14 days of the date of the court's order, he must pay it to the prescribed charity, namely The Access to Justice Foundation (the payment to be sent to PO Box 64162, London WC1A 9AN).

Disposition

[30] I would make orders on this appeal accordingly. I would also ask Mr de Waal to draft an appropriate form of order giving effect to the court's decision, having first provided a copy to Mr Gill for his comments.

Giving the second judgment, **Lloyd LJ** said:

[31] I agree with Rimer LJ that the issue on this appeal (apart from the point about interest) turns on whether plaster on the ceiling and the walls of the flat is part of the structure of the flat for the repair of which Mr Gill as landlord is responsible under clause 2.3.1 of the tenancy agreement and section 11(1)(*a*) of the 1985 Act.

[32] The only decision on the point is that of Mr Recorder Thayne Forbes QC (as he then was) in *Irvine*, from which Rimer LJ has quoted the relevant passages. In *Niazi*, to which he has also referred, the Court of Appeal, at p1257G, described the point as being "not free from difficulties" and understandably refrained from deciding it, since it was unnecessary for the outcome of that case. The point does require thought, and we have less material than would be ideal on which to come to our conclusion, but we cannot avoid the point, as the Court of Appeal has been able to on previous occasions.

[33] Like Rimer LJ, I would respectfully differ from Mr Recorder Thayne Forbes on this point, not as regards the first passage (quoted in [20] above) in which he sought to identify a principle on which to decide the point, but rather as regards his application of that principle to the particular case of plaster, in the passage set out in [21] above. As Rimer LJ says, in the case of the use of plaster in ways that are now perhaps historic rather than current (but of which examples are certainly still to be found) such as lath-and-plaster construction, the plaster is clearly part of the structure. The expert's second report suggests (as quoted in [16] above) that this technique may have been used in parts of the premises relevant to this case.

[34] I would not limit my reasoning, however, to cases where the wall or ceiling is of lath-and-plaster or similar construction. I agree that plaster as applied to even a solid wall or ceiling is not "in the nature of a decorative finish", as Mr Recorder Thayne Forbes said, and that it is to be regarded as a part of the wall or ceiling on or to which a decorative finish, of whatever kind, may be applied. Accordingly, I would hold, as a general proposition, that plaster forming part of or applied to walls and ceilings is part of the structure of the relevant premises.

[35] I therefore agree with Rimer LJ that, to the extent of the damage caused by the failure to repair the plasterwork, Mr Gill was liable for the whole of that loss, not for only a small percentage of it.

[36] I also agree with his assessment, on the very limited materials that we have, of the appropriate figure for the damages that should have been awarded to the appellant, taking account of this factor.

[37] Likewise, I agree with what he says about interest and about costs, for the reasons that he gives.

Thomas LJ said:

[38] I agree with both judgments. The reasoning and conclusion of Mr Recorder Thayne Forbes in respect of wall plaster quoted in [21] of Rimer LJ's judgment are not correct. Plaster forming part of or applied to walls and ceilings is part of the structure of the relevant premises.

Appeal allowed.

K/S Victoria Street *v* House of Fraser (Stores Management) Ltd and others

COURT OF APPEAL

[2011] EWCA Civ 904

27 July 2011

LORD NEUBERGER MR AND THOMAS AND ETHERTON LJJ

Estates Gazette 13 August 2011

[2011] 32 EG 56

Landlord and tenant — Assignment — Guarantee — Landlord and Tenant (Covenants) Act 1995 — Lease of commercial premises — Obligations of tenant guaranteed by parent company — Agreement for lease requiring onward assignment of lease by tenant to sister company with further guarantee by parent company — Whether requirement for further guarantee frustrating operation of section 24(2) of 1995 Act so as to be void under section 25(1)(a) — Whether assignee entitled to effect immediate reassignment to original tenant

Pursuant to an agreement dated January 2006, the first defendant sold a department store to the claimant and immediately took a leaseback of the premises for a term of 35 years at an initial rent of £2.25m, with its obligations as tenant guaranteed by the third defendant, its parent company. The third defendant had chosen the first defendant to act as vendor and the initial tenant under the leaseback for tax reasons, but it was not a satisfactory tenant from the claimant's point of view. In order to provide the claimant with a tenant of acceptable covenant strength, clause 3.5 of the agreement required the first defendant, within three months, to assign the lease to another company in the same group, which was to be the second defendant if no other choice was made, and the third defendant to guarantee the assignee's liabilities.

A The first defendant did not assign the lease and the claimant brought a claim for specific performance of clause 3.5. In those proceedings, it was held that, although the claimant could in principle require an assignment to the second defendant, it could not require the third defendant to act as the second defendant's guarantor since, on the authority of *Good Harvest Partnership LLP* v *Centaur Services Ltd* [2010] EWHC 330 (Ch); [2010] 1 EGLR 29; [2010] 14 EG 114, that requirement would frustrate the operation of section 24(2) of the Landlord and Tenant (Covenants) Act 1995 and was therefore void under the anti-avoidance provisions of section 25: see [2010] EWHC 3006 (Ch); [2010] 44 EG 125 (CS).

B In a further judgment, it was held that the second defendant, if required to take an assignment of the lease, could not immediately effect a reassignment to the first defendant. It was held that para 3.15(F) of the agreement, on which the defendants sought to rely, did not confer a stand-alone right to assign to another group company but was subject to the requirements of clause 3.15(E) regarding the financial standing of the assignee: see [2010] EWHC 3344 (Ch); [2010] PLSCS 320. The claimant appealed from the first judgment and the defendants appealed from the second. The appeal court considered the extent and effect of the *Good Harvest* decision.

***Held*: The appeals were dismissed. (1) The requirement in**
C clause 3.5 for the third defendant to guarantee the assignee's liabilities was void by reason of section 25(1)(a) of the 1995 Act. That was so even though clause 3.5 was a sensible quid pro quo for the claimant's agreement to accept the first defendant as the initial tenant under the lease for a short period, in order to assist the defendants' tax affairs. The validity of a provision, for the purposes of section 25(1), is determined by the objective effect of that provision, not the subjective reasons for its existence. The thrust of section 24(2) is that a party should not remain liable under a tenancy after the tenant with whose liability it is associated has been released from that liability. Any contractual arrangement contained
D in a tenancy or in a prior agreement that obliges an existing or prospective guarantor of the tenant's liabilities to guarantee the liabilities of a future assignee should be void under section 25 as frustrating the operation of section 24(2). The same will be true of a contractual arrangement contained in a later document, since section 25 applies to agreements "whether or not" they were made in or prior to a tenancy. (2) Section 25(1) will invalidate any agreement that involved a guarantor of a tenant guaranteeing that tenant's assignee, regardless of whether the agreement was entered into at the insistence of the landlord or was freely offered by the assignor and guarantor. An important qualification to that
E rule arises in cases involving an authorised guarantee agreement (AGA) under section 16. The guarantor of an assignor can validly guarantee the liability of the assignee by being party to a valid AGA under section 16. A landlord whose consent is required for an assignment can, where it is reasonable to do so, require the tenant to enter into an AGA guaranteeing the performance of the lease covenants by the assignee and, since section 24(2) releases the tenant's guarantor only "to the same extent as the tenant", the guarantor can guarantee the tenant's continued liability under the AGA. Moreover, if a guarantor of the tenant is released from its guarantee on an assignment of the tenancy, nothing prevents it
F from becoming a guarantor again on a further assignment. (3) The second defendant could not, on taking an assignment of the lease, immediately reassign it to the first defendant. The right to assign in clause 3.15(F) had to be read cumulatively with the requirements as the financial standing of the assignee contained in clause 3.15(E). That construction accorded with the commercial context of the sale and leaseback and the purpose of clause 3.5.

The following cases are referred to in this report.

Arbuthnott v *Fagan* [1995] CLC 1396, CA

Avonridge Property Co Ltd v *Mashru*; *sub nom London Diocesan Fund* v *Phithwa*; *Mashru* v *Avonridge Property Co Ltd*; *Avonridge Property Co Ltd* v *London Diocesan Fund*; *London Diocesan Fund* v *Avonridge Property Co Ltd* [2005] UKHL 70; [2005] 1 WLR 3956; [2006] 1 All ER 127; [2006] 1 P&CR 25; [2006] 1 EGLR 15; [2006] 01 EG 100

Good Harvest Partnerships LLP v *Centaur Services Ltd* [2010] EWHC 330 (Ch); [2010] Ch 426; [2010] 1 EGLR 29; [2010] 14 EG 114

K/S Victoria Street v *House of Fraser (Stores Management) Ltd* [2010] EWHC 3006 (Ch); [2010] 44 EG 125 (CS), [2010] EWHC 3344 (Ch); [2010] PLSCS 320

This was an appeal by the claimant, K/S Victoria Street, from a decision of Mr John Randall QC, sitting as a deputy High Court judge, in proceedings against the defendants, House of Fraser (Stores Management) Ltd, (House of Fraser (Stores) Ltd and House of Fraser Ltd, for specific performance of obligations under an agreement for lease; and by the defendants from a decision of Mann J on a further issue in the same proceedings.

Anthony Speaight QC (instructed by Stockler Brunton) appeared for the claimant; Jonathan Seitler QC and Nicholas Taggart (instructed by Lawrence Graham LLP) represented the defendants.

Giving judgment, **Lord Neuberger MR** said:

[1] This is the judgment of the court, to which all members have made substantial contributions.

[2] The judgment relates to two appeals, each of which raises an issue arising out of an agreement made on 26 January 2006 (the agreement). The agreement provided for the sale of the freehold of a property on terms that it was immediately leased back by the buyer to the seller. In due course, the sale was completed and the lease (the lease) was granted.

[3] The first issue is whether (as Mr John Randall QC, sitting as a deputy High Court judge, held) part of clause 3.5 of the agreement (clause 3.5), which stipulated that the lease should be assigned to a specific assignee on terms that required the guarantor of the tenant's liability to guarantee the liability of the assignee, is invalid. The second issue is whether, if the lease was assigned in accordance with clause 3.5, it could (contrary to what Mann J subsequently held) be freely assigned back to the tenant.

[4] The first issue touches on a question of some general significance in the law of leases, namely the ambit of section 25(1) of the Landlord and Tenant (Covenants) Act 1995 (the 1995 Act), and whether the reasoning of Newey J in *Good Harvest Partnership LLP* v *Centaur Services Ltd* [2010] EWHC 330 (Ch); [2010] Ch 426* is correct. The second issue is rather more specific to the facts and documents in this case, and turns on the interpretation of clause 3.15 of the lease (clause 3.15). Before turning to those issues, it is necessary to set out the relevant factual and procedural history, and the background and relevant sections of the 1995 Act. All references hereafter to sections are to sections of that Act.

Factual and procedural history

[5] By the agreement, a Danish partnership, K/S Victoria Street (Victoria), agreed to purchase the freehold of a large department store in Wolverhampton (the premises) from the then freehold owner. That freehold owner was a subsidiary of House of Fraser plc (HoF, which later that year ceased to be a publicly listed company, and became House of Fraser Ltd). The subsidiary was called House of Fraser (Stores Management) Ltd (Management), and it had recently acquired the premises from another subsidiary of HoF, James Beattie Ltd (Beatties). The price payable for the freehold under the agreement was £46m.

[6] The agreement provided that, immediately upon completion of the purchase, Victoria would grant, and Management would accept, the lease, with HoF acting as guarantor of Management's liabilities, in accordance with the provisions of schedule 3 to the lease. The term of the lease was to be 35 years and the rent was to be £2.25m pa, subject to upward-only reviews every five years.

[7] HoF had selected Management from among its subsidiaries, to be the seller under the agreement and the initial tenant under the lease, for tax reasons. However, in the light of Management's financial

* Editor's note: Also reported at [2010] 1 EGLR 29

situation, it was not a satisfactory tenant so far as Victoria was concerned. Accordingly, the parties agreed that, within three months, the lease would be assigned to another company within the HoF group, failing which it would be assigned to a specific HoF subsidiary, namely House of Fraser (Stores) Ltd (Stores), a company with a healthy balance sheet. To reflect this arrangement, clause 3.5 of the agreement, provided as follows (the division of the clause into three paragraphs being added for convenience):

> 3.5 (i) [Management] agrees to assign the Lease to an assignee (being a Group Company of [HoF] being of equal or greater covenant strength to [Beatties], and
>
> (ii) if a company is not chosen by 20 April 2006 then the assignee shall be Stores and Stores agrees to take that assignment by no later than 26 April 2006, and
>
> (iii) [HoF] agrees to enter into a deed of guarantee of that assignee's liabilities as surety in the form set out in schedule 3 to the Lease.

[8] The agreement was completed, and, in accordance with its terms, the lease was granted by Victoria to Management, with HoF acting as Management's guarantor.

[9] The only provision of the lease that is relevant for present purposes is clause 3.15, which contained the tenant's obligations in respect of "alienation". The first six paragraphs of the clause were concerned with assignment, and were in these terms:

> (A) Not to assign, charge, underlet, hold upon trust for another or part with or share possession or occupation of the whole or any part of the Premises except as provided in this sub-clause.
>
> (B) Not to assign the Premises nor to underlet the whole or any part to a person entitled to claim diplomatic or Sovereign immunity.
>
> (C) Not to assign the whole of the Premises unless either:
>
> (i) the Tenant demonstrates that the Net Profits of the assignee in each of the three Accounting Periods ending immediately before the date of the assignment exceed in each of those Accounting Periods the figure equal to three times the principal yearly rent; or
>
> (ii) on or before completion of the assignment the Tenant enters into an authorised guarantee agreement with the Landlord in accordance with section 16 of the [1995 Act] in such form as the Landlord may lawfully require and any surety of the Tenant guarantees in such form as the Landlord reasonably requires the Tenant's obligations under such authorised guarantee agreement.
>
> (D) Not to assign the whole of the Premises without first:
>
> (i) obtaining the consent of the Landlord (such consent not to be unreasonably withheld);
>
> (ii) procuring that such sureties as the Landlord reasonably requires covenant by deed directly with the Landlord as principal debtors or covenantors in such form as the Landlord reasonably requires to pay to the Landlord all losses, costs and expenses arising out of or incidental to any failure by such assignee to comply with its obligations to the Landlord from time to time.... .
>
> (E) Not to assign the whole of the Premises to a Group Company of the Tenant unless the Group Company is of the same or better financial standing than the Tenant or has offered a guarantor or guarantors which when considered with the Group Company are of the same or better financial standing than the Tenant and [HoF] taken together.
>
> (F) Notwithstanding the provisions of this clause where the Tenant is [Management] or any other Group Company of [HoF] consent shall not be required to an assignment of the whole to another Group Company of [HoF] provided [HoF] acts as surety to the assignee Group Company.

The remaining paragraphs of clause 3.15 of the lease were concerned with underletting or parting with possession and occupation.

[10] Subsequently, the lease was not assigned in accordance with clause 3.5(i), and it remains vested in Management. Victoria contended that it should be assigned to Stores in accordance with clause 3.5(ii), with HoF renewing its guarantee pursuant to its apparent obligation to do so under clause 3.5(iii). Management and HoF contended that clause 3.5(iii) was unenforceable, and refused to effect the assignment to Stores. Accordingly, Victoria began these proceedings, seeking an order that the lease be assigned to Stores with HoF as guarantor of Stores' liability. This was resisted by Stores, Management and HoF (the HoF defendants) on various grounds, and the matter came before Mr Randall.

[11] A number of issues were canvassed before him, but the only relevant issues for present purposes were whether: (a) clause 3.5(iii) of the agreement was void because of the provisions of the 1995 Act: and, if so, (b) clause 3.5(i) and (ii) were none the less enforceable. Mr Randall decided that (a) clause 3.5(iii) was void, but (b) it was severable from the rest of clause 3.5, so that clause 3.5(i) and (ii) were, at least in principle enforceable: see [2010] EWHC 3006 (Ch)*.

[12] The HoF defendants then raised a new argument, namely that it would be pointless to order specific performance of clause 3.5, because, in the light of clause 3.15(F) of the lease, if it was assigned by Management to Stores, Stores could immediately assign it back to Management without any guarantee other than that of HoF, which was insolvent on a balance sheet basis. At a further hearing, Mann J rejected that contention: see [2010] EWHC 3344 (Ch)†.

[13] Victoria appeals against Mr Randall's decision that clause 3.5(iii) of the agreement is void (but not against his conclusion that, if it is void, the remainder of the clause is none the less enforceable), and the HoF defendants appeal against Mann J's decision that, if the lease were assigned to Stores, it could not immediately be assigned back to Management.

Background and relevant provisions of 1995 Act

[14] According to well-established common law principles, an original tenant was bound by all the tenant's obligations in the lease for the duration of a tenancy, even after it had assigned it (unless there was a provision to the contrary, which was very rare). In the case of many, indeed most, assignments, assignees would similarly covenant with the landlord to observe the tenant's covenants for the duration of the tenancy. Such liability for the duration of the tenancy after, often long after, the original tenant (or an assignee) had assigned it, was perceived in many quarters as being unfair, especially in a time of inflation and after a rent review. In a report in 1988, *Landlord & Tenant Law: Privity of Contract and Estate*, Law Com No 174, the Law Commission recommended that this state of affairs be radically changed in respect of future tenancies.

[15] Parliament accepted this recommendation and enacted the 1995 Act, which implemented the proposals of the Commission with amendments. The long title of the Act describes its purpose as being that "persons bound by covenants of a tenancy... be released from such covenants on the assignment of the tenancy". In *London Diocesan Fund* v *Phithwa* [2005] UKHL 70; [2005] 1 WLR 3956‡, in [39], Baroness Hale of Richmond said:

> The mischief at which the Commission's recommendations were aimed was the continuation of a liability long after the parties had parted with their interests in the property to which it related.

[16] This aim is centrally achieved through section 5, subsection (2)(a) of which provides, in simple and clear terms, that if a "tenant assigns the whole of the premises demised to him,... he is released from the tenant covenants of the tenancy... as from the assignment". Sections 6, 7 and 8 contain some rather more complex and conditional provisions for the release of a landlord on the assignment of the reversion to a tenancy. All these sections apply only to a "new tenancy", that is, a tenancy entered into after the 1995 Act came into force, 17 July 1995.

[17] Section 24(2) is concerned with the position of any party, other than a tenant, which is bound by any of the tenant's covenants in a tenancy. It states that where:

> (a) by virtue of this Act a tenant is released from a tenant covenant of a tenancy, and
>
> (b) immediately before the release another person is bound by a covenant of the tenancy imposing any liability or penalty in the event of a failure to comply with that tenant covenant,

* Editor's note: Reported at [2010] 44 EG 125 (CS)
† Editor's note: Reported at [2010] PLSCS 320
‡ Editor's note: Also reported at [2006] 1 EGLR 15

A then, as from the release of the tenant, that other person is released from the covenant mentioned in paragraph (b) to the same extent as the tenant is released from that tenant covenant.

The expression "tenant covenant" is defined in section 28(1) as "a covenant falling to be complied with by the tenant of premises demised by the tenancy".

[18] Section 25 states as follows:

> (1) Any agreement relating to a tenancy is void to the extent that
> (a) it would apart from this section have effect to exclude, modify or otherwise frustrate the operation of any provision of this Act, or
> B (b) it provides for
> (i) the termination or surrender of the tenancy, or
> (ii) the imposition on the tenant of any penalty, disability or liability,
> in the event of the operation of any provision of this Act, or
> (c) it provides for any of the matters referred to in paragraph (b)(i) or (ii) and does so (whether expressly or otherwise) in connection with, or in consequence of, the operation of any provision of this Act.
> (2) To the extent that an agreement relating to a tenancy constitutes a covenant (whether absolute or qualified) against the assignment, or parting with the possession, of the premises demised by the tenancy or any part of them
> (a) the agreement is not void by virtue of subsection (1) by reason only of the fact that as such the covenant prohibits or restricts any such assignment or parting with possession; but
> C (b) paragraph (a) above does not otherwise affect the operation of that subsection in relation to the agreement (and in particular does not preclude its application to the agreement to the extent that it purports to regulate the giving of, or the making of any application for, consent to any such assignment or parting with possession).
> (3) In accordance with section 16(1) nothing in this section applies to any agreement to the extent that it is an authorised guarantee agreement; but (without prejudice to the generality of subsection (1) above) an agreement is void to the extent that it is one falling within section 16(4)(a) or (b).
> (4) This section applies to an agreement relating to a tenancy whether or not the agreement is
> (a) contained in the instrument creating the tenancy; or
> D (b) made before the creation of the tenancy.

[19] Section 16, which is referred to in section 25(3), is concerned with "authorised guarantee agreements" or "AGAs", and applies only to new tenancies. The section was included to mitigate the effect of section 5(2), and to confer a degree of protection on landlords when a tenancy is assigned. AGAs are agreements that a tenant can, in some circumstances, be required to enter into to guarantee the obligations of its assignee until the tenancy is assigned again. Section 16 provides as follows:

> (1) Where on an assignment a tenant is to any extent released from a tenant
> E covenant of a tenancy by virtue of this Act…, nothing in this Act shall preclude him from entering into an authorised guarantee agreement with respect to the performance of that covenant by the assignee.
> (2) For the purposes of this section an agreement is an authorised guarantee agreement if
> (a) under it the tenant guarantees the performance of the relevant covenant to any extent by the assignee; and
> (b) it is entered into in the circumstances set out in subsection (3); and
> (c) its provisions conform with subsections (4) and (5).
> (3) Those circumstances are as follows
> (a) by virtue of a covenant against assignment (whether absolute or qualified) the assignment cannot be effected without the consent of the landlord
> F under the tenancy or some other person;
> (b) any such consent is given subject to a condition (lawfully imposed) that the tenant is to enter into an agreement guaranteeing the performance of the covenant by the assignee; and
> (c) the agreement is entered into by the tenant in pursuance of that condition.

Validity of clause 3.5(iii): the effect of section 25(1) of the 1995 Act

[20] The HoF defendants' contention is that clause 3.5(iii) is rendered void by section 25(1)(a) because it would "frustrate the operation" of section 24(2)(b). That argument proceeds as follows. On the assignment of the lease by Management to Stores in accordance with clause 3.5(ii), Management will be released from all further liability under the lease, by virtue of section 5(1), so section 24(2)(a) is satisfied; and, as HoF is "another person" who is "bound by [the] covenant[s] of the [lease]", section 24(2)(b) also applies. Accordingly, it is the effect and intention of section 24(2) that, "as from the release of [Management]", that is, on the assignment to Stores, HoF should be released from its liabilities as guarantor under the lease. Any provision, such as clause 3.5(iii), that stipulates in advance that HoF must reassume precisely that liability as a term of the assignment, would therefore "frustrate" the operation of section 24(2)(b), and it is therefore rendered void by section 25(1)(a).

[21] In our judgment, this argument is correct. If a landlord could: (a) when granting a tenancy, impose an obligation on the tenant's guarantor to guarantee the liability of the assignee in the event of an assignment; and (b) on an assignment by a tenant, enforce that obligation, it would, as a matter of ordinary language, "frustrate" the operation of section 24(2). If it were otherwise, it would mean, for instance, that a landlord, when granting a tenancy, could require a guarantor of the tenant's liabilities, on every assignment of the tenancy, to guarantee the liability of each successive assignee. Such an obligation (a renewal obligation) would plainly be wholly contrary to the purpose of section 24(2), because it would enable a well-advised landlord to ensure that any guarantor was in precisely the position in which it would have been before 1995 Act came into force.

[22] Were a renewal obligation enforceable, in many cases, for instance where the prospective occupier of commercial property was a member of a group of companies, a landlord could also effectively avoid the effect of section 5(2) by requiring a subsidiary company in the group to take the tenancy, with the parent company acting as guarantor on terms that included a renewal obligation. Indeed, it could go further than this: as a matter of logic, if such a liability could lawfully be imposed on and enforced against a guarantor of the original tenant, it is hard to see why it could not be imposed on the original tenant.

[23] Given the plain purpose of the 1995 Act, and the widely expressed terms of section 25(1)(a) and 25(4), any contractual arrangement contained in the tenancy (or in a prior agreement), which imposes an obligation, on an existing or prospective guarantor of the tenant's liabilities, to guarantee the liabilities of a future assignee should be void. That conclusion is supported by what was said in *London Diocesan Fund*, in [14] and [18], by Lord Nicholls of Birkenhead. He described section 25 as "a comprehensive anti-avoidance provision", which was "to be interpreted generously, so as to ensure the operation of the 1995 Act is not frustrated, either directly or indirectly".

[24] For the avoidance of doubt, this conclusion would also apply to a contractual arrangement contained in a later document, for instance a renewal obligation imposed on a guarantor of an assignee's liabilities in an assignment or a licence to assign. Subsection (4) makes it clear that section 25 applies to agreements "whether or not" they are made in, or prior to, a tenancy. Accordingly, if, as we consider, a renewal obligation imposed on a guarantor of the original tenant is void, it must follow that the imposition of a renewal obligation on the guarantor of an assignee would also be void.

[25] When it comes to frustrating the effect of section 24(2), clause 3.5(iii) is *a fortiori* the example of a guarantor of the original tenant entering into a renewal obligation. That is because clause 3.5(iii) is not merely an undertaking by a guarantor of the liability of the original tenant, namely Management, to guarantee the liability of a future assignee, namely Stores: it is an undertaking to do so in respect of an assignment that Victoria, the landlord, is entitled to insist on taking place.

[26] For Victoria, Mr Anthony Speaight QC raised two arguments against this conclusion. The first was that the purpose of section 24(2) was solely to ensure that the release of the tenant under section 5(2) was effective. This was on the basis that, if the tenant's guarantor is not released under a provision such as section 24(2), he would be able to seek through subrogation an indemnity from the tenant as the principal creditor, even though the tenant had been released from primary liability. On that basis, ran the argument, a provision such as clause 3.5(iii), or indeed a renewal obligation, does not "frustrate" the purpose of section 24(2).

[27] We are by no means satisfied of the correctness of the premise of this argument, but we are prepared to assume, without deciding, that the statutory release of the tenant without releasing a guarantor would not have deprived the guarantor of its right of subrogation. However, even on that basis, we would reject Mr Speaight's first argument, because we do not accept that section 24(2) has the limited purpose for which he contended. Its language does not suggest such a limited purpose, and the whole thrust of section 24(2), indeed of the 1995 Act itself, is that a person should not remain liable under a tenancy after the tenant with whose liability he is associated has been released from his liability. Further, paras 4.53 and 4.54 of the Law Commission report (*op cit*), which were relied on by Mr Speaight, do not support the notion that the section has such a limited purpose. The former paragraph states that when the lessee's liability ceases, "that would automatically end the responsibility of [his] guarantor". Paragraph 4.54 says that if a guarantor remains liable after the lessee is released, it "would undermine the thrust and purpose of [the Commission's] recommendations".

[28] Second, Mr Speaight contended that, on the unusual facts of this case, section 25(1)(a) should not apply. He pointed out that neither party to the agreement had any intention of evading the provisions of the 1995 Act, and that clause 3.5 was included as part of an arrangement intended to benefit the HoF defendants (because it was HoF that wanted the lease granted to Management for a short initial period for tax purposes).

[29] We have considerable sympathy with that argument, given that clause 3.5 was a perfectly sensible quid pro quo from the HoF defendants in return for Victoria's agreement to accept Management as the initial tenant under the lease for a short period to assist the HoF group's tax affairs. However, it seems to us that it would be wrong to accede to the argument, both as a matter of principle and as a matter of practice. It would be wrong in principle, because section 25(1)(a) applies to any provision that would "have effect to… frustrate the operation of", *inter alia*, section 24(2). In other words, it is the objective effect of a particular provision, and not the subjective reasons for its existence, which determine its validity for the purpose of section 25(1). It would be wrong in practice to adopt the approach inherent in the submission, because it would introduce an element of uncertainty, and the opportunity for disputes of fact, into an area where it is important to have predictability.

Section 25(1): clause 3.15(F) and the decision in Good Harvest

[30] The foregoing discussion is enough to dispose of Victoria's appeal against the decision of Mr Randall. However, it is appropriate to consider the reasoning and decision of Newey J in *Good Harvest*, for two reasons. First, the decision seems to have led to some controversy, and, because it is a first instance decision, to some uncertainty, in the field of landlord and tenant. Second, it is common ground in this case that the proviso to clause 3.15(F) is valid, and that might appear questionable, especially in the light of the wide way in which Mr Jonathan Seitler QC put his primary argument as to the validity of clause 3.5(iii).

[31] The effect of the reasoning of Newey J in *Good Harvest* appears to be that, on an assignment of a lease from the existing tenant (the assignor) to an assignee, section 25(1) invalidates a guarantee of the assignee's liabilities by the person who was the guarantor of the assignor's liabilities. This is based on the point that the agreement by the guarantor to guarantee the liabilities of the assignee, with effect from the date on which the assignor assigns the lease, "frustrates" the effect of section 24(2), which requires the guarantor to be released from such liability with effect from the assignment.

[32] In *Good Harvest*, the tenancy was granted to Chiron, whose liability was guaranteed by Centaur, and Chiron wished to assign to Total. The alienation covenant, clause 5.9, was (as in the present case) a fairly complex provision (see [4)]; by clause 5.9.3, the tenant covenanted not to assign without landlord's consent (not to be unreasonably withheld), clause 5.9.6.1 contained a requirement that the tenant and its guarantor enter into an AGA on any assignment, and clause 5.9.6.6 included a term that a guarantor would be provided for an assignee to be approved by the landlord (such approval not to be unreasonably withheld).

[33] The tenancy was assigned to Total pursuant to an agreement under which both Chiron (the tenant) and Centaur (its guarantor) guaranteed the liability of Total. Total failed to pay the rent, and the landlord sued Centaur under its guarantee in the agreement pursuant to which the assignment was made. In an admirably clear and concisely expressed judgment, Newey J held that Centaur's covenant in that agreement was rendered void by section 25(1) as it frustrated the effect of section 24(2). He seems to have taken a wide view of section 25(1), holding that it "serves to bar a guarantor [of the assignor] from giving a guarantee for an assignee" : see [22(viii)] and also [22(iv) and (ix)].

[34] For the reasons already discussed, an agreement that requires a guarantor to provide a further guarantee in the future falls foul of section 25(1), because it involves a guarantor, at a time that it is, or is agreeing to become, the tenant's guarantor, committing itself to reassume its liabilities on a future assignment, when the plain purpose of section 24(2) is to ensure that he is released from its liabilities with effect from the assignment. What is less clear, however, is how much further the ambit of section 25(1) goes. In *Good Harvest*, Newey J appears to have given an interpretation (interpretation (i)) to section 25(1), which meant that it invalidated any agreement that involves a guarantor of the assignor guaranteeing that assignor's assignee. An alternative interpretation (interpretation (ii)), which was raised before us, is that section 25(1) invalidates such an agreement only if it was entered into at the insistence of the landlord.

[35] We shall begin by considering the issue without regard to sections 16 and 25(3) that is, ignoring the provisions relating to AGAs.

[36] Where a lease contains a tenant's covenant against assignment in the normal form, namely that it requires landlord's consent, which is not to be unreasonably withheld, the landlord may often reasonably refuse consent to a particular assignment unless a suitable guarantor of the assignee is provided. If interpretation (i) is correct, the assignor could not offer its own guarantor, because a guarantor of the present tenant is, in effect, absolutely precluded from providing a subsequent guarantee of that tenant's assignee. Interpretation (i) would thus appear to give the 1995 Act an unattractively limiting and commercially unrealistic effect.

[37] As Newey J accepted, interpretation (i) would mean that, even where it suited the assignor, the assignee and the guarantor that the assignee should have the same guarantor as the assignor (because, for instance, the assignor and the assignee had the same parent company, or shared a common bank, which was the guarantor), they could not offer that guarantor. It would also appear to mean that the lease could not be assigned to the guarantor, even where both tenant and guarantor wanted it. Lord Nicholls said, in *London Diocesan Fund*, in [16], that section 5 was "intended to benefit tenants… . That is [its] purpose. That is how [it is] meant to operate." So, too, section 24(2) is meant to benefit guarantors. It can therefore be argued that, where the assignor and the guarantor who want the guarantor to guarantee an assignee, or who want the lease to be assigned to the guarantor, such a renewal, or such an assignment, would not "frustrate the operation of any provision of [the 1995 Act]".

[38] All this provides some support for the contention that a guarantee of the assignee by the assignor's guarantor may not fall foul of section 25(1) if it is freely offered by the assignor and guarantor. On the other hand, there is much to be said for interpretation (i). It leads to a clear and simple position. It avoids argument, after, possibly long after, the subsequent guarantee has been given, as to whether it really had been insisted on by the landlord or freely offered by the assignor. Such problems could be particularly acute where the landlord's interest had been assigned after the subsequent guarantee had been obtained, and the current landlord therefore may have no knowledge of the circumstances in which it had been obtained.

[39] Interpretation (ii) is, in many ways, more attractive in terms of commercial sense, as it is consistent with the idea that that section 25 is to be treated as an anti-avoidance provision that (in a case involving

A an assignment of the tenancy rather than the reversion) protects the assignor and its guarantor, but does not work to their disadvantage.

[40] Thus, in a case of a covenant against assignment subject to landlord's consent, interpretation (ii) would prevent the landlord insisting on the assignor's guarantor guaranteeing an assignee, as that would involve requiring an arrangement in the future whereby the present guarantor was not released. However, it would not prevent the landlord, if it was reasonable to do so, from requiring a guarantor of the assignee, and the assignor then proposing the present guarantor for that role. Such a course would involve the landlord imposing a requirement that did not fall foul of section 25(1), but the reason for the subsequent
B guarantee from the existing guarantor would be because the assignor offered it freely (that is, following a reasonable and unobjectionable request from the landlord), which, on interpretation (ii), would result in the subsequent guarantee not being invalidated by section 25(1).

[41] However, as already mentioned above, section 25(1) appears to be concerned with "the effect" of a particular agreement, which suggests that what it aims to invalidate is the objective effect of an agreement, rather than its subjective purpose. It does appear that interpretation (ii) runs rather counter to that approach, as it looks to the subjective reason for the guarantor guaranteeing the assignee, rather than its objective effect.

C [42] Interpretation (ii) also carries with it the problem that, unlike interpretation (i), it would not be enough to look at the documents of title to see whether section 25 applies. The fact that the assignment included a guarantee from the existing guarantor may mean that section 25 is engaged, but it would be necessary to look at the negotiations leading to that guarantee in order to see if it had been insisted on by the landlord, or freely offered by the assignor and/or guarantor. Such an exercise may raise difficult points both of principle and on the facts, as the negotiations may have been conducted orally as well as in writing, and there may have been nuances in the negotiations that render it difficult to decide whether the landlord was insisting on
D the subsequent guarantee or it was being freely offered.

[43] So we have to choose between interpretation (i) and (ii). (It is right to add that, in addition to interpretations (i) and (ii), we have considered whether section 25(1) should be given a rather narrower effect, so that its ambit is limited to agreements that precede the subsequent guarantee, but it could not invalidate any subsequent guarantee. It seems to us that, although it has its attractions, this interpretation cannot be justified. First, subsections (1), (2) and (4) of section 25 are concerned with agreements, not conditions or imposed terms; second, the exclusion in section 25(3) of an AGA, as opposed to a requirement that an assignor enters into an AGA, seems inconsistent
E with this interpretation; third, section 16(1) also envisages that an AGA, rather than a requirement for an AGA, would, in the absence of the section, be ruled out by section 25, as would an agreement that, although similar to an AGA, for some reason does not comply with the requirements of section 16.)

[44] In our view, interpretation (i) is correct, albeit subject to a very important qualification, which relates to AGAs under section 16. It appears to us that the principled objections to interpretation (ii) are correct. Further, the qualification we have mentioned removes most of the unsatisfactory commercial consequences to which interpretation (i) appears otherwise to give rise.

F [45] As already mentioned, in almost all cases (save relatively freely assignable long leases at low rents), leases are assignable only with the landlord's consent, not to be unreasonably withheld. Where consent is not required, the lease is normally pretty freely assignable or wholly unassignable, so the 1995 Act is unlikely to present any problem. In the normal case, where the landlord's consent (not to be unreasonably withheld) is required for an assignment, the landlord can, if it is reasonable to do so, require the tenant to enter into an AGA: see section 16(3). That plainly qualifies the extent to which the assignor is released from its liability in accordance with section 5(2), as sections 16(1) and 25(3) recognise, because, instead of being released from liability under the lease, such liability is immediately reimposed on it.

[46] By section 24(2), on any assignment, a guarantor of the assignor is required to be released only "to the same extent as the tenant". Accordingly, if, where section 16(2) applies, the landlord is entitled to require the assignor to reassume liability under an AGA, it does not appear to us to be inconsistent with section 24(2), and hence it would not be void under section 25(1), for the landlord in such a case to require the guarantor to guarantee the liability of the tenant under the AGA. Where an assignor is lawfully required to enter into an AGA, then, when it assigns the lease, it is released from its obligations under the lease, save to the extent that it reassumes those obligations under the AGA. There appears to be nothing inconsistent with section 24(2) if the assignor's guarantor is required to guarantee the assignor's liability under the AGA: the guarantor is released to precisely the same extent as the assigning tenant.

[47] (It may well be that the guarantor could simply act as a coguarantor under the AGA with the assignor, as might have been the position in *Good Harvest* see [5] rather than being a guarantor of the assignor's liability, under the AGA. However, that issue is one that we should not determine because it involves going much further than we need on any view, although there seems to us to be much force in Mr Randall's point that, if the assignor's guarantor can guarantee the assignor's liability under an AGA, it is hard to see why that guarantor should not be able "to do something to the same substantial effect, different only in form": see [41].

[48] The conclusion that the assignor's guarantor can be required to guarantee the liability of the assignor under an AGA appears to comply with commercial sense and to avoid many of the problems that are otherwise thrown up by interpretation (i). It is also consistent with the purpose of an AGA. Take the facts of the present case: Victoria granted the lease to Management, an impecunious subsidiary of a substantial company, HoF, taking HoF as a guarantor. On an assignment to another company outside the HoF group, it would be of very little value to Victoria to have the right to ask for an AGA from Management, unless it could require HoF to guarantee Management's liability under the AGA. It would therefore be positively surprising if section 16 extended to an assignor but not to the assignor's guarantor.

[49] Newey J appears to have left open the question of whether the assignor's guarantor could validly guarantee the liability of the assignor under an AGA: see [22(vii)]. However, for the reasons we have given, in the light of the terms of sections 5(2), 16(1), (2), 24(2) and 25(1), although the guarantor of an assignor cannot normally validly guarantee the liability of the assignee, it can validly do so by being party to a valid AGA.

[50] We would add that, if (which we doubt) the reasoning of Newey J was predicated on the ground that a provision such as clause 5.9.6.1, which entitled the landlord to require the tenant and its guarantor to enter into an AGA on any assignment, was invalid, we do not agree. Section 16(3) may or may not preclude a landlord relying on a provision in an alienation covenant that purports to entitle it to insist on an AGA as a matter of right on any assignment (we do not need to decide the point and should not do so). However, if it does preclude the landlord having such an absolute right, so that it is ineffective, that would not prevent the landlord insisting on an AGA, if the alienation covenant also contained a provision that consent to an assignment could not be unreasonably withheld (like clause 5.9.6.6 in *Good Harvest*), and it was reasonable to require an AGA.

[51] This conclusion does not mean that the proviso to clause 3.15(F) of the lease in this case is unenforceable in respect of the contemplated assignment by Stores back to Management. It is true that it would involve HoF, the guarantor of the original tenant, Management, which would have been released on the assignment to Stores under clause 3.5(ii) (because clause 3.5(iii) was unenforceable), becoming liable again under the lease, as guarantor of Stores' liability. If a tenancy, granted to a tenant whose obligations are guaranteed by a guarantor, is assigned, and thereafter the assignee assigns it back to the tenant, whose obligations are again guaranteed by the guarantor, it seems unlikely that section 25(1) could possibly have been intended to invalidate the guarantor's renewed guarantee or the tenant's renewed assumption of

liability under the tenancy. In such a case, the original tenant and the original guarantor are released from liability under the tenancy on the first assignment, and the fact that they choose subsequently, namely on a further assignment, to reassume liability under the lease cannot be said to "frustrate" their release on the first assignment.

[52] However, it is only right to add that a provision such as the proviso to clause 3.15(F) would, on our analysis, appear to have a rather capricious effect if it was invoked to effect an assignment between companies in the same group. If, on the present facts, and ignoring clause 3.5, Management wished to assign to Stores, it seems to us that Victoria could not require HoF to guarantee Stores' liability, because that would fall foul of section 25(1) for the reasons discussed. Whether that would undermine Management's ability to rely on clause 3.15(F) we do not need to decide. However, if it did not, and Management then assigned to Stores, without HoF guaranteeing Stores' liability, and Stores then wished to assign on to another HoF company (such as Management or Beatties), in reliance on clause 3.15(F), Victoria could invoke the proviso and insist on HoF guaranteeing the new assignee's liabilities.

[53] Reverting to the general effect of the 1995 Act on existing guarantors, our conclusion, as explained above is that the effect of sections 24 and 25 is that:

(i) an existing or contracting guarantor of a tenant cannot validly be required to commit itself in advance to guarantee the liability of a future assignee;

(ii) subject to (iii) and (iv), a guarantor of an assignor cannot validly guarantee the liability of the assignor's assignee;

(iii) such a guarantor can validly do so by being party to an AGA that otherwise complies with section 16; and

(iv) such a guarantor can in any event validly guarantee the liability of an assignee on a further assignment.

We would hope that those responsible for drafting leases are aware of these conclusions, and that, as a result, the 1995 Act should not lead to many practical difficulties of the sort discussed above.

[54] Accordingly, we largely agree with Newey J's analysis in *Good Harvest*, save in respect of AGAs (point (iii) in the previous paragraph), which he appears to have left open. We have been much assisted by his analysis of the effect of the 1995 Act, as well as by Mr Randall's discussion of that analysis in the judgment under appeal.

Interpretation of clause 3.15(F)

[55] Turning now to the second issue, the HoF defendants' contention is that clause 3.15(F) is a "stand-alone" provision, which operates independently of the other parts of the clause, so that an assignment is lawful if the requirements of clause 3.15(F) alone are fulfilled. If that is right, then, even if the lease is assigned to Stores pursuant to clause 3.5(ii), it could freely be reassigned straight back to Management, and so, argue the HoF defendants, it would therefore be pointless to order specific performance of clause 3.5(ii).

[56] Mann J did not accept that clause 3.15(F) was a "stand-alone" provision. He concluded that the removal in clause 3.15(F) of the need for consent means that assignments within the HoF group do not need to comply with paras (D) or (C)(ii) of that clause, but that they did, however, need to comply with para (E) and, arguably, with para (C)(i). He did not have to resolve whether para (C)(i) would apply because, on the facts, the requirements of para (E) could not be satisfied: Stores is a valuable company with a good balance sheet, whereas HoF has a negative balance sheet and Management's covenant is not and has never been worth having. The judge concluded, therefore, that the threatened reassignment would be a breach of the provisions of clause 3.15.

[57] Mr Seitler attacked the judge's analysis on several grounds. He made a general criticism that the judge had subjected the wording of clause 3.15 to examination under a forensic microscope (adapting the language of Sir Thomas Bingham MR in *Arbuthnott* v *Fagan* [1995] CLC 1396), resulting in a strained and uncommercial result. Mr Seitler's overriding point was that the judge had failed to give proper weight to the opening words of para (F) of clause 3.15 "Notwithstanding the provisions of this clause" which, he said, showed that the paragraph is intended to operate independently of all the preceding provisions of G
the clause and is to be "free-standing". Mr Seitler drew attention to similar wording in clause 3.15(N) (relating to parting with possession and occupation), and said that the literal meaning of the words could not be ignored. He submitted that, consistently with that literal meaning, "consent" in para (F) refers to each of paras (C) to (E) of clause 3.15, all of which relate to what is required to be satisfied when consent is necessary.

[58] Ignoring para (B) of clause 3.15, which, as both parties agree, is an absolute prohibition applicable to all assignments, Mr Seitler adopted the judge's view that paras (C) and (D) are cumulative and must be read together; the requirements of para (D) are excluded from H
an assignment within para (F) because para (D) is expressly dealing with consent; and the requirements of paras (C)(ii) are similarly excluded because, as is made clear in section 16(3)(a) of the 1995 Act, an AGA arises only in the context of a covenant not to assign without the landlord's consent. Mr Seitler submitted that, since para (C)(ii) has no application to an assignment within para (F), then nor can para (C)(i) since they are stated to be alternatives. Furthermore, he contended, the fact that both paras (C) and (F) are dealing with the same subject matter, namely guarantees and expressly in the case of para (C) and implicitly in the case of para (F) financial standing, shows that their requirements are not intended to be cumulative but separate. J

[59] Turning to para (E), Mr Seitler advanced a number of reasons supporting his argument, based on his overarching point on the opening words of para (F), that it is not cumulative with para (F). He pointed out that, unlike para (F), para (E) does not begin with the words "Notwithstanding the provisions of this clause", thereby both showing that para (E) is cumulative with para (D), and, at the same time, that para (F) is not cumulative with para (E). He also pointed out that both paras (E) and (F) are dealing with the same subject-matter, namely assignments within a group of companies. He said that para (F) is, therefore, naturally to be read as giving a special concession to the HoF Group, reducing the requirements that would otherwise apply to K
such assignments. He further submitted that, as in the case of paras (F) and (C), on the one hand, and paras (E) and (C), on the other hand, the fact that para (F) is dealing with the same subject matter as para (E), namely guarantees and (explicitly or implicitly) financial standing, also shows that their requirements were not intended to be cumulative but alternatives.

[60] The HoF defendants also say that, if and so far as necessary, the restrictions in clause 3.15 should be construed in favour of the tenant since they deprive the tenant of what would otherwise be an important incident of the normal right of a tenant to turn the tenancy to account by alienation. They submit that the presumption that it was not L
intended to deprive the tenant of that right should have been applied by the judge once he had concluded that the clause was unsatisfactorily drafted and unclear.

[61] Mr Seitler submitted that the judge's interpretation produces a result that is contrary to commercial common sense. Instead of transfers within the HoF group being subject to a lower level of restriction on alienation than transfers outside the HoF group, which was the manifest intention, to read para (F) as cumulative with paras (C) and (E) would make such assignments subject to more onerous requirements and so more difficult.

[62] We agree with the judge and Mr Seitler that para (F) necessarily M
excludes the requirements of paras (D) and (C)(ii) since those covenants are directly and exclusively concerned with the landlord's consent to an assignment. The critical issue is whether the requirements of para (F) stand alone or are cumulative with those in paras (C)(i) and (E).

[63] Both paras (E) and (F) are dealing with transfers within a group of companies. Although it is true that para (E), like para (F), addresses the issue of guarantors in different terms, there is nothing inherently impossible about reading its requirements as cumulative with those in para (F). Furthermore, to do so would still leave para (F) conferring an advantage on the HoF group when compared with assignments within other groups of companies. As Mr Speaight pointed out in his admirably succinct oral submissions, the effect of para (F) is to preclude

A the landlord from objecting to an assignment within the HoF group on matters not addressed in para (C) or para (E), such as the fact that the assignee is a foreign subsidiary or there are outstanding breaches of covenant, such as disrepair.

[64] So far as the language of clause 3.15(F) is concerned, two points are worth making. First, we do not accept that the opening words of clause 3.15(F) naturally mean that all the preceding paragraphs of the clause must be disregarded. Although they could have that meaning, what they more naturally imply is that, in so far as any express provision of para (F) is inconsistent with what is in a previous paragraph, the terms of para (F) prevail. Further, if the opening words of para (F) had the
B meaning contended for by Mr Seitler, it is hard to see why the paragraph goes on to refer to the absence of any need for consent. Second, while the concept of "consent" mentioned in the paragraph could refer, as Mr Seitler contended, to everything in the preceding paragraphs, it seems to us that the word is at least as apt to be naturally read as referring to the consents mentioned to in paras (D)(i) and (C)(i).

[65] Critically, we agree with the judge that the commercial context in which the lease was granted eliminates any doubt on the point. The agreement and the lease are composite parts of a single commercial transaction, namely the sale and leaseback of the premises. The annual rent of £2.25m (subject to five yearly rent reviews) was a critical
C commercial aspect of the transaction under which Victoria paid the purchase price of £46m. The financial ability of the tenant to pay that rent would obviously be a matter of the central concern to Victoria, as would the financial standing of the tenant's guarantor. That concern is reflected in clause 3.5. Management was known to be a company with a worthless covenant at that time, whereas Stores and HoF Ltd were financially strong. HoF wanted Management to be the first tenant in order to assist HoF's tax position. Victoria agreed, but only on the basis that clause 3.5 was included in the agreement. It would make commercial nonsense of that vital part of the deal if, following the assignment from Management to Stores, the purpose of clause 3.5 could
D be rendered futile by an immediate reassignment of the lease back to Management without Victoria's consent. The commercial objective underlying clause 3.5 would be achieved if paras (E) and (F) were read cumulatively; but it would be frustrated if they were not.

[66] Mr Seitler submitted that there is a conflict between clause 3.5 and clause 3.15(F), and that conflict must be resolved in favour of para (F) since, once the sale was completed and the lease was granted, the agreement was spent. We cannot accept that contention. In the first place, the argument assumes a conflict, and the anterior question is whether there is a conflict. In any event, clause 3.5 was intended to operate precisely because and while the lease is on foot. Accordingly, the provisions of both the lease and clause 3.5 are to be interpreted in a way that gives effect to the commercial intention of the parties underlying both of them, if it is possible to do so consistently with their language.

[67] Mr Speaight submitted that, in addition to para (E), both limbs of para (C) are to be read cumulatively with para (F). As we have said, the judge did not consider he had to resolve that issue, and nor do we. It is sufficient if, as we find, (E) is to be read cumulatively with (F). However, it is right to say that, as at present advised, we see considerable force in the argument that para (C)(i) is difficult to reconcile with para (F), and that, if para (C)(i) is not applicable to an assignment falling within para (F), then para (C)(ii) should not be applicable either.

[68] There is a respondent's notice on behalf of Victoria seeking to uphold the judge's decision on the further ground that the *contra proferentem* rule should apply, so that any uncertainty concerning the interpretation of para (F) should be resolved in favour of Victoria, since the paragraph was introduced at the request of the HoF defendants and it favours them. Like the judge, we consider that there is no need to resort to the rule in order to resolve the issue of interpretation in favour of Victoria: such rules are rarely, if ever, of any assistance when it comes to construing commercial contracts. Quite apart from raising abstruse issues as to who is the *proferens* (and, in particular, whether the issue turns on the precise facts of the case or hypothetical analysis), "rules" of interpretation such as *contra proferentem* are rarely decisive as to the meaning of any provisions of a commercial contract. The words used, commercial sense, and the documentary and factual context, are, and should be, normally enough to determine the meaning of a contractual provision.

[69] Accordingly, for substantially the same reasons as Mann J in his careful judgment, we consider that clause 3.15(E) could be relied on by Victoria if, after taking an assignment of the lease, Stores sought to invoke clause 3.15(F) and assign the lease back to Management.

Conclusions

[70] For these reasons, we would dismiss the appeal from: (a) the decision of Mr Randall that clause 3.5(iii) of the agreement is void; and (b) the decision of Mann J that, where 3.15(F) of the lease applies, the application of clause 3.15(E) is not excluded.

Appeals dismissed.

Hashemi *v* Gladehurst Properties Ltd

COURT OF APPEAL

[2011] EWCA Civ 604

19 May 2011

CARNWATH AND PATTEN LJJ AND BARON J

Estates Gazette 23 July 2011

[2011] 29 EG 90

Landlord and tenant — Housing Act 2004 — Tenancy deposit scheme — Assured shorthold tenancy — Landlord failing to deal with tenant's deposit in accordance with authorised scheme — Tenancy ending — Section 214(3) and (4) of 2004 Act — Whether tenant entitled to three times deposit amount as sanction for landlord's failure to comply with obligations under section 213 — Whether sanction ceasing to be available on termination of tenancy

The respondent was one of two tenants of a flat let to them by the appellant in 2007 on an assured shorthold tenancy. The tenants paid an initial deposit of £6,240 on signing the tenancy agreement. The appellant did not deal with the deposit in accordance with an authorised scheme, as required by section 213 of the Housing Act 2004, but retained it in its own bank account until October 2008, when the tenants vacated the flat. The appellant retained the sum of £1,123.99 from the deposit for various breaches of the tenants' cleaning and repairing covenants, as it was entitled to do under the terms of the tenancy agreement, and repaid the remainder to the tenants.

Thereafter, the respondent brought a claim against the appellant for payment of three times the deposit sum as the sanction, under section 214(4) of the 2004 Act, for its failure to comply with the initial requirements of an authorised deposit scheme or to provide the prescribed information in that regard. He also claimed the sum of £618 representing certain disputed items of the appellant's deposit retention. The respondent sued both on his own behalf and for the other tenant.

A district judge struck out the claim under section 214(4) on the ground that section 214 had ceased to apply once the tenancy had come to an end. The respondent's appeal against that decision was allowed by a judge, who reinstated the claim and gave judgment on it for the respondent. The judge made an order for the return of the deposit under section 214(3), on terms that recited the amounts already returned, those validly retained and those still in dispute, and further ordered payment of three times the deposit amount under section 214(4). The appellant appealed.

***Held*: The appeal was allowed. The court could no longer make an order under section 214(3) and (4) once the tenancy had come to an end. The initial requirements of an authorised scheme are matters to be dealt with at the inception of the lease and not later than the expiry of the term. It is up to the tenant to bring proceedings under section 214 to enforce the requirements imposed on the landlord by section 213. Any default by the landlord in that respect may be remedied and, if it is not, the court is required by section 214(3) either to order the repayment of the deposit to the tenant or to order the landlord to pay it into an authorised scheme. For the court to have a genuine discretion to exercise, both alternatives must be available. Section 214(1)(a) should therefore be read as meaning that the tenant may apply to the court where the initial requirements of an authorised scheme have not been, but could still be, complied with. That is the only meaning that ties in with the two alternatives in section 214(3) continuing to be available. Accordingly, the grounds for a section 214 application will cease to exist once the lease expires and no order under either section 214(3) or (4) can be made after that date.**

The following case is referred to in this report.

Tiensia v *Vision Enterprises Ltd (t/a Universal Estates); Honeysuckle Properties* v *Fletcher* [2010] EWCA Civ 1224; [2011] 1 All ER 1059; [2010] 3 EGLR 53; [2010] 49 EG 80; [2011] HLR 10

This was an appeal by the appellant, Gladehurst Properties Ltd, from a decision of HH Judge Cryan, sitting in Clerkenwell and Shoreditch County Court, allowing an appeal by the respondent, Farid Hashemi, against an order of Judge Stary striking out a claim, under section 214 of the Housing Act 2004, for three times the amount of the deposit on an assured shorthold tenancy.

Laura West (instructed by Bude Nathan Iwanier) appeared for the appellant; Kevin Gannon (instructed by Osbornes Solicitors) represented the respondent.

Giving judgment, **Patten LJ** said:

[1] Gladehurst Properties Ltd (Gladehurst) is the landlord of flat 8, 238 City Road, London EC1V 2PQ, which it let to Mr Farid Hashemi and Mr Matthew Johnson under an assured shorthold tenancy agreement dated 8 August 2007. The tenancy was granted for a fixed term of one year commencing on 6 September 2007 subject to a break clause under which either the landlord or the tenants could terminate the lease after six months by giving two months' notice in writing.

[2] The rent payable under the tenancy was £2,080 per calendar month. It was payable in advance on the sixth day of each month. Under clause 1 of the general tenancy conditions, the tenants covenanted to keep the interior of the flat in a good and clean state of repair and condition and to deliver up the premises in that state at the end of the term. They also paid a deposit of £6,240 that was held by the landlord as stakeholders on the following terms:

> The Tenant shall pay to the Landlord's Agent as stakeholder on the signing of this Agreement the amount of the Deposit and the first payment of Rent provided always that the Landlord may, at the end of the tenancy, howsoever determined apply to the stakeholder to deduct and pay over from the Deposit whatever monies may be due to him from the Tenant either against the Inventory or by virtue of any breach by the Tenant of any of his obligations under this Agreement including rent owed to the Landlord including rent in advance which has fallen due and any sum expected by the Landlord in remedying any breaches of this Agreement by the Tenant.
>
> As soon as practicable after the determination of the tenancy (howsoever the same may be determined) the Stakeholder shall retain (and account to the Landlord) such part of the Deposit as the Landlord's Agent shall deem necessary to enable the Landlord as at the date of such determination to make good any breach or non-compliance by the Tenant with his obligations hereunder and pay all costs, charges and expenses incurred in connection therewith including the Landlord's Agent's reasonable fees and disbursements and shall account to the Tenant for any balance of such sum. If the deposit shall be insufficient for the purposes aforesaid the Tenant shall pay to the Landlord forthwith on

demand such further sum as shall in the opinion of the stakeholder be required for such purposes.

[3] The Housing Act 2004 (the 2004 Act) introduced provisions for the creation of tenancy deposit schemes, which were designed to put an end to complaints by residential tenants that their deposits had been unreasonably withheld by their landlords at the termination of the lease or, in some cases, had even been misappropriated. Until 2000, the taking of deposits (which are a common feature of short residential tenancies) was entirely unregulated. In March 2000, a voluntary scheme was set up by the Independent Housing Ombudsman but it was not widely used and subsequently closed down. The government then decided to legislate.

[4] The purpose of tenancy deposit schemes is to safeguard the deposit during the term of the lease and then to provide a machinery to resolve any disputes that arise about the proportion of the deposit that the landlord is entitled to retain to meet the tenant's liability for rent or any breaches of covenant. To this end, section 212 of the Act requires the appropriate national authority to arrange for one or more tenancy deposit schemes to be made available. The schemes must comply with the requirements of Schedule 10 to the Act, which contains detailed provisions concerning the contents of the schemes and what they must provide for on the termination of the tenancy.

[5] I shall return later to some of those provisions but, for the moment, it is enough to record that tenancy deposit schemes must take one of two forms: either a custodial scheme under which the deposit is paid to the landlord and then transferred to a designated account held by the scheme administrator or an insurance scheme under which the landlord retains the deposit but pays over to the scheme administrator at the end of the lease any part of the deposit that is in dispute between the landlord and its tenant. The administrator is then responsible for paying out the disputed sums under the scheme, subject to a right of reimbursement from the landlord, which is backed by insurance: see para 1 of Schedule 10.

[6] The Act requires all deposits to be dealt with in accordance with an authorised scheme. Section 213 provides that:

> (1) Any tenancy deposit paid to a person in connection with a shorthold tenancy must, as from the time when it is received, be dealt with in accordance with an authorised scheme.
>
> (2) No person may require the payment of a tenancy deposit in connection with a shorthold tenancy which is not to be subject to the requirement in subsection (1).
>
> (3) Where a landlord receives a tenancy deposit in connection with a shorthold tenancy, the initial requirements of an authorised scheme must be complied with by the landlord in relation to the deposit within the period of 14 days beginning with the date on which it is received.
>
> (4) For the purposes of this section "the initial requirements" of an authorised scheme are such requirements imposed by the scheme as fall to be complied with by a landlord on receiving such a tenancy deposit.
>
> (5) A landlord who has received such a tenancy deposit must give the tenant and any relevant person such information relating to
>
> (a) the authorised scheme applying to the deposit,
>
> (b) compliance by the landlord with the initial requirements of the scheme in relation to the deposit, and
>
> (c) the operation of provisions of this Chapter in relation to the deposit,
>
> as may be prescribed.
>
> (6) The information required by subsection (5) must be given to the tenant and any relevant person
>
> (a) in the prescribed form or in a form substantially to the same effect, and
>
> (b) within the period of 14 days beginning with the date on which the deposit is received by the landlord.
>
> (7) No person may, in connection with a shorthold tenancy, require a deposit which consists of property other than money.
>
> (8) In subsection (7) "deposit" means a transfer of property intended to be held (by the landlord or otherwise) as security for
>
> (a) the performance of any obligations of the tenant, or
>
> (b) the discharge of any liability of his,
>
> arising under or in connection with the tenancy.
>
> (9) The provisions of this section apply despite any agreement to the contrary.
>
> (10) In this section
>
> "prescribed" means prescribed by an order made by the appropriate national authority;
>
> "property" means moveable property;
>
> "relevant person" means any person who, in accordance with arrangements made with the tenant, paid the deposit on behalf of the tenant.

[7] Clause 8.1(i) of the tenancy agreement provided that:

> The Tenant agrees to pay the Landlord's costs of and in connection with:
>
> (i) The sum of £176.25 being £150 + VAT for the set up costs of this tenancy including the preparation and completion of this contract together with the checking of the said Inventory at the commencement of the tenancy. A further £100 + VAT will become payable if the tenant wishes the Landlord to register the deposit with the government deposit scheme.

[8] But the deposit that the tenants paid was never registered or paid into a tenancy deposit scheme. Instead, it was retained by Gladehurst in its own bank account until October 2008 when the tenants vacated the flat. There were various breaches of the cleaning and repairing covenants outstanding at the end of the lease and Gladehurst arranged for an independent inventory clerk to inspect the flat and to assess the cost of cleaning and making good the disrepair. He recommended the deduction of £1,123.99 from the deposit. This was therefore retained by Gladehurst and the balance of £5,116.01 was repaid to the tenants via a bank transfer to their nominated account.

[9] In November 2008, Mr Hashemi contacted Gladehurst and requested receipts and a breakdown of the amount deducted from the deposit. At the same time, he put the landlord on notice that he would issue a claim for three times the deposit as provided for under section 214 of the 2004 Act. Section 214 provides that:

> (1) Where a tenancy deposit has been paid in connection with a shorthold tenancy, the tenant or any relevant person (as defined by section 213(10)) may make an application to a county court on the grounds
>
> (a) that the initial requirements of an authorised scheme (see section 213(4)) have not, or section 213(6)(a) has not, been complied with in relation to the deposit; or
>
> (b) that he has been notified by the landlord that a particular authorised scheme applies to the deposit but has been unable to obtain confirmation from the scheme administrator that the deposit is being held in accordance with the scheme.
>
> (2) Subsections (3) and (4) apply if on such an application the court
>
> (a) is satisfied that those requirements have not, or section 213(6)(a) has not, been complied with in relation to the deposit, or
>
> (b) is not satisfied that the deposit is being held in accordance with an authorised scheme,
>
> as the case may be.
>
> (3) The court must, as it thinks fit, either
>
> (a) order the person who appears to the court to be holding the deposit to repay it to the applicant, or
>
> (b) order that person to pay the deposit into the designated account held by the scheme administrator under an authorised custodial scheme,
>
> within the period of 14 days beginning with the date of the making of the order.
>
> (4) The court must also order the landlord to pay to the applicant a sum of money equal to three times the amount of the deposit within the period of 14 days beginning with the date of the making of the order.
>
> (5) Where any deposit given in connection with a shorthold tenancy could not be lawfully required as a result of section 213(7), the property in question is recoverable from the person holding it by the person by whom it was given as a deposit.
>
> (6) In subsection (5) "deposit" has the meaning given by section 213(8).

[10] Both parties accept that the tenancy came to an end when the respondents vacated the premises in October 2008. The principal issue on this appeal is whether the tenants are entitled to a judgment in a sum equal to three times the deposit, which the court is obliged to order the landlord to pay under section 214(4). A claim form was issued in the names of both tenants and served on 16 February 2009. The defence filed referred to the tenants not having taken up the option of registering the deposit, which is referred to in clause 8(i) of the tenancy agreement, but it is now accepted that this is not a defence to a breach by a landlord of its obligations under section 213 of the Act.

[11] The claim was considered on paper by District Judge Manners on 2 April 2009. She struck out the claim on the ground that it had been made after the tenancy had come to an end and that the provisions of section 214 no longer therefore applied. The tenants applied to set that order aside and their application was heard by District Judge Stary on 9 July 2009 when Mr Hashemi appeared in person. The district judge dismissed the application in so far as it related to the claim for three times the deposit but allowed the claim to stand in respect of the sum of £618 out of the £1,123.99 retained by Gladehurst from the deposit.

[12] The £618 represents the cost to the landlord of repainting the flat and of various other items of repair. These are disputed by the tenants. However, they accept that the landlord was entitled to deduct unpaid water rates; unpaid inventory charges and the cost of cleaning the premises, all of which totalled £505.99. The district judge ordered that the dispute about the £618 should continue as a small claim and gave directions for a new defence to be filed in respect of that.

[13] The tenants appealed against the striking out of their claim for three times the amount of the deposit. This was heard by HH Judge Cryan, who allowed the appeal; restored the claim under section 214(4) of the Act; and gave judgment for the tenants in respect of the deposit and for a further sum of £18,720 representing the section 214(4) claim. He also ordered that the balance of the claim in respect of the sum of £618 should be stayed for a period of 28 days and struck out if no application to restore it was made by the tenants within that period.

[14] Gladehurst now appeals against the judge's order with the permission of Arden LJ. The principal ground of appeal relates to the order for payment of the £18,720 but the landlord raises two further issues that I propose to deal with before coming to the substance of the appeal. These relate to Mr Hashemi's authority to bring the claim in the name of himself and Mr Johnson and the correctness of the judge's order for the return of the deposit without deduction.

Authority to sue

[15] The claim form was issued in the name of both tenants but did not contain a statement of truth signed by them. Ms Laura West, who appeared for Gladehurst, submitted to Judge Cryan that Mr Johnson appeared to have taken no part in the proceedings and that Mr Hashemi was conducting them on his behalf with no apparent authority to do so. Mr Johnson had not instructed the solicitor that acted for them; had not filed a witness statement or other documents in the proceedings and had had no direct communication with the court or with her clients. The same applied to the appellant's notice filed in respect of the order of District Judge Stary.

[16] The issue about Mr Johnson's involvement in the proceedings and the appeal was important because it is, I think, common ground that any remedy that the tenants are entitled to either under the lease or under section 214 is enjoyed by them as joint tenants and can be enforced only by both of them. In these circumstances, CPR 19.3(1) requires both of them to be parties to the action unless the court orders otherwise.

[17] At the hearing of the appeal, Judge Cryan gave leave for Mr Hashemi to rely on a witness statement from Mr Johnson dated 6 December 2009 in which he says that he was consulted by Mr Hashemi about the claim and agreed that they should issue the proceedings. He confirms that he is content for Mr Hashemi to pursue the appeal on the basis that they will split between them any damages that are awarded.

[18] The judge was satisfied on the basis of this evidence that Mr Hashemi did have authority to instruct a solicitor to issue the claim in the names of both tenants. But to correct any technical defect in the proceedings, he ordered that Mr Hashemi should have permission under CPR 19.6 to continue the proceedings on behalf of himself and Mr Johnson.

[19] Gladehurst does not seek to challenge the veracity of Mr Johnson's evidence on this appeal but Ms West submitted that the judge erred in making the representation order. Neither Mr Hashemi nor Mr Johnson had applied for such an order and the making of the order G led to unfairness.

[20] I do not think that there is anything in this point. Like Ms West, I think that the effect of Mr Johnson's evidence was in fact to confirm that he was always a party to the action. He had authorised the commencement of the proceedings in his name and his failure to sign a statement of truth on the claim form or the appellant's notice does not invalidate that. It was therefore unnecessary for the judge to make a representation order. However, it has caused the landlord no prejudice because the effect of such an order is to allow the court to make orders (including orders for costs) that are binding on both the representative party and on those whom he represents. I can see no reason therefore to H set aside the judge's order on this point.

Return of the deposit

[21] This is more complicated. As mentioned earlier, the parties were agreed that of the £1,123.99 deducted from the deposit, only £618 is in dispute. The claim form did not include a claim for the return of the deposit but the district judge allowed the tenants to proceed with a claim for the £618 in the small claims court. On the first appeal, Judge Cryan accepted that only this sum was in dispute out of the original deposit and that £5,116.01 had already been repaid to the respondent. However, he expressed doubts as to whether the orders that he regarded himself as bound to make under section 214(3) and (4) could be subject to a J set-off in respect of what was due contractually to the landlord out of the deposit.

[22] The order that was made therefore recites that: (i) the £5,116.01 has already been paid to the tenants; (ii) they accept that £505.99 of the deposit is due to the landlord; and (iii) the landlord is to retain the £618 pending the resolution of the tenants' claim for that sum. But it then goes on and orders Gladehurst to pay both the £18,720 and the £6,240 to the tenants and merely stays the tenants' claim for the £618 on the terms referred to above.

[23] Ms West submitted that the orders made are inconsistent with the recitals and highlight the artificiality of the court being required K to order the refund of the deposit after the lease has come to an end and the landlord has repaid to the tenant what he accepts is due. Mr Kevin Gannon, for the tenants, said that the order was simply the obvious way of producing a practical solution for the circumstances of this case. The recitals make it clear that the tenants cannot seek to enforce the order for the repayment of the whole of the deposit and that the £618 remains to be litigated (if at all) in further proceedings.

[24] Although Mr Gannon is right about the effect of the order, it is, I think, overcomplicated. Even if the landlord is not entitled to a set-off in respect of its contractual claims, I cannot see why the agreed retention of the £505.99 and the agreed repayment of the £5,116.01 L should not be reflected by a reduction in the order for repayment of the £6,240. The order as drawn also fails to make it clear that if the tenants fail to pursue their small claim for the £618, this sum should also be credited to Gladehurst. The order should therefore be amended to that extent as least regardless of the outcome of the main ground of appeal.

Section 214

[25] The mandatory provisions of section 214(3) and (4) apply if two conditions are satisfied. First, the court must be seized of an application by "the tenant" under section 214(1); and second, the court must be satisfied on that application either that the initial M requirements of an authorised scheme (as defined by section 213(4)) have not been complied with or that section 213(6)(a) (the landlord's obligation to give the tenant the prescribed information relating to the authorised scheme applicable to the deposit and his compliance with the initial requirements of the scheme) has not been complied with.

[26] The grounds on which a tenant may make a section 214 application include the landlord's failure to observe and comply with the initial requirements as defined and the provisions of section 213(6)(a). One might therefore suppose that once the 14-day time limit set by section 213(3) had passed, the tenant's cause of action was, so to speak, complete. But in *Tiensia* v *Vision Enterprises Ltd (t/a*

Universal Estates) [2010] EWCA Civ 1224; [2011] HLR 10*, this court held (Sedley LJ dissenting) that the sanction provisions contained in section 214 are focussed on the date of the hearing so that, where the landlord has by then complied with the initial requirements and with section 213(6)(a), the court has no power to make the orders described in section 214(3) and (4). The position (as summarised by Rimer LJ in [36] of his judgment) is that:

> the pre-condition of a tenant's application to the court under section 214 is not a failure by the landlord to comply with the "initial requirements" or the notification thereof to the tenant within the 14-day period specified in section 213. It is the failure to comply with either of those obligations at all. It follows in my judgment that if, therefore, the landlord is late in complying with his dual section 213 obligations, but he nevertheless duly does so before any section 214 proceedings are brought by the tenant, the tenant will have no cause of action under section 214 and any claim he brings under it will fall to be dismissed.

[27] The cut-off date for compliance is the hearing date. In a case where compliance by the landlord post-dates the issue of the proceedings but occurs prior to the hearing date, the court will likewise have no power to make a section 214 order but the tenant will be able to recover his costs: see [44]:

> As I have said, the objective of the legislation is not the punishment of landlords but the achieving of proper protection of tenants' deposits. The legislation should not be interpreted in a sense that implicitly encourages the ambushing of landlords by tenants who have grounds for believing that the landlords have not complied with their section 213 obligations. It should be interpreted in a way that avoids litigation. Litigation will or should be avoided if, following a letter before claim, the landlord promptly puts his house in order. If the landlord declines or fails to do so, then of course it is open to the tenant to pursue his section 214 claim. If the landlord later (before the hearing) repents and remedies his defaults, the claim will still fail, although the tenant will ordinarily recover his costs. He may not recover his full costs, but there is nothing unusual about a claimant not doing so. The tenant will bring his claim knowing of that risk.

[28] The particular point decided in *Tiensia* does not assist the landlord in this case. Gladehurst has never complied with the initial requirements or with section 213(6)(a) and cannot now do so. But its very inability to comply post the termination of the lease and the return of the deposit is relied on by Ms West to support her argument that it cannot have been the intention of the legislature to require compliance with section 213 after the lease has expired and where in the majority of cases the deposit will have been returned safely to the tenant subject to any appropriate deductions.

[29] In terms of statutory construction, this argument can succeed only if the tenant's right to make an application under section 214(1) or the court's power to make an order under section 214(3) and (4) are to be treated as conditional on the continuing existence of the tenancy as at the relevant time. Subsections (3) and (4) are the mandatory consequence of the court being satisfied on a section 214 application of the matters specified in section 214(2). Even applying the construction of that provision adopted in *Tiensia*, it is common ground that those conditions were satisfied in this case. Section 214(2) does not include a further express requirement that the lease should be subsisting at the date of the hearing so that the issue has to be whether the application that the respondents made and that came before Judge Cryan on appeal was "such an application" within the meaning of section 214(2): that is, an application made under section 214(1).

[30] Ms West therefore submitted that the words "the tenant" in section 214(1) should be construed to mean a tenant under a subsisting lease. "Tenant" is not a defined term under the Act and, as the judge pointed out, it is used in paras 4 and 4A of Schedule 10 in connection with the arrangements under a custodial scheme for the resolution of any disputes about retentions. Those disputes can arise only after the tenancy has come to an end and their resolution is one of the principal objects of the scheme. Schedule 10 is therefore a clear illustration of "the tenant" being used in that context to describe the former tenant who had paid the deposit under the now expired lease.

[31] Ms West's argument has therefore to concentrate on the context in which "the tenant" is used in section 214(1). Unlike paras 4 and 4A of Schedule 10, which relate in terms to the termination of tenancies where the deposit is held under a custodial scheme, section 214(1) looks to the time, she submitted, when the deposit is paid at the commencement of the lease and continues to be held by the landlord otherwise than in accordance with an authorised scheme. The principal purpose of section 214 is to remedy a landlord's continued failure to comply with section 213 by registering the deposit with or paying it to the administrator of an authorised scheme. This was recognised by the decision in *Tiensia*, where Rimer LJ, in [40], said:

> That interpretation of the legislation means that late, but nevertheless due, compliance by the landlord with his dual obligations under section 213(3) and (6) will furnish him with a complete defence to any claim by the tenant under section 214. Such interpretation appears to me to be not only firmly supported by what I would regard as the carefully chosen statutory language, it is also a properly precise, or strict, one to apply to legislation such as section 214 that is manifestly penal in intent. Moreover, it is an interpretation that is consistent with the purpose of the legislation. That purpose is to achieve the due protection of deposits paid by tenants, ideally within the 14-day period but, if not, then later. It cannot be its purpose to punish landlords who may for example, for innocent reasons, be just a day late in securing such protection.

[32] But once the lease has terminated, the landlord cannot comply with the initial requirements of a scheme. None of the authorised schemes that have been set up intend a deposit to be protected under the scheme once the contractual term has come to an end. Moreover, if the respondents are right and it remains open to a former tenant to take proceedings under section 214 after the termination of the lease, their landlords could be faced with proceedings for three times the amount of the deposit even years after the lease has come to an end and the initial deposit has been returned.

[33] As an alternative to her main argument, Ms West submitted that section 214(4) should be interpreted so as to read "the amount of the deposit" as meaning the amount of the deposit still retained by the landlord at the date of the hearing or the order. This would reduce the landlord's liability under section 214(4) in this case from £18,720 to £1,834. I am not persuaded by this. "Deposit" is a defined term and has the meaning set out in section 213(8). It clearly means the deposit as paid by the tenant at the commencement of the lease. Section 214(4) is a penal provision, set in amount, which does not create a kind of sliding scale depending on how much of the original deposit the landlord happens to retain at the relevant time. That would be entirely arbitrary and it is unsupported by the language and structure of Chapter 4 of the Act.

[34] The strongest argument against the landlord's construction of the Act is that it would operate as an incentive to landlords to avoid complying with section 213. A failure to deal with the deposit in accordance with an authorised scheme would carry no penalty (except perhaps as to costs) if the tenant did choose to make a section 214 application during the term of the lease. The landlord would have until the hearing date in which to comply. However, if the tenant delayed in making an application (as in this case) until after the termination of the lease, the landlord would never need to comply with section 213 and would have a complete answer to the claim.

[35] Mr Gannon, for the respondents, supported the judge's view that the protection of a deposit under an authorised scheme was intended to operate as much after the lease has terminated as it did before it expired. The provisions of Schedule 10 that I have referred to are an obvious example of this. Without a scheme in place, the tenant will encounter all the former problems of having to bring proceedings in the county court to secure the return of his deposit. These were intended to be avoided by the compulsory machinery of an authorised scheme. The unscrupulous landlord, by contrast, could, on Ms West's argument, sit on its hands and hope for the best knowing that it could always avoid a section 214 order by delay.

* Editor's note: Also reported at [2010] 3 EGLR 53

[36] Mr Gannon submitted that it is nothing to the point that the scheme rules do not allow deposits to be protected after the end of the tenancy. The provisions of section 213 require the deposit to be placed into an authorised scheme at the beginning of the tenancy and to remain so protected during the whole of the term. The landlord is not legally entitled to take a deposit on any other basis. Its obligation therefore falls to be performed at a time when the deposit can be protected as required and if it fails to comply with its legal duties at the time when they fall to be performed, the consequences prescribed by section 214(3) and (4) follow.

[37] The point is not an easy one, but I have come to the conclusion that the power of the court to make an order under section 214(3) and (4) is no longer exercisable once the tenancy has come to an end. Although section 213 makes it unlawful for a landlord to require the payment of a deposit that is not to be dealt with in accordance with an authorised scheme and requires the landlord within 14 days of receipt of a deposit to comply with the initial requirements of such a scheme, it is important to note that no criminal penalty is imposed for non-compliance with these provisions. Instead, they are made enforceable at the option of the tenant under section 214. It is entirely a matter for him as to whether it chooses to take advantage of the provisions in section 213, which were created for its benefit.

[38] The answer therefore to the argument that the construction of section 214 contended for by Gladehurst will encourage landlords not to comply with their legal obligations under section 213 is the same as applies to any breach by a landlord of its covenants or other obligations under the lease. The tenant always has it within his hands to secure their enforcement by the taking of proceedings. That is the remedy prescribed by section 214 of the Act and it is up to the tenant to make use of it.

[39] Looked at in this way, it is entirely understandable if some temporal limits apply to the exercise of the right of enforcement under section 214. The initial requirements of an authorised scheme are, as Mr Gannon submitted, matters to be dealt with at the inception of the lease and not later than the expiry of the term. Section 214(1), as interpreted in *Tiensia*, speaks in terms of these requirements not yet having been complied with and therefore carries the strong implication that the default can still be remedied. This impression is confirmed by section 214(3), which requires the court either to order the repayment of the deposit to the applicant or to order the landlord to pay it into an authorised scheme. For the court to have a genuine discretion to exercise both alternatives must be available. On the facts of this case, neither was. Although Judge Cryan treated the £618 as part of the deposit retained by Gladehurst, it was artificial to do so. The tenants had agreed to Gladehurst retaining the £618 pending the prosecution of their claim for its return that it was for them to pursue and make out. The retention was therefore consensual.

[40] The issue raised by the landlord has also to be resolved in a case where the landlord has returned the entirety of the deposit on the termination of the lease. Mr Gannon accepted that if this were the case, no section 214 claim could be made. Section 214 envisages that it must still be open to the court to make both a section 214(3) and a section 214(4) order. Where this is not the case neither applies.

[41] However, that argument would mean that a defaulting landlord that nevertheless was scrupulous in dealing with the deposit at the end of the lease would be in a worse position with a defaulting tenant than it would be with a tenant who observed his own covenants to the letter. If the landlord was able to refund the deposit in its entirety, it would escape the consequences of section 214(4) but not if it had legitimate grounds for retaining part of it to pay for repairs.

[42] Anomalies of this kind are avoided by reading section 214(1)(a) as meaning that the initial requirements of an authorised scheme have not been but are still capable of being complied with. This is consistent with the decision in *Tiensia* and is the only meaning that ties in with the two alternatives in section 214(3) continuing to be available. In practice, this means that the grounds for a section 214 application will cease to exist once the lease expires and no order under either section 214(3) or (4) can therefore be made after that date. From that moment on, the application will cease to be "such an application" as is described in section 214(2).

[43] It follows that the district judge was right to reject the section 214(4) claim in this case. I would therefore allow this appeal and set aside the orders for repayment of the deposit and for the payment of the £18,720.

Baron J said:

[44] I agree.

Carnwath LJ said:

[45] I also agree.

Appeal allowed.

Kahlon *v* Isherwood

COURT OF APPEAL

[2011] EWCA Civ 602

19 May 2011

RIX, STANLEY BURNTON AND PATTEN LJJ

Estates Gazette 16 July 2011

[2011] 28 EG 76

Landlord and tenant — Housing Act 1988 — Assured tenancy — Possession claim by landlord — Mediation resulting in agreement whereby landlord to waive rent arrears and tenant to accept change from assured tenancy to assured shorthold tenancy — Agreement reflected in schedule to* Tomlin *order — Whether tenant continuing to hold assured tenancy by reason of non-compliance with notice requirements of para 7(2)(a) of Schedule 2A to 1988 Act — Whether* Tomlin *order constituting notice to substantially same effect as prescribed notice

Possession proceedings brought by the respondent landlord against the appellant assured tenant were compromised in 2008 after a successful mediation. A mediation agreement, signed by the parties and the mediator, provided that the appellant's assured tenancy would be replaced with an assured shorthold tenancy (AST) for a term of 12 months from June 2008, while the respondent would repair items of disrepair for which the appellant had counterclaimed and would waive the arrears of rent and the benefit of a costs order. The parties signed a *Tomlin* order drawn up by the appellant's solicitor with a schedule setting out the agreed terms. A tenancy agreement headed "Assured shorthold tenancy" was duly executed.

The respondent subsequently served notice on the appellant, under section 21 of the Housing Act 1988, to terminate the AST at the end of the 12-month term. The appellant failed to vacate on the required date. He asserted that he still held an assured tenancy by reason of section 19A of the Housing Act 1988 owing to non-compliance with the notice requirements imposed by para 7(2)(a) of Schedule 2A in cases where an existing assured tenant was downgraded to an AST from the same landlord. He relied on the absence of any notice served by him in the form prescribed by the Assured Tenancies and Agricultural Occupancies (Forms) Regulations 1997 (form 8), setting out the tenant's acknowledgment that the tenancy was to be an AST, with all the consequences for security of tenure that that entailed.

The respondent applied to the county court for a possession order. Granting the order, the recorder held that the schedule to the earlier *Tomlin* order, although not identical to form 8, was to "substantially the same effect" within the meaning of para 2 of the 1997 Regulations, in circumstances where the appellant had been advised by a solicitor throughout the relevant proceedings and the subsequent mediation and must have been advised of, and been aware of, the effects of the change to his tenancy. The appellant appealed.

Held: **The appeal was allowed. The relevance or materiality of a departure from the prescribed form of notice must be assessed by reference to the purpose of the notice. The purpose of the notice to be prepared and served by the tenant under para 7(2)(a) of Schedule 2A to the Housing Act 1988 is to give to the tenant the information that para 4 of form 8 contains and to provide the landlord with an acknowledgment of that matter. If, on that approach, a notice does not comply with para 7(2)(a), the tenancy will continue to be an assured tenancy; there is no provision for the court to dispense with the requirements of para 7(2)(a). The schedule to the *Tomlin* order was not to "substantially the same effect" as the prescribed form 8 since it did not contain the bullet-pointed information that form 8 includes as to the tenant's rights and there was nothing corresponding to para 4 of the form, which sets out the tenant's understanding and acceptance of the consequences of a change from an assured to an AST in terms of security of tenure. Those were matters of substance. Where a provision in the prescribed form is part of the substance of the notice, it is no answer to its omission to say that the information that it conveys was well known to the tenant at the relevant time. The principles relevant to the construction of inaccurate particulars in a notice cannot fill gaps in the notice: either it contains all the necessary matters of substance or it does not. Accordingly, the possession order should be set aside and the case remitted to the county court to reconsider the claim for possession on the basis that an assured, rather that an assured shorthold, tenancy had been created. It might also be necessary to address any arguments as to the continued effectiveness of the other terms of the mediated compromise.**

Per curiam: **It is doubtful whether a *Tomlin* order, even if it contains wording in the prescribed form, can fulfil the role of a para 7 notice. Para 7 requires a notice to be served before the tenancy is entered into, in order to give the tenant an opportunity to consider fully the consequences of what is proposed and to resile if he changes his mind. That is not possible where the execution of a *Tomlin* order binds him to take a new tenancy by creating a specifically enforceable contract for the grant of a new tenancy in a given form. Similar problems arise with regard to mediation agreements. Therefore, the requirement for service of a para 7 notice in such cases should be dealt with by the tenant serving a notice in the prescribed form immediately prior to the execution of the mediation agreement.**

The following cases are referred to in this report.

Manel v *Memon* [2000] 2 EGLR 40; [2000] 33 EG 74; (2001) 33 HLR 24

Mannai Investment Co Ltd v *Eagle Star Life Assurance Co Ltd* [1997] AC 749; [1997] 2 WLR 945; [1997] 3 All ER 352; [1997] 1 EGLR 57; [1997] 24 EG 122; [1997] 25 EG 138, HL

Osborn & Co Ltd (B) v *Dior; sub nom Osborn & Co Ltd* v *Dior; Marito Holdings SA* v *Deneche* [2003] EWCA Civ 281; [2003] 05 EG 144 (CS); [2003] HLR 45

Tegerdine v *Brooks* (1977) 36 P&CR 261; [1978] 1 EGLR 33; 245 EG 51, CA

York v *Casey* [1998] 2 EGLR 25; [1998] 30 EG 110; (1999) 31 HLR 209, CA

This was an appeal by the appellant, Andrew Isherwood, from a decision of Mr Recorder Willetts, sitting in Coventry County Court, allowing a claim by the respondent, Jasbir Kahlon, for possession of a residential property on the ground that the appellant held an assured shorthold tenancy that had been terminated by notice under section 21 of the Housing Act 1988.

Naomi Winston (instructed by Blakemores Solicitors, of Birmingham) appeared for the appellant; Andrew Maguire (instructed by Heer Manak Solicitors, of Coventry) represented the respondent.

Giving judgment, **Patten LJ** said:

[1] This is an appeal by Mr Andrew Isherwood against an order for possession that was made against him by Mr Recorder Willetts in Coventry County Court on 15 November 2010. The order relates to a property at 74 Sunningdale Road, Coventry (the property) that Mr Isherwood has occupied as a tenant since around 1994. The respondent to the appeal, Mrs Jasbir Kaur Kahlon, is the freehold owner of the property.

[2] Since 2000, there have been three separate sets of proceedings for possession not counting the action in which the order under appeal was made. All three claims have been based on arrears of rent. The first claim was compromised in 2003 on terms that Mr Isherwood and his wife remained in possession. The second action was withdrawn in 2005. The third action was commenced in 2007, alleging arrears of rent and a breach of a covenant to allow access to the property. Mr Isherwood counterclaimed in those proceedings for damages for breach of the landlord's repairing covenants.

[3] In May 2008, the parties submitted their dispute to mediation, which was successful in that it concluded with an agreement signed by the parties and the mediator on 8 May that provided for the appellant's solicitor, Mr Baxendale, to prepare a *Tomlin* order in the proceedings under which:

> An assured shorthold tenancy agreement for the period of 12 months in the form annex will be completed between parties to the commencement date will be 2nd June 2008.

[4] The agreement also provided for the arrears of rent alleged in the action and the benefit of a costs order to be waived by the respondent and for her to arrange for certain specified repairs to the property to be carried out within four months of the commencement of the assured shorthold tenancy (AST).

[5] The *Tomlin* order was drawn up and signed by the parties on 16 May 2008. The schedule was in the following terms:

> The Claimant and the First Defendant have agreed Full and Final Settlement of all matters in dispute between them in the Present Proceedings upon the following terms:
>
> 1. The Claimant and the First Defendant shall execute an Assured Shorthold Tenancy Agreement for the period of 12 months commencing 2nd June 2008 in the form annexed hereto, to be signed by the parties not later than 2nd June 2008.
>
> 2. The Rent reserved by the said Assured Shorthold Tenancy Agreement shall be £140 per week subject to the proviso at paragraph 3 below.
>
> 3. All alleged rent arrears in relation to the Premises shall be waived.
>
> 4. It is hereby provided and agreed between the parties that the Claimant will not call for payment of any rental or take any steps or enforcement in relation to any payment or alleged non-payment of rental by the First Defendant in excess of rent paid by the Housing Benefit office.
>
> 5. The parties shall take all reasonable steps to co-operate in regard to the claim by the First Defendant for Housing Benefit.
>
> 6. The Claimant and the First Defendant shall enter into an appropriate Consent Order whereby the First Defendant shall agree to set aside and not enforce the Costs Order amounting to £2,500 in proceeding with No: 5CV03452 in the Coventry County Court.
>
> 7. The Claimant Undertakes to instruct an appropriate Contractor to attend the premises and to rectify items 3, 4, 5, 6, 16, 17, 18, 19 and 20 of the Joint Expert's Report dated 11th January 2008 not later than 2nd October 2008.
>
> 8. The First Defendant agrees to allow reasonable access on reasonable notice to enable the Claimant to discharge her agreement under paragraph 6 above.

[6] On 21 May 2008, the parties executed a tenancy agreement headed "Assured Shorthold Tenancy Agreement" under which Mr Isherwood was granted a lease of the property for a term certain of 12 months from 2 June 2008. Clause 2 of the agreement provided that the landlord may terminate the tenancy at any time before the expiry of the term by giving to the tenant not less than two months' written notice that she requires possession of the property.

[7] An ASTgives to the tenant very limited security of tenure. If the fixed term has come to an end and the landlord has served on the tenant not less than two months' notice in writing stating that he requires possession of the demised premises, the court must make an order for possession: see section 21(1) of the Housing Act 1988 (the HA 1988). The notice may be served either before or on the day on which the tenancy comes to an end: see section 21(2).

[8] By contrast, an assured tenancy (as defined in section 1 of the HA 1988) can be brought to an end only by an order for possession made on one of the grounds set out in section 7 of the HA 1988. An

assured tenancy granted for a fixed term will continue after the expiry of the term as a periodic tenancy subject to the same protection unless it is surrendered: see section 5(2) of the HA 1988.

[9] The change from an assured tenancy to an AST has therefore serious consequences for the tenant and the HA 1988 contains specific provisions designed to ensure that the tenant has been put on notice of what those consequences are. In the case of ASTs granted before 28 February 1997, section 20 of the HA 1988 required the landlord to serve on the tenant a notice in prescribed form before the tenancy was entered into stating that the assured tenancy was to be an assured shorthold tenancy. Section 20(3) also prohibited the grant of an AST by an existing landlord to an existing tenant where the existing tenancy was an assured tenancy.

[10] However, all assured tenancies entered into after 28 February 1997 are now ASTs unless they fall within certain excluded categories set out in Schedule 2A: see section 19A of the HA 1988. These include the case where the landlord serves a notice on the proposed tenant stating that the tenancy is not to be an AST (para 1) and cases where the new tenancy is to be granted to an existing assured tenant. The previous embargo on granting ASTs to existing assured tenants is removed but for the new tenancy to be an AST (as opposed to an assured tenancy) the conditions set out in para 7 of Schedule 2A must be complied with. The tenancy will not therefore qualify as an AST if it is:

> 7(1) An assured tenancy which
>
> (a) is granted to a person (alone or jointly with others) who, immediately before the tenancy was granted, was the tenant (or, in the case of joint tenants, one of the tenants) under an assured tenancy other than a shorthold tenancy ("the old tenancy"),
>
> (b) is granted (alone or jointly with others) by a person who was at that time the landlord (or one of the joint landlords) under the old tenancy, and
>
> (c) is not one in respect of which a notice is served as mentioned in sub-paragraph (2) below.
>
> (2) The notice referred to in sub-paragraph (1)(c) above is one which
>
> (a) is in such form as may be prescribed,
>
> (b) is served before the assured tenancy is entered into,
>
> (c) is served by the person who is to be the tenant under the assured tenancy on the person who is to be the landlord under that tenancy (or, in the case of joint landlords, on at least one of the persons who are to be joint landlords), and
>
> (d) states that the assured tenancy to which it relates is to be a shorthold tenancy.

[11] In respect of some types of notice, such as notices of proceedings of possession under section 8 of the HA 1988, the court is given the power to dispense with the requirements for such a notice if it considers it just and equitable to do so. However, parliament has not included any such power in the case of para 7.

[12] The prescribed form of notice is contained in the Assured Tenancies and Agricultural Occupancies (Forms) Regulations 1997 (SI 1997/194) (the 1997 Regulations). Paragraph 2 of the Regulations provides that:

> In these Regulations any reference to a section or Schedule is a reference to a section of, or Schedule to, the Housing Act 1988 and any reference to a numbered form is a reference to the form bearing that number in the Schedule to these Regulations, or to a form substantially to the same effect.

[13] The relevant form referred to in para 7 of Schedule 2A to the HA 1988 is form 8, which is set out in full in the appendix to this judgment.

[14] The principal difference between para 7 of Schedule 2A and the provisions of section 20 that govern the grant of pre-Housing Act 1996 tenancies is the requirement that the notice should be served by the tenant on the landlord rather than vice versa. As explained earlier, there is nothing comparable to section 19A and para 7 in section 20 because, until February 1997, it was not possible to grant an AST to an existing assured tenant. But the form prescribed under section 20(2)(a) was contained in the Assured Tenancies and Agricultural Occupancies (Forms) Regulations 1988 (SI 1988/2203), which contained an identical regulation 2 requiring the notice to be in the prescribed form or "a form substantially to the same effect".

[15] On 31 March 2009, Mrs Kahlon served on Mr Isherwood G
a notice under section 21 of the HA 1988 stating that she required possession of the property on 2 June 2009. When Mr Isherwood failed to vacate, Mrs Kahlon issued proceedings against him for possession on 12 June 2009. These were dealt with under the accelerated procedure for AST possession proceedings adopted in the county court because, in most such cases, the tenant will have no answer to the claim. However, in this case, Mr Isherwood took the point that no notice in the prescribed form had been served by him prior to the execution of the tenancy agreement on 21 May 2008 with the consequence that he was an assured rather than an assured shorthold tenant. Mrs Kahlon could not therefore obtain an order for possession without making out one of H
the statutory grounds specified in section 7 of the HA 1988.

[16] The recorder rejected this defence and made an order for possession. He held that the schedule to the *Tomlin* order, although not identical in form to form 8 in the 1997 Regulations, was in a form substantially to the same effect. He observed that Mr Isherwood had the services of a solicitor (Mr Baxendale) throughout the relevant proceedings and the subsequent mediation and must, in the absence of evidence to the contrary, be assumed to have been advised and to have known about the effect of the agreed change from an assured to an assured shorthold tenancy. This knowledge, coupled with the parties' signature to the *Tomlin* schedule, was sufficient to comply J
with the notice formalities required as a prerequisite to the grant of an assured shorthold tenancy. The tenancy agreement therefore took effect according to its terms and Mr Isherwood had no defence to the action.

[17] The defence raised by the appellant was entirely technical in nature and stands or falls with our view as to whether the *Tomlin* schedule complies with para 7(2) of Schedule 2A. The point is in many ways deeply unattractive given that it was Mr Baxendale who was to draft the *Tomlin* schedule in accordance with the mediation agreement and Mr Isherwood's undoubted agreement at the time to accept an AST in place of his existing assured tenancy as part of the settlement. Mrs Kahlon has agreed to waive significant sums in respect of rent and K
costs as part of a bargain on which Mr Isherwood is now seeking to renege. If he is right in his argument concerning the lack of a compliant notice, Mrs Kahlon will doubtless wish, if she is able, to revisit the enforceability of the remaining terms of the agreement.

[18] But confining myself, as I must, to the issue under appeal, I am in no doubt that the recorder was wrong to reject Mr Isherwood's defence on the grounds that he did. The question of whether a notice is in a form substantially to the same effect as the prescribed form under the relevant regulations has been considered by this court on a number of previous occasions in respect of section 20 of the HA 1988. This is, however, the first time when it has been considered in the context of L
section 19A and para 7 of Schedule 2A.

[19] In *Manel* v *Memon* [2000] 2 EGLR 40, the section 20 notice omitted the four bullet points on the prescribed form, including one that advised the tenant to go to a solicitor or the Citizens' Advice Bureau before signing the tenancy agreement if he did not understand anything. The bullet point was very similar to the one contained in form 8. This court held that the bullet points were part of the substance of the notice so that their omission was fatal to its validity. At p42D-E, Nourse LJ said that:

> in 1988, an assured shorthold tenancy was both novel in concept and notably less advantageous to tenants than a tenancy protected by the M
> Rent Acts, which had for many years been the means of tenure by which residential properties of low rateable value were generally held. Moreover, the persons who were likely to be offered tenancies of such properties were also likely to be unable to assess for themselves the full legal and practical consequences of taking them. In the circumstances, the three bullet points on whose omission Mr Buttimore primarily relies, in particular, I would say, the exhortation to get prior legal advice from a solicitor or a Citizens' Advice Bureau, and the statement that the giving of the notice by the landlord does not commit the tenant to take the tenancy, can only be treated as part of the substance of the notice. They do not fall into the same category as the notes that were omitted from the section 25 notice in *Tegerdine* v *Brooks*. Without them, a section 20 notice is not in substance to the same effect as a notice in Form 7.

A [20] *Tegerdine* v *Brooks* (1977) 36 P&CR 261* was a decision under Part II of the Landlord and Tenant Act 1954 about the validity of a section 24 notice served by the landlord. In the notice terminating the tenancy, the landlord stated that he would not oppose the grant of a new tenancy. The tenant failed to serve a counter-notice but then contended that the section 4 notice was invalid because it omitted some of the notes on the prescribed form. The relevant regulations permitted that notice to be as prescribed or "substantially to the same effect" and the Court of Appeal held that the omissions were immaterial because the notes in question were made irrelevant by the landlord's stated willingness to accept the grant of a new tenancy.

B [21] Relevance or materiality has to be assessed by reference to the purpose of the notice. But where the provision in the prescribed form is clearly part of the substance of the notice as found in *Manel*, it is no answer to its omission to say that the information it conveys was well known to the tenant at the relevant time. This was the basis on which the recorder in this case justified any significant discrepancies between form 8 and the *Tomlin* schedule. But that approach is impermissible. Precisely the same argument was deployed and rejected in *Manel*. At p42F-G, Nourse LJ said that:

> In dealing with this point, Judge Sich said:
> "It is suggested that if she had seen that warning (about seeking independent
> C advice) she would have heeded it and would have declined to sign the document until she had been to see the Citizens' Advice Bureau. I have already made the point that those two documents in conjunction, the Tenancy Agreement and the Section 20 Notice, when taken together, were clearly inconsistent with what she understood the Agreement to be. If she was prepared to sign that I cannot really accept that the inclusion of that note would have made any difference. It is clearly an important and formal legal document and I do not think that it can, or the suggestion that that would or might have made any difference is really no more than speculative."
>
> It thus appears that the judge dealt with the point as if it were one of prejudice. However, as Mr Brett accepts, that is not the correct approach. If a notice is not in a form substantially to the same effect as the prescribed form, it
> D is defective irrespective of whether the defect has caused the tenant prejudice or not.

[22] A comparison between form 8 and the *Tomlin* schedule discloses, I think, a number of material omissions. The bullet points are missing and there is nothing in the schedule that corresponds to para 4 of form 8 setting out the tenant's understanding and acceptance of the consequences in terms of security of tenure of a change from an assured to an AST. These are clearly matters of substance. I am inclined to think that the contents of para 2 of form 8 can be inferred from the reference in the *Tomlin* schedule to the present proceedings, which, on
E examination, were clearly concerned with the termination of an assured tenancy. However, para 4 is an essential part of the notice and there is nothing in the *Tomlin* schedule that even remotely corresponds to it. The obvious purpose of the notice is to give to the tenant (even though it is he who prepares and serves the notice) the information that para 4 contains and to provide the landlord with an acknowledgement of that. The schedule is not therefore a form substantially to the same effect as form 8.

[23] Mr Andrew Maguire ultimately, I think, accepted that he could not support the recorder's reasoning if the authorities I have referred to apply in this case. In his skeleton argument, he made reference to the
F well-known decision of the House of Lords in *Mannai Investment Co Ltd* v *Eagle Star Life Assurance Co Ltd* [1997] AC 749†, which set out the new approach to the construction of a notice exercising a break clause in a lease that specified the wrong date of termination. In *York* v *Casey* (1999) 31 HLR 209‡, this court accepted that the same approach to construction was relevant to a section 20 notice that contained inaccurate particulars of the commencement date of the proposed tenancy and the identity of the landlords. However, although that approach may operate to correct obvious factual mistakes in the notice, it cannot, in my view, operate to fill gaps in the form of notice of the kind we are concerned with on this appeal. Either the notice contains all the necessary matters of substance or it does not.

[24] Mr Maguire's final submission was that the section 20 decisions have no application to a case governed by section 19A. He pointed to the fact that the notice is served by the tenant and that the landlord has no need for much of the information that the prescribed form of notice contains, including the contents of para 4. However, the requirement that the tenant should serve the notice is not determinative, in my view, of its purpose and this argument fails to recognise the intention behind the changes introduced by section 19A and Schedule 2A. It is clear from the structure of form 8 that the primary purpose of the notice remains that of informing the tenant in the document that he has to serve of the consequences of what he is proposing. The requirement that the tenant should prepare and serve the notice was obviously intended to guarantee that he is fully aware of the effect of his intended change of status and can be seen as the quid pro quo for removing the earlier bar on the grant of a shorthold tenancy to an existing assured tenant. Therefore, although it is in form a notice to the landlord, it retains a dual purpose of warning the tenant of the legal consequences of the grant to him of an AST. If anything, the need to ensure that the tenant is fully informed about these matters is far greater in a case under para 7 than it is in a case under section 20 where he is not exchanging his existing security of tenure for a lesser form of protection. Consistently with this, the reasoning in cases such as *Manel* must apply with equal force to a case under section 19A. The notice remains the means of providing a clear record that the tenant has been appraised of his rights and the loss of security that the new arrangements will entail.

[25] That is sufficient to dispose of the appeal. However, I should mention out of completeness a further point raised by Ms Naomi Winston in argument about the adequacy of the *Tomlin* schedule as a para 7 notice. She submitted that although there was no reason in principle why a schedule to an order could not constitute a notice in a prescribed form as required under para 7(2)(a) of Schedule 2A, it could satisfy that description only if it was recognisable as a notice. This submission is based on a passage in the judgment of Arden LJ in *B Osborn & Co Ltd* v *Dior* [2003] EWCA Civ 281* (another case about a section 20 notice) where she says:

> 39. Parliament clearly attributed importance to the formality of a section 20 notice. The primary legislation requires that the notice must be in a particular form, which we now know as Form 7. The regulations permit that form to be substantially to the same effect, but the notice must still be in a form which is Form 7 or recognisable as such. This requirement therefore goes to the form of the notice. The notice cannot be given orally or informally in a letter. It must be in the prescribed form or a form substantially to that effect. Those latter words permit some formal defects, but the essence of the form must remain.

[26] In practice, the requirement that the notice should be in a form substantially to the same effect as the prescribed form is likely to ensure that a notice contained in the schedule to an order is recognised as such. In the present case, the point therefore adds very little in terms of whether the condition in regulation 2 of the 1997 regulations is satisfied. However, it has led us to focus on a slightly different but related question, which is whether the *Tomlin* schedule that forms a contract between the parties for the grant of an AST can also fulfil the role of a para 7 notice that needs to be served before the assured tenancy is entered into.

[27] It is, I think, important to note that para 7(2)(b) refers in terms to an assured tenancy and not to an AST. This is because the assured tenancy that the parties are proposing to enter into will be a shorthold tenancy only if all of the para 7 conditions are complied with. If the form of the notice is defective, an assured rather than an AST will have been entered into. However, the terms of para 7(2)(b) mean that if the document under which the parties contract to enter into the assured

* Editor's note: Also reported at [1978] 1 EGLR 33
† Editor's note: Also reported at [1977] 1 EGLR 57
‡ Editor's note: Also reported at [1998] 2 EGLR 25

* Editor's note: Reported at [2003] HLR 45

tenancy is also to constitute the notice then that condition is not satisfied. The assumption behind the requirement for a gap in time between the service of the notice and the entering into the tenancy agreement must be that the tenant should be given an opportunity to consider fully the consequences of what is proposed in the notice and to be able to resile if he has a change of mind. That is not possible if his execution of the *Tomlin* schedule binds him to take a new tenancy.

[28] It is said that no such danger can exist because the lack of a gap between the service of the notice and the contract for a new lease will (for the reasons just given) always prevent the new tenancy from being an assured shorthold. But that is only because para 7(2)(b) requires the service of the notice and the grant of the assured tenancy to be separated in point of time in every case. It seems to be common ground that para 1 of the *Tomlin* schedule did create a specifically enforceable contract for the grant of a new tenancy in the form of the tenancy agreement annexed to the schedule and this contract was subsequently completed by the execution of the agreement. In these circumstances, the condition set out in para 7(2)(b) is not, in my view, satisfied in the present case.

[29] The same argument could, I think, be deployed in respect of clause 1 of the mediation agreement that undoubtedly preceded the giving of any notice but it would, I think, be most unfortunate if one of the consequences of this appeal was to discourage the use of mediation to resolve disputes about assured tenancies. The object of any mediation must be to reach a concluded and binding agreement that effectively compromises the issues between the parties. The requirement for the service of a para 7 notice prior to the entry into an assured tenancy could and should have been catered for in this case by the tenant preparing and serving a notice in the prescribed form on the landlord immediately prior to the execution of the mediation agreement. It would then have been open to the parties to enter into the mediation agreement without any risk that the creation of an AST might be jeopardised by the coming into effect of an equitable lease.

[30] I would therefore allow the appeal and set aside the order for possession. The case will have to be remitted to the county court for the judge to reconsider the claim for possession on the basis that the agreement executed on 21 May 2008 created an assured rather than an AST. Any argument as to whether all of the terms of the mediated compromise remains effective may also have to be addressed.

Stanley Burnton LJ said:

[31] I agree.

Rix LJ said:

[32] I also agree.

Appendix to judgment

Specified date: 28 February 1997: see reg 1.

Form No 8

Housing Act 1988 Schedule 2A, paragraph 7(2) as inserted by Schedule 7 to the Housing Act 1996 F

Tenant's notice proposing that an Assured Tenancy be replaced by an Assured Shorthold Tenancy

"Please write clearly in black ink.

"Please cross out text marked with an asterisk (*) that does not apply.

"This notice should only be used by an assured tenant. You should only use this notice to notify your landlord that you wish your assured tenancy to be replaced by an assured shorthold tenancy.

" This notice must be served by a tenant on a landlord before an assured tenancy can be replaced by an assured shorthold tenancy.

"You should be aware that by serving this notice, you will be giving up your right to stay in the property after the first six months of the assured shorthold tenancy or, if you agree a fixed term with your landlord, after the end of the fixed term. G

"You do not have to complete this form even if your landlord has asked you to do so. Your existing security of tenure as an assured tenant will be unaffected if you do not complete it.

"If you are in any doubt about whether to complete this form, take it immediately to a citizens' advice bureau, housing advice centre, a law centre or a solicitor.

"Once you are clear that you wish to issue this notice, complete the form and send it to your landlord.

1 To:......

Name(s) of landlord(s) H

2 I/We*, the tenant(s) of:......

Address of premises

give notice that I/we* propose that the assured tenancy to which this notice relates should be replaced by a shorthold tenancy.

3 I/We* propose that the new shorthold tenancy should commence on:

...... /...... /...... (day/month/year)

"The new shorthold tenancy cannot commence until after the date this notice is served on the landlord.

4

(a) I/We* understand that under my/our* existing tenancy, I/we* can only be required to give up possession in accordance with the grounds set out in Schedule 2 to the Housing Act 1988, whereas under the new shorthold tenancy, the landlord(s) will be able to recover possession of the premises without being J required to prove a ground for possession, after the first six months of the assured shorthold tenancy, or, if there is a fixed term for longer than 6 months, at the end of that fixed term, subject to two months' notice.

Signed...... Date......

To be signed and dated by the tenant. If there are joint tenants each tenant must sign.

(b) Name and address of tenant.

Name(s) (Block Capitals)......

Address......

TelephoneDaytime...... Evening......

NOTES

Initial Commencement K

Specified date

Specified date: 28 February 1997: see reg 1.

Appeal allowed.

LAND REGISTRATION

Baxter *v* Mannion

COURT OF APPEAL

[2011] EWCA Civ 120

22 February 2011

MUMMERY, JACOB AND TOMLINSON LJJ

Estates Gazette 21 May 2011

[2011] 20 EG 114

Land registration — Rectification of register for mistake — Land — Registration Act 2002 — Registration of title to land obtained on ground of adverse possession under Schedule 6 to 2002 Act — Necessary period of adverse possession not occurring — Owner of paper title failing to serve counternotice objecting to registration within prescribed time limit — Whether paper owner entitled to be reinstated as registered proprietor by rectification on ground of mistake under para 5(a) of Schedule 4 — Extent of jurisdiction to rectify for mistake

In 2005, the appellant applied to the Land Registry to be registered as the owner of a field in place of the respondent on the ground of 10 years' adverse possession pursuant to Schedule 6 to the Land Registration Act 2002. The respondent was served with notice of the application and a blank NAP form to be completed and returned if he wanted to object to the application or serve a counternotice. He failed to return the form within the prescribed period of 65 working days after the date of the notice. The appellant was registered as the proprietor of the field shortly after the deadline had passed.

The respondent applied to be re-registered as the proprietor by means of rectification of the title to correct a mistake within para 5(a) of Schedule 4 to the 2002 Act. The appellant objected to the application, which was accordingly referred to a deputy adjudicator. The adjudicator found that the appellant's registration as proprietor had been a "mistake" because he had not been in adverse possession of the field for the requisite 10-year period; she therefore granted rectification. The appellant appealed to the High Court. The judge dismissed the appeal and upheld the adjudicator's decision. He concluded that it would be unjust not to alter the register such that the requirements of para 6(2) of Schedule 4, in respect of rectification of the title of land of which the proprietor was in possession, were met: see [2010] EWHC 573 (Ch); [2010] 1 WLR 1965. The appellant appealed.

***Held*: The appeal was dismissed. An application to be registered with title to land under Schedule 6(1) to the 2002 Act may be made only by a person who has been in adverse possession of an estate for 10 years. A person who has not been in adverse possession is not entitled to apply and a registration obtained by such a person will be mistaken. Putting the register back in the condition it was in prior to the application will therefore involve correcting a mistake within the meaning of para 5 of Schedule 4. There is no reason to limit the meaning of correcting a mistake to one that was caused through some official error in the course of examining the application. To hold otherwise would invite fraud by dishonest applicants claiming to have been in adverse possession in circumstances where they knew that the owner was away or otherwise unable or unlikely to submit an NAP form in time. The true owner should not lose his land for want of serving a form in time (there being no provision for extensions of time) where a registration application looks in order F on its face but is in fact erroneous. That applies not only where a mistake is induced by fraud but also in the case of a merely wrong application. It would be contrary to the policy of the 2002 Act, which aims to make it more difficult to obtain land by adverse possession where a registered proprietor could be at risk of losing his land to a person who had never been in adverse possession. The deputy adjudicator's finding that the appellant had not been in adverse possession was sound. It would be unjust not to reinstate the respondent as the registered proprietor because he would otherwise lose his land to someone who had not been in adverse possession and who had made an unjustified attempt to acquire G title. The respondent's failure to return the NAP form in time did not affect the balance of justice in the case.**

The following case is referred to in this report.

Baxter v *Mannion* [2010] EWHC 573 (Ch); [2010] 1 WLR 1965; [2011] 1 P&CR 6

This was an appeal by the appellant, Steven Baxter, from a decision of Henderson J, sitting in the Chancery Division of the High Court, dismissing an appeal against a determination of a deputy adjudicator of the Land Registry allowing an application by the respondent, Thomas Mannion, to reinstate him as the registered proprietor of land by H rectifying the title to correct a mistake pursuant to Schedule 4 to the Land Registration Act 2002.

Helen Galley (instructed by MA Law LLP) appeared for the appellant; Thomas Weekes (instructed by Taylor Vinters, of Cambridge) represented the respondent.

Giving the first judgment at the invitation of Mummery LJ, **Jacob LJ** said:

[1] Mummery LJ gave permission for this second appeal because he considered that it raises an important question of principle. And so it does. It is this: can a man who has got his name registered as the J proprietor of a parcel of registered land by wrongly claiming that he had been in adverse possession for 10 years hang on to that title if the original proprietor, within 65 days of its being posted to him, failed to fill in and return a form posted to him by the Land Registry? Or can the original proprietor apply to the registrar to have the register of title rectified by "correcting a mistake"? Does the machinery of the Land Registration Act 2002 (the 2002 Act) allow a party to take someone else's land by operation of a bureaucratic machinery that trumps reality?

[2] The appeal is from a judgment of Henderson J of 18 March 2010, [2010] EWHC 573 (Ch)*, upholding a decision of a deputy adjudicator to the Land Registry, Ms Ann McAllister, of 5 February K 2009. It concerns a small field in Chatteris, a little market town near Ely in Cambridgeshire. The dispute is between Mr Thomas Mannion, the original registered proprietor of the field, and Mr Steven Baxter, who successfully invoked the procedure under the 2002 Act to get himself registered as the proprietor instead of Mr Mannion. Having found out about this, Mr Mannion applied to the registrar to have his name put back as the owner and that of Mr Baxter removed. Mr Baxter in turn

---

* Editor's note: Reported at [2010] 1 WLR 1965

applied for registration of a right of way over a yard leading to the field. Nothing turns on this later application and I say no more about it.

Legislation

[3] Rather than clutter up this judgment with the relevant text of the legislation, the 2002 Act, I have set it out in an annex.

[4] The Act was the result of much work by and ultimate recommendations of the Law Commission and HM Land Registry. The Commission and the Land Registry published a consultative report in 1998, *Land Registration for the Twenty-First Century: A Consultative Document* (Law Com No 254) and a final report, *Land Registration for the Twenty-First Century: A Conveyancing Revolution* (Law Com No 271). The final report was accompanied by a draft bill. The text of the relevant provisions of the 2002 Act is the same as in that draft bill. It is not surprising, therefore, that both Ms Helen Galley (for Mr Baxter) and Mr Thomas Weekes (for Mr Mannion) took us to parts of both reports, each contending that certain parts supported their respective cases.

[5] It should be noticed that both the consultative document and the final report were the responsibility of Mr Charles Harpum, who was a Law Commissioner for England and Wales from 1994-2001, head of the property and trust law team at the Commission and a consultant to the Land Registry during the passage of the Act through parliament.

What happened in this case

[6] Mr Mannion bought the field in 1996 and was duly registered as its proprietor. Over the years, Mr Baxter made some use of the field by way of keeping horses on it, although the amount of use was not, the deputy adjudicator found, either continuous or as extensive as Mr Baxter claimed.

[7] In August 2005, Mr Baxter made an application to the Land Registry for registration of the field in his name. The application was made under para 1(1) of Schedule 6. Mr Baxter claimed that he had been in adverse possession of the field since 1985, supporting the application with a statutory declaration to that effect. The Registry examined the application. It seemed in order and indeed on its face it apparently was. So, on 1 February 2006, the registry sent a notice to Mr Mannion as required by para 2(1). The notice contained the para 2(2) warning, stating in bold type that if Mr Mannion objected or wanted to give a counternotice to the Chief Land Registrar he must do so before 12 noon on 8 May 2006 (that is, 65 working days after the date of the notice, that being the period prescribed by rules pursuant to para 3(2)). The notice was accompanied by detailed explanatory notes and a blank form (NAP) for Mr Mannion to complete if he wanted the application to be dealt with under para 5.

[8] The form was simple and would have been very easy to complete. Mr Mannion did receive it. However, he did not complete it and send it back in time. There were reasons: he had first consulted a solicitor, who said this was not his kind of expertise. Before Mr Mannion could consult another solicitor terrible things happened: his only brother died (in Ireland, where Mr Mannion went to be with him for his last few weeks) and then his grandson of only three weeks died of meningitis. In addition, Mr Mannion went to the US for a short period to attend his godson's wedding.

[9] It was not until 8 September, some four months too late, that Mr Mannion's new solicitor wrote. It asked for an extension of time to deal with the notice. Actually, there is no power under the legislation to extend time, whether prospectively or retrospectively. This is a not unimportant matter when one considers whether matters can be put right out of time. In any event, the Registry had already acted on Mr Baxter's application, doing so on 8 May soon after the deadline had been passed.

[10] The Registry replied saying that it was too late and that it had already acted on the application. It added, however, that if Mr Mannion wanted to pursue the matter, he would need to lodge an application for rectification. It is noteworthy that this was the official view: for it indicates that the remedy of rectification could be invoked after a title had been registered pursuant to the Schedule 6 procedure. Whether that view is correct is in issue before us.

[11] Mr Mannion took that advice and duly lodged an application for rectification pursuant to Schedule 4. He sought an alteration of the registration of title by way of substituting his name for that of Mr Baxter. He said that this would be an alteration *for the purpose of correcting a mistake* within the meaning of para 5(a) of Schedule 4.

[12] Mr Baxter challenged the application, so the matter was referred to the deputy adjudicator. She heard the evidence from both sides (which, on Mr Mannion's side, included a number of neighbours) and held that Mr Baxter had not been in exclusive possession during the relevant period (August 1996 to August 2006) and that he had not had the necessary intention to exclude the world at large. She considered that both limbs of para 6(2) of Schedule 4 were satisfied and ordered the rectification sought.

[13] Mr Baxter appealed. His principal point was that there was no "mistake" to be corrected: that the Registry had acted correctly on the material before it. The registration of Mr Baxter as a proprietor was a result of the failure of Mr Mannion to object by filling in and returning the NAP form, not any mistake by anyone.

[14] Henderson J rejected that contention. He said:

> [39] … I can see no good reason to confine the jurisdiction of the registrar under paragraph 5(a) to the correction of mistakes of a procedural nature. In my judgment there is a mistake in the register, which the registrar has power to correct, if any statutory condition which is a prerequisite for registration is shown not have been satisfied.

A little later in the same paragraph:

> The precondition refers to a factual test (ten years' adverse possession) which has to be satisfied, and upon which a squatter's right to apply for registration is predicated. The paragraph does not say that a squatter may apply for registration if he can produce prima facie evidence of adverse possession…

[15] Mr Baxter advanced other points before Henderson J, namely that the deputy adjudicator had applied the wrong onus of proof and had wrongly considered that Mr Baxter was not at the time of the application for rectification in possession of the land so that the safeguard of Schedule 4, para 6(2) applied. He also pursued an appeal on the facts.

[16] As to the facts, Henderson J, not surprisingly since it involved an overall evaluation of these, rejected the argument. It is resurrected before us only as part of an argument concerning the burden of proof. As to the onus point, Henderson J accepted that the adjudicator had apparently got that wrong. However, he held that it made no difference. As to the "in possession" point, he again accepted that the adjudicator had erred. He also held that she had erred in simply saying, shortly, that both limbs of para 6(2) were satisfied. He thought in particular that a finding on limb (a) (fraud or lack of proper care) required a clear pleading and proper investigation, which had not happened. So he reassessed the position under limb (b) for himself, concluding that it would be unjust for the alteration not to be made.

Points on this appeal

[17] Ms Galley took three points:

(a) Henderson J was wrong about *correcting a mistake*;

(b) Henderson J was wrong about the consequences of his decision concerning onus; and

(c) Henderson J was wrong when he held that it would be unjust not to make the alteration.

(a) Meaning of "correcting a mistake"

[18] Ms Galley submitted that *correcting a mistake* has limited ambit. She did not put it quite as it had been put by leading counsel below, namely that the ambit was limited to procedural mistake. She submitted that the whole purpose of the new procedure was to do away with the previous law under which difficult questions of fact could arise where a squatter claimed the benefit of the Limitation Act 1980. Whether the squatter had been in exclusive possession for 12 or more years was a notoriously difficult question to try and created uncertainty in dealings in land that might be subject to "squatter's rights". The

inquiry of the deputy adjudicator in this case was just the sort of dispute the Act was intended to do away with.

[19] She submitted that the essence of the new scheme was simplicity, so as to give certainty in dealing with ownership of land. The principle is notice after an initial factual examination of the application by the registrar. There is a limited time period to challenge. (She described it as "generous", but it should be borne in mind that the 65-day period is provided only by the rules which cannot be used to construe the statute). Failing such challenge, title would be registered and everyone would know where they were.

[20] Ms Galley supported her contentions by reference to a number of passages in both the consultative and final reports of the Commission and the Registry. I go only to her best ones, to a passage in the consultative document and the two passages in the final report:

> 10.11 (of the consultative document)… If a system of registered title is to be effective, those who register their titles should be able to rely upon the fact of registration to protect their ownership except where there are compelling reasons to the contrary. All that should be required of them is to keep the Registry informed of their address for service. As we explain below, the land registration system enables registered proprietors to be protected against adverse possession in ways that would be very much harder to achieve where title is unregistered.
>
> 2.74 (of the final report) The essence of the new scheme in the Bill is that it gives a registered proprietor one chance, but only one chance, to terminate a squatter's adverse possession. In summary, a squatter will be able to apply to be registered as proprietor after 10 years' adverse possession. The registered proprietor and certain other persons (such as a chargee) who are interested in the property will be notified of the application. If any of them object, the squatter's application will be rejected, unless he or she can establish one of the very limited exceptional grounds which will entitle him or her to be registered anyway.
>
> 14.34 (of the final report). Where a notice is served by the registrar (as explained in paragraph 14.32) and no counter-notice is served on him within the time prescribed, the registrar *must* approve the squatter's application to be registered as proprietor of the land in place of the existing proprietor. We explain the effect of such registration below, at paragraph 14.71.

[21] She submitted that the intention was clear: a once-and-for-all system by way of notice to be followed by a counternotice, failing the latter, registration. That produced the intended clarity, certainty and simplicity

[22] Ms Galley also showed us the views of the learned authors Ruoff and Roper in *The Law and Practice of Registered Conveyancing* 46.009 (the sheets of the loose-leaf work are dated December 2007). They say:

> The court has power to order the alteration of the register for the purpose of correcting a mistake. The mistake may consist of a mistaken entry in the register or the mistaken omission of an entry which should have been made. Whether an entry in the register is mistaken depends upon its effect at the time of registration. So the entry of an estate or interest purportedly arising under a void disposition is a mistake. The entry made in the register does not reflect the true effect of the purported disposition when the entry was made. However, the entry of a person as having acquired an estate or interest under what proves to be a voidable disposition is not a mistake. Unless it had been rescinded at the time of registration, the disposition would be valid and it would not be a mistake to enter the disponee as the proprietor of the estate or interest under it. An entry cannot retroactively become a mistake. It cannot be argued therefore that the rescission of a voidable transaction retroactively makes the entry which recorded the disposition a mistake. That would undermine the policy of the Act.

It is implicit from this, Ms Galley suggested, that all that matters for the purpose of deciding whether there was a mistake is the position at the time of registration.

[23] I am unable to accept these submissions for a variety of reasons that I think are compelling.

[24] I start with the language of the Act. Schedule 6(1) says that "a person may apply to the registrar to be registered… if he has been in adverse possession of an estate". That surely indicates that a person who has not in fact been in adverse possession is simply not entitled to apply. Parliament cannot have intended that such a person could obtain registered title. A registration obtained by a person not entitled to apply for it would be mistaken. So, putting the register back in the condition it was in prior to the application would be correction of a mistake within the meaning of Schedule 4(1) and (5)

[25] I can see no reason for limiting the *correction of a mistake* to a mistake through some official error in the course of examination of the application, for which Ms Galley in part contended.

[26] Second, like the judge, I think the proposed construction would be an invitation to fraud. A dishonest applicant (perhaps knowing that the registered proprietor would be away or otherwise unable or unlikely to send in a NAP form in time or at all) could falsely claim that he had been in adverse possession for 10 years. His application would succeed because on its face it looked in order and the true owner would lose his land. The fact that there is no possibility of extending the prescribed time means that parliament either intended that the rectification power could cover such a case or that the true owner could lose his land for want of a form in time. The latter is wholly improbable.

[27] Ms Galley felt the force of that. She first suggested that the difficulty could be overcome because a fraudulent applicant would be met by a counternotice. However, that will not cover the case where the counternotice is not sent back in time which could include a case where the fraudster knew the registrar's letter would not be received or, if received, would not be dealt with in time.

[28] Her second submission on this point involved a concession: that a registration pursuant to a fraudulent application could amount to a mistake, whereas a registration pursuant to an innocent but mistaken application could not. Thus, she accepted that a registration obtained by the use of a forged conveyance could be rectified as *correcting a mistake*. Further, when she was pressed by the case of an application made when the applicant knew that the landowner would not respond in time, for example because he or she would be away or for some other reason other than mental incapacity specifically provided for in para 8 of Schedule 6 she accepted that that too would be a mistake within the meaning of Schedule 4. Her concession was made by a general appeal to the old adage "fraud unravels everything".

[29] The insuperable difficulty with this submission is that it is impossible to draw any rational distinction between a mistake induced by fraud and a mistake induced by a wrong application. The reason for the mistake that the registrar was given false information is the same in both cases.

[30] Third, her submission goes against the policy of the Act with regard to obtaining land by adverse possession. That policy was to make it *more* difficult to do that. It was set out in the consultation document thus:

> 10.2 We then put forward proposals for a wholly new substantive system of adverse possession that would apply only to registered land, and which is consistent with the principles of title registration. This would offer much greater security of title for a registered proprietor than exists under the present law, and would confine the acquisition of land by adverse possession to cases where it was necessary either in the interests of fairness or to ensure that land remained saleable.

As the judge observed, "it would be very strange if a registered proprietor could… be at risk of irrevocably losing his land to a squatter who had never in fact been in adverse possession".

[31] Fourth, her reliance on Ruoff and Roper is misplaced. Their suggestion that there is a distinction to be drawn between a void and a voidable transaction, interesting though it is, sheds no light on an application made by someone not entitled to apply. I would add that I would reserve my position as to whether the authors are right in their proposed distinction: it is difficult to see why, for instance, a transaction induced by a fraudulent misrepresentation (which would be only voidable) could not be corrected once the victim had elected to treat it as void.

[32] Fifth, I am not persuaded by her references to passages in the consultation and final reports. True it is that they refer to the need for certainty and simplicity. The normal operation of the provisions will indeed lead to that a 10- or more year squatter who wants to claim title

A can invoke the Schedule 6 procedure and the landowner will generally respond. If he objects to registration on receipt of the notice, it will be refused. However, if he then does nothing about it, the squatter can come back after two years and get title. The normal procedure brings things to a head one way or another.

[33] What the passages do not do is say anything about what happens when the application is made by someone who has not been in adverse possession. Nor is there anything in either report suggesting that a special limited meaning should be given to *correction of a mistake*. Indeed, if anything, to the contrary. The discussion of the bill's provisions, which became Schedule 4, says:

B

> 10.61… Rectification is confined to cases where a mistake is to be corrected. This will not include every case which is at present treated as rectification. It will not therefore cover cases where the register is altered to give effect to rights that have been acquired over the land since it was registered, or where the register was originally correct, but subsequent events have made it incorrect.

This is far from suggesting any special meaning of *mistake to be corrected.*

[34] Nor can I see any answer to the point accepted by the judge, that if land could be lost in this way the landowner would have lost it without compensation with a consequential breach of Article 1 to the C first Protocol of the Convention on Human Rights.

[35] Finally, although we were referred to other passages of Ruoff and Roper and to passages of the 7th edition of *Megarry & Wade: The Law of Real Property* (2008), whose editors include Mr Harpum, none of the passages was directly in point no bull's eye for either side. However, one passage in *Registered Land: Law and Practice under the Land Registration Act 2002* (2004) co-authored by Mr Harpum is to use a mixed metaphor squarely in point. It is a footnote to para. 30.1, and says:

> If an applicant was registered under the scheme provided by the LRA 2002 D and it then transpired that he had not in fact been in adverse possession for 10 years, his registration would be a mistake, and there would, therefore, be grounds for an application for rectification of the register: see LRA 2002, Sch 4, paras 2(1)(a), 5(a).

So the opinion of the architect of the Act is dead against Ms Galley's contentions. That clearly has considerable weight in view of Mr Harpum's deep connection with the Act and its drafting.

[36] Accordingly, I unhesitatingly reject the main point on this appeal.

(b) Onus point

E [37] It was common ground that, as the judge said, the adjudicator had "by implication misdirected herself on the burden of proof". The judge went on to consider whether the error arguably invalidated her findings of fact. If so, he would have to remit. He held that the decision on the facts did not turn on the onus and, accordingly, there should be no remission. Ms Galley submitted that the judge was wrong to hold that the wrong approach to onus could not have made any difference.

[38] I do not agree. The legal onus of proof lay on Mr Mannion. He put in evidence prima facie showing that Mr Baxter's claim to have been in exclusive possession for upwards of 10 years was wrong. That shifted the evidential burden to Mr Baxter. On the facts, he failed to F prove that he had been in adverse possession.

[39] Actually the findings simply did not turn on the onus of proof. The deputy adjudicator had "no hesitation in accepting the evidence given by and on behalf of Mr Mannion". She rejected part of Mr Baxter's evidence (for example: "I reject the assertion that the horses were on the Field every day"; "I do not accept that Mr Baxter seeded the land every year nor that he sprayed it regularly"). These holdings were not as a result of any question of onus of proof, they were a result of the adjudicator seeing and hearing key witnesses and evaluating the effect of this evidence along with the written evidence, which had been accepted without challenge.

[40] The judge was right to reject this point.

(c) "Unjust point"

[41] It is common ground that the adjudicator overlooked the fact that Mr Baxter had, by the time of the application, assumed possession of the land. So Schedule 4(6)(2) applied. Putting aside subpara (a), the question that the judge assessed for himself was this: would it be unjust not to put Mr Mannion back as the registered title holder. He held that it would be, saying that it was a matter of "simple justice". So it was. Mr Baxter had made an unjustified attempt to get himself title. Mr Mannion would otherwise lose his property.

[42] The only factor to which Ms Galley could point to suggest that it would be unjust to alter the register now was that Mr Mannion had failed to return the NAP form when he could have done. Mere failure to operate bureaucratic machinery is as thistledown to Mr Mannion losing his land and Mr Baxter getting it when he had never been in adverse possession. There is nothing in this point.

[43] So I would dismiss this appeal.

Tomlinson LJ said:

[44] I agree.

Mummery LJ said:

[45] I also agree.

Appeal dismissed.

Annex: Relevant provisions of Land Registration Act 2002

Alteration of the register

s65 Schedule 4 (which makes provision about alteration of the register) has effect.

Disapplication of periods of limitation

s96(1) No period of limitation under section 15 of the Limitation Act 1980 (c58) (time limits in relation to recovery of land) shall run against any person, other than a chargee, in relation to an estate in land or rentcharge the title to which is registered.

(2) No period of limitation under section 16 of that Act (time limits in relation to redemption of land) shall run against any person in relation to such an estate in land or rentcharge.

(3) Accordingly, section 17 of that Act (extinction of title on expiry of time limit) does not operate to extinguish the title of any person where, by virtue of this section, a period of limitation does not run against him.

Registration of adverse possessor

s97 Schedule 6 (which makes provision about the registration of an adverse possessor of an estate in land or rentcharge) has effect.

Schedule 4: Alteration of the Register

Introductory

1 In this Schedule, references to rectification, in relation to alteration of the register, are to alteration which

(a) involves the correction of a mistake, and

(b) prejudicially affects the title of a registered proprietor.

…

Alteration otherwise than pursuant to a court order

5 The registrar may alter the register for the purpose of

(a) correcting a mistake,

…

6(1) This paragraph applies to the power under paragraph 5, so far as relating to rectification.

(2) No alteration affecting the title of the proprietor of a registered estate in land may be made under paragraph 5 without the proprietor's consent in relation to land in his possession unless

(a) he has by fraud or lack of proper care caused or substantially contributed to the mistake, or

(b) it would for any other reason be unjust for the alteration not to be made.

(3) If on an application for alteration under paragraph 5 the registrar has power to make the alteration, the application must be approved, unless there are exceptional circumstances which justify not making the alteration.

Schedule 6

Registration of adverse possessor

Right to apply for registration

1(1) A person may apply to the registrar to be registered as the proprietor of a registered estate in land if he has been in adverse possession of the estate for the period of ten years ending on the date of the application.

...

Notification of application

2(1) The registrar must give notice of an application under paragraph 1 to

(a) the proprietor of the estate to which the application relates,

...

(2) Notice under this paragraph shall include notice of the effect of paragraph 4.

Treatment of application

3(1) A person given notice under paragraph 2 may require that the application to which the notice relates be dealt with under paragraph 5.

(2) The right under this paragraph is exercisable by notice to the registrar given before the end of such period as rules may provide.

4 If an application under paragraph 1 is not required to be dealt with under paragraph 5, the applicant is entitled to be entered in the register as the new proprietor of the estate.

5(1) If an application under paragraph 1 is required to be dealt with under this paragraph, the applicant is only entitled to be registered as the new proprietor of the estate if any of the following conditions is met.

(2) The first condition is that

(a) it would be unconscionable because of an equity by estoppel for the registered proprietor to seek to dispossess the applicant, and

(b) the circumstances are such that the applicant ought to be registered as the proprietor.

(3) The second condition is that the applicant is for some other reason entitled to be registered as the proprietor of the estate.

(4) The third condition is that

(a) the land to which the application relates is adjacent to land belonging to the applicant,

(b) the exact line of the boundary between the two has not been determined under rules under section 60,

(c) for at least ten years of the period of adverse possession ending on the date of the application, the applicant (or any predecessor in title) reasonably believed that the land to which the application relates belonged to him, and

(d) the estate to which the application relates was registered more than one year prior to the date of the application.

(5) In relation to an application under paragraph 1(2), this paragraph has effect as if the reference in sub-paragraph (4)(c) to the date of the application were to the day before the date of the applicant's eviction.

Right to make further application for registration

6(1) Where a person's application under paragraph 1 is rejected, he may make a further application to be registered as the proprietor of the estate if he is in adverse possession of the estate from the date of the application until the last day of the period of two years beginning with the date of its rejection.

(2) However, a person may not make an application under this paragraph if

(a) he is a defendant in proceedings which involve asserting a right to possession of the land,

(b) judgment for possession of the land has been given against him in the last two years, or

(c) he has been evicted from the land pursuant to a judgment for possession.

7 If a person makes an application under paragraph 6, he is entitled to be entered in the register as the new proprietor of the estate.

Restriction on applications

8(1) No one may apply under this Schedule to be registered as the proprietor of an estate in land during, or before the end of twelve months after the end of, any period in which the existing registered proprietor is for the purposes of the Limitation (Enemies and War Prisoners) Act 1945 (8&9 Geo 6 c16)

(a) an enemy, or

(b) detained in enemy territory.

(2) No-one may apply under this Schedule to be registered as the proprietor of an estate in land during any period in which the existing registered proprietor is

(a) unable because of mental disability to make decisions about issues of the kind to which such an application would give rise, or

(b) unable to communicate such decisions because of mental disability or physical impairment.

Effect of registration

9(1) Where a person is registered as the proprietor of an estate in land in pursuance of an application under this Schedule, the title by virtue of adverse possession which he had at the time of the application is extinguished.

LEASEHOLD ENFRANCHISEMENT

41-60 Albert Palace Mansions (Freehold) Ltd *v* Craftrule Ltd

COURT OF APPEAL

[2011] EWCA Civ 185

24 February 2011

SIR ANDREW MORRITT CVO C, AND SMITH AND BLACK LJJ

Estates Gazette 14 May 2011

[2011] 19 EG 98

Leasehold enfranchisement — Leasehold Reform, Housing and Urban Development Act 1993 — Collective enfranchisement — Application to acquire part of building containing 20 flats — That part divisible into two sets of 10 flats — Whether property a "self-contained part of a building" within section 3 — Whether that term denoting part incapable of further subdivision into smaller self-contained parts

The respondent was the nominee purchaser acting in a collective enfranchisement claim by the tenants of 10 flats under the Leasehold Reform, Housing and Urban Development Act 1993. The flats formed part of a five-storey, terraced building in London SW11. The tenants' initial notice, served on the appellant freeholder under section 13 of the Act, sought to acquire the freehold of that part of the building containing flats 41-60. This consisted of a vertically divided section of the building, itself divided into a left- and right-hand half, each containing 10 flats and each having separate entrances and common parts. The flats of three of the participating tenants were in the left-hand section while the remainder were located in the right-hand half.

The appellant did not serve a counternotice. However, it disputed the tenants' right to enfranchise on the ground that the property did not qualify as a "self-contained part of a building" for the purposes of sections 3 and 4 of the Act. It contended that the statutory right to collective enfranchisement was exercisable only in respect of a part of a building that could not be further subdivided into smaller self-contained parts and that each half of the property, containing flats 41-50 and 51-60 respectively, itself qualified as a self-contained part.

The county court judge rejected that contention and made an order, under section 25(1) of the Act, that the participating tenants were entitled to enfranchise the entire property specified in the initial notice on the terms therein proposed. That decision was upheld by the High Court, which held that section 3 did not, either expressly or by necessary implication, require that a self-contained part of a building should be indivisible into smaller self-contained parts: see [2010] EWHC 1230 (Ch); [2010] 2 EGLR 45; [2010] 31 EG 64. The appellant appealed.

***Held*: The appeal was dismissed. On the natural meaning of the words, the expression "self-contained part of a building", in section 3 of the Act, includes a self-contained part that is itself capable of being divided into smaller self-contained units. The words are neither ambiguous nor obscure and such a meaning does not lead to absurdity. That interpretation is supported by section 13(8), (9) and (10), which envisages that a notice that specifies a self-contained part of an entire building might later be replaced by another notice specifying a different self-contained part within that covered by the first notice. Those provisions contemplate that an enfranchisement claim can validly be made in respect of a larger self-contained part that can be divided into smaller self-contained parts, provided that sufficient qualifying tenants agree to participate. Section 4(3A), excluding enfranchisement of premises where different parties own the freehold of different parts of them and any of those parts is self-contained, likewise implies that a self-contained part of a building may itself contain two or more self-contained units. There is no justification for putting a gloss on the clear statutory words so as to require that a self-contained part must be the smallest possible self-contained part.** F

The following cases are referred to in this report.

Attorney-General v *Lamplough* (1878) LR 3 Ex D 214 G
Denetower Ltd v *Toop* [1991] 1 WLR 945; [1991] 3 All ER 661; [1991] 1 EGLR 84; [1991] 20 EG 194, CA
41-60 Albert Palace Mansions (Freehold) Ltd v *Craftrule Ltd*; *sub nom Craftrule Ltd* v *41-60 Albert Palace Mansions (Freehold) Ltd* [2010] EWHC 1230 (Ch); [2010] 1 WLR 2046; [2010] 3 All ER 952; [2010] 2 EGLR 45; [2010] 31 EG 64
Holding & Management (Solitaire) Ltd v *Finland Street 1-16 RTM Co Ltd* [2008] 1 EGLR 107; [2008] 02 EG 152, LT
Kay-Green v *Twinsectra Ltd (No 1)* [1996] 1 WLR 1587; [1996] 4 All ER 546; [1996] 2 EGLR 43; [1996] 38 EG 136, CA
Long Acre Securities Ltd v *Karet* [2004] EWHC 442 (Ch); [2005] Ch 61; [2004] 3 WLR 866; [2005] 4 All ER 413; [2004] 2 EGLR 121
Majorstake Ltd v *Curtis* [2008] UKHL 10; [2008] 1 AC 787; [2008] 2 WLR 338; [2008] 2 All ER 303; [2008] 2 P&CR 2; [2008] 1 EGLR 44; [2008] 14 EG 102 H
Oakwood Court (Holland Park) Ltd v *Daejan Properties Ltd* [2007] 1 EGLR 121, CC
Pepper (Inspector of Taxes) v *Hart* [1993] AC 593; [1992] 3 WLR 1032; [1993] 1 All ER 42, HL

This was an appeal by the appellant, Craftrule Ltd, from a decision of HH Judge Madge, sitting in Central London County Court, allowing a claim by the respondent, 41-60 Albert Palace Mansions (Freehold) Ltd, to acquire the freehold of part of a building by collective enfranchisement under the Leasehold Reform, Housing and Urban Development Act 1993. J

Kenneth Munro (instructed by Olswang LLP) appeared for the appellant; Philip Rainey QC (instructed by Butcher Burns LLP) represented the respondent.

Giving judgment, **Smith LJ** said:

Introduction

[1] This is an appeal from the order of Henderson J made on 27 May 2010 (see [2010] EWHC 1230 (Ch)*) dismissing the appeal of Craftrule Ltd against the order of HH Judge Madge, sitting in Central London County Court. It is therefore a second appeal. It raises a point of K statutory construction as to the meaning of the words "a self-contained part of a building" in sections 3 and 4 of the Leasehold Reform, Housing and Urban Development Act 1993 (the 1993 Act). Rimer LJ gave permission to appeal since he considered that the issue raised an important point of principle that this court ought to consider.

[2] The appellant, Craftrule Ltd, is the owner of the freehold reversionary interest in the premises known as 41-60 Albert Palace

* Editor's note: Reported at [2010] 2 EGLR 45

A Mansions, Lurline Gardens, London SW11 4DQ (the property). The respondent is a limited company set up for the purpose of purchasing the freehold of the property pursuant to the enfranchisement provisions within the 1993 Act. It is controlled by the leasehold tenants of the flats comprising the property and is their nominee purchaser for the purpose of their claim for enfranchisement.

[3] Part 1 of Chapter 1 of the 1993 Act provides for the collective enfranchisement of flats by qualifying tenants. For the purposes of this appeal, the essentials are that, pursuant to section 13 of the Act, a group of long leaseholding tenants holding not less than half the flats within the premises concerned may serve a notice on the landlord B seeking to acquire the freehold. The premises concerned must be a self-contained building or part of a building, as defined in section 3 of the 1993 Act. Section 3 is headed "Premises to which this Chapter applies" and provides:

> (1) Subject to section 4, this Chapter applies to any premises if
> (a) they consist of a self-contained building or part of a building…;
> (b) they contain two or more flats held by qualifying tenants; and
> (c) the total number of flats held by such tenants is not less than two-thirds of the total number of flats contained in the premises.
> (2) For the purposes of this section a building is a self-contained building if it is structurally detached, and a part of a building is a self-contained part of C a building if
> (a) it constitutes a vertical division of the building and the structure of the building is such that that part could be redeveloped independently of the remainder of the building; and
> (b) the relevant services provided for occupiers of that part either
> (i) are provided independently of the relevant services provided for occupiers of the remainder of the building, or
> (ii) could be so provided without involving the carrying out of any works likely to result in a significant interruption in the provision of any such services for occupiers of the remainder of the building;…

[4] Section 4 of the 1993 Act provides for certain exclusions from D the right to enfranchisement, but it is common ground that none of them applies in the present case.

[5] The property forms part of a terraced building comprising 160 flats, built in the late 19th or early 20th century. The building consists of eight "handed pairs". Each "pair" comprises 20 flats arranged on five floors. The property comprises one such pair consisting of 20 flats, nos 41-60, situated roughly in the middle of the terrace. Each half of the pair has a separate entrance leading to separate common parts and staircases. There is a dividing wall between each half (and the adjacent parts of the building) that is vertically continuous from the footings to roof level. Each half has its own drainage system, E mains water riser, electricity supply and entry phone. Telephone and cable TV services are provided to individual flats, although the cables are bundled together. There are no other communal services. Each half of the pair could be redeveloped independently of the other half of the pair and/or of the rest of the building. All the flats in the property save for nos 41 and 50 are demised under long residential leases for a term of more than 21 years. The service charge provisions in the leases are defined by reference to the entire property (flats 41-60).

[6] Save in one respect, it is not disputed that the entire property (flats 41-60) satisfies all the requirements of section 3 as set out above. The only issue that the appellant raises is that the property is F not "a vertical division of the building". It contends that it comprises two vertical divisions of the building. However, it is agreed that each half of the pair (41-50 and 51-60) would independently satisfy all the requirements of section 3. I interpose to say that, in my view, the expression "vertical division" is rather odd. A division is a line of no thickness. It seems to me that what section 3(2)(a) means is that the premises must be a vertical slice of the building.

[7] Section 13(1) of the 1993 Act provides that a claim for enfranchisement is to be made by the giving of a notice to the landlord. By section 13(2), the notice must be given by the qualifying tenants (viz tenants with long leases) of not less than half the total number of flats within the relevant premises. In March 2008, a group of 10 long leasehold tenants (the participating tenants) gave notice, pursuant to section 13, claiming the right to acquire the freehold reversionary interest in the entire property (flats 41-60). Seven of "the participating tenants" lived in the 51-60 half and only three lived in the 41-50 half. The landlord failed to issue a counternotice as it could have done pursuant to section 21 of the 1993 Act. The result was that, unless the notice were to be declared invalid, "the participating tenants", through their nominee purchaser, would be entitled to purchase the freehold on the terms set out in the notice.

Proceedings below

[8] A dispute arose between the parties as to the validity of the notice and the tenants commenced proceedings in Central London County Court, seeking a declaration, pursuant to section 25 of the Act, that their notice was valid and that they were entitled to acquire the freehold on the terms proposed. It was common ground that the notice was valid subject to the one issue that remains for determination on this appeal. The landlord contended that it was not open to the tenants to seek to enfranchise the entire property, that is all 20 flats, by the issue of one notice. It argued that the right to enfranchisement attaches only to the smallest possible self-contained part of a building. Since each half of the pair is a self-contained part of a building, each half must be the subject of a separate claim and notice must be given by qualifying tenants of at least half the flats in that half of the pair.

[9] Judge Madge rejected that submission and held that the notice in respect of the entire property was valid. The same arguments were deployed before Henderson J on the first appeal. He dismissed the appeal, taking the same view for much the same reasons as those given by Judge Madge.

[10] The basis of Henderson J's decision was that the statutory language of section 3 was clear and unambiguous. It permitted the enfranchisement of any self-contained part of a building (subject to the exclusions in section 4, which did not apply). There was nothing in the section that suggested that the right attached only to the smallest possible self-contained part. Had parliament intended to oblige tenants to claim the smallest part of the building to satisfy the requirements of section 3(2), it would have said so. He was satisfied that this construction was consistent with the statutory purpose of the scheme, as identified in *Majorstake Ltd* v *Curtis* [2008] UKHL 10; [2008] 1 AC 787*, in [21] and [23]. Moreover, this construction of section 3 derived support from other provisions of the Act, in particular subsections 13(8), (9) and (10) and section 4(3A). He rejected the submission advanced by Mr Kenneth Munro, for the landlord, that there was authority, which although not directly in point was of assistance. The judge regarded those cases as irrelevant. Further, he rejected Mr Munro's invitation to have regard to certain passages from the speeches of Lord Strathclyde taken from *Hansard*. The judge said that there was no ambiguity in the statutory words to justify such a course. In any event, having read the passages *de bene esse*, the judge did not think that they were capable of resolving any supposed ambiguity.

[11] Rimer LJ gave permission for this second appeal because the issue is of some general importance on which there is no direct authority.

Appeal to this court: Submissions

[12] In outline, Mr Munro, (counsel for the landlord in this court and below) submitted that Henderson J had erred in several respects. He had been wrong to reject the authorities cited as irrelevant. He should not have refused to have regard to the *Hansard* extracts. He had failed to deal with the submissions advanced as to the implications of the decision. Mr Munro submitted that the construction accepted by the judge would have a most undesirable effect. It would enable a group of tenants to enfranchise a part of a building in which they had no financial interest against the wishes of the tenants in that part.

[13] Mr Philip Rainey QC (counsel for the tenants here and below) submitted that the judge had been right in all respects for the reasons

---

* Editor's note: Also reported at [2008] 1 EGLR 44

that he gave. In particular, he had been right to rely on subsections (8), (9) and (10) of section 13 and section 4(3A).

Discussion

[14] I will begin by considering subsections (8), (9) and (10) of section 13. These provide:

> (8) Where any premises have been specified in a notice under this section, no subsequent notice which specifies the whole or part of those premises may be given under this section so long as the earlier notice continues in force.
>
> (9) Where any premises have been specified in a notice under this section and
>
> (a) that notice has been withdrawn…, or
>
> (b) …
>
> no subsequent notice which specifies the whole or part of those premises may be given under this section within the period of twelve months beginning with the date of the withdrawal…
>
> (10) In subsections (8) and (9) any reference to a notice which specifies the whole or part of any premises includes a reference to a notice which specifies any premises which contain the whole or part of those premises; and in those subsections and this "specifies" means specifies under subsection (3)(a)(i).

[15] These provisions appear to envisage that a notice that specifies a self-contained part of the building (the whole of the relevant premises, which I will call X) may be later replaced by a notice that specifies a different self-contained part of the building (which I will call Y) where Y is only a part of X. In that event, Y is necessarily a smaller part of the building than X. Therefore, if notices in respect of both X and Y are potentially valid, it follows that a self-contained part of a building (for the purposes of section 3) cannot be limited to the smallest possible self-contained part.

[16] Mr Munro argued in response that the expression "part of those premises" in section 13(8) is intended to relate to a part of the premises, such as appurtenant garages or walkways, that may have been included in the "larger" self-contained part of the building but that may be hived off and excluded from the specified premises in a later notice. This, he said, is intended to give tenants the choice whether they enfranchise just the flats themselves or whether they include associated premises such as garages and walkways. The power of choice appears to derive from section 1(5) of the 1993 Act, which provides that qualifying tenants may exercise their right of enfranchisement in respect of any premises despite the fact that those premises are less extensive than the entirety of the premises in respect of which those tenants are entitled to exercise the right.

[17] I would accept that that subsection gives the tenants the choice to which Mr Munro referred. However, it also enables them to choose whether to specify a larger or smaller self-contained part of a building, provided that they would be entitled to seek enfranchisement of either. The smaller part must be self-contained if it is to support a claim, but the provisions plainly envisage that the claim may be made in respect of a larger self-contained part of the building, provided that there are sufficient qualifying tenants who agree to participate. In my judgment, Mr Munro's submission, although not incorrect in itself, does not take the argument any further. It appears to me that section 13(8), (9) and (10) provides powerful support for Mr Rainey's argument.

[18] The other statutory provision on which Henderson J relied is section 4(3A), which was introduced into the 1993 Act by amendment in 1996. Prior to the amendment, section 3(1)(a) contained an additional requirement that the freehold of the whole of the building or self-contained part of the building should be owned by the same person. It was found that some landlords were avoiding enfranchisement by hiving off the freehold of small parts of the premises to another landlord, usually an associated company. So, in 1996, that additional requirement was repealed and a new exclusion clause was inserted in section 4, as follows:

> 4 Premises excluded from right.
>
> …
>
> (3A) Where different persons own the freehold of different parts of premises within subsection (1) of section 3, this Chapter does not apply to the premises if any of those parts is a self-contained part of a building for the purposes of that section. G

[19] Mr Rainey's submission in respect of this subsection was that this additional exclusion would have been wholly unnecessary if a valid notice could be served only in respect of an indivisible self-contained part of the building. The freehold of that indivisible self-contained part would inevitably be held by one landlord. The implication of this provision is that a self-contained part of a building may well contain two or more self-contained parts.

[20] Mr Munro made two submissions on this point. First, he submitted that it was not permissible to have regard to this amended provision in construing the words of the original statutes: see H *Attorney-General* v *Lamplough* (1878) LR 3 Ex D 214. Second, if that were wrong, section 4(3A) does no more than show that parliament wanted to clarify the position in respect of blocks of flats where the freehold is divided between owners and says nothing about the position where (as here) the freehold of the relevant premises is in single ownership. In my view, it is permissible to have regard to this amendment, which became necessary once the additional requirement (which had been abused) was removed from section 3 in order that it could be made clear that enfranchisement was available only in respect of a self-contained part of a building, the freehold of which was in single ownership. I share the view of Henderson J that this provision J provides support for the tenants' contention.

[21] In my judgment, sections 4(3A) and 13(8), (9) and (10) taken together are conclusive of the true construction of the expression "self-contained part of a building" within section 3(1). I can see no justification for putting a gloss on the clear statutory words so as to require that a self-contained part must be the smallest possible self-contained part.

[22] I can deal with Mr Munro's other arguments quite briefly. I reject his submission that assistance is to be derived from *Oakwood Court (Holland Park) Ltd* v *Daejan Properties Ltd* [2007] 1 EGLR 121. There, a claim for enfranchisement of several flats (being part K of a building) failed; the part was not self-contained because it shared the services of a boiler house with the other flats. The case was not concerned with the divisibility of a self-contained part of a building and said nothing of assistance to the present issue. Similarly, *Holding & Management (Solitaire) Ltd* v *Finland Street 1-16 RTM Co Ltd* [2008] 1 EGLR 107 is of no assistance. It concerned the right of tenants to manage part of a building in which their flats were located pursuant to the Commonhold and Leasehold Reform Act 2002. The statutory provisions are similar to those of the 1993 Act, and the question arose as to whether the part of the building in question complied with the requirement of the equivalent of section 3(2)(a) of the 1993 Act. The L problem was that the dividing line between the premises and the rest of the building was not vertical the whole way down; part of the dividing line was lateral. The case was not concerned, as we are, with the divisibility of a self-contained part where the verticality of the division is not in issue.

[23] Mr Munro also sought to rely on *Denetower Ltd* v *Toop* [1991] 1 WLR 945*, *Kay-Green* v *Twinsectra Ltd (No 1)* [1996] 1 WLR 1587† and *Long Acre Securities Ltd* v *Karet* [2004] EWHC 442 (Ch); [2005] Ch 61‡. These were all cases under the Landlord and Tenant Act 1987 relating to a tenant's right of first refusal to purchase the freehold that the landlord wanted to sell. The statutory provisions are similar but not M identical to those of the 1993 Act, but in any event the issues decided in those cases bore no relation to the issue in the present case.

[24] I would also reject Mr Munro's submission that the construction favoured by the judge would have wide implications, including undesirable and unintended consequences. His complaint is that if a self-contained part of a building can include two or more smaller self-contained parts, the right to enfranchisement will or might be

---

* Editor's note: Also reported at [1991] 1 EGLR 84

† Editor's note: Also reported at [1996] 2 EGLR 43

‡ Editor's note: Also reported at [2004] 2 EGLR 121

exercisable by tenants who do not have any financial interest in the other part. Take, for example, the situation in the present property and a proposal to claim enfranchisement of the entire property comprising 20 flats. If all the tenants of flats 51-60 were in favour of the proposal and all those in 41-50 were against it, those in favour could impose their will on the others, even though they had no financial interest in flats 41-50. Even worse, said Mr Mr Munro, if the whole building of 160 flats had but one freehold owner (which in fact it does not), it would be possible for the tenants of flats 1-80 to impose their will on those in flats 81-160, even though all the latter were opposed. I must say that these circumstances seem unlikely to occur, but I accept that hypothetically they are possible. I cannot see any reason why such a situation should be regarded as so undesirable that one should infer that parliament cannot have intended it to occur. It is obvious that parliament envisaged that 50% of the tenants within any self-contained part of the building might be unwilling to join in the enfranchisement. That could be so even if Mr Munro were correct that the right of enfranchisement attaches only to the smallest possible self-contained part of the building. I cannot see that there is anything more undesirable in imposing a new landlord on 10 unwilling tenants out of 20 than there is in imposing it on five tenants out of 10. In my view, it is clear that parliament thought that there were real advantages to tenants if they were able to enfranchise and that those advantages should not be lost merely because there were some unwilling tenants. Those advantages are explained in the passage from *Majorstake* to which Henderson J referred.

[25] I come finally to Mr Munro's submission that the expression "a self-contained part of a building" is ambiguous and that we should resolve the ambiguity by reference to the speech of Lord Strathclyde during a debate in Committee in the House of Lords on 9 March 1993. We are entitled to look at *Hansard* as an aid to construction only if the legislation is either ambiguous or obscure or leads to an absurdity: see *Pepper (Inspector of Taxes)* v *Hart* [1993] AC 593. In my judgment, on the ordinary natural meaning of the words, the expression "self-contained part of a building" in section 3 includes a self-contained part of a building that is itself capable of being divided into smaller self-contained parts of a building. The words are neither ambiguous nor obscure. Nor does such a meaning lead to absurdity. I would hold that we are not entitled to have regard to Lord Strathclyde's speech.

[26] In the course of argument, we examined, *de bene esse*, the extract from *Hansard* on which Mr Munro sought to rely. I do not propose to quote it. It does not appear to me that these passages would resolve the ambiguity, if one existed.

[27] Lord Strathclyde was responding to three quite technical amendments. Amendment 13 proposed that the criterion for establishing viability in respect of the provision of services should be changed. In the bill, a self-contained part of a building could be enfranchised only if it were possible to separate services from the remainder of the block without *significant disruption* to the rest of the block. The amendment proposed would prevent separate parts of the building from being enfranchised unless separation of services could be achieved without *significant costs*. In explaining why the government would not want to adopt that amendment, Lord Strathclyde said that it was intended that enfranchisement should be exercisable over the smallest viable unit. He feared that the proposed amendment might make this more difficult because it might mean that a larger unit had to be purchased. He said nothing to suggest that tenants who chose to enfranchise a larger unit would not be permitted to do so, provided of course that there were enough of them within that unit who wanted to do so.

[28] In respect of amendments 14 and 15, the concern raised appears to have been for the position of those tenants who were left outside the enfranchised part of the building, although this is not entirely clear. Lord Strathclyde again said that the bill provided that the smallest viable unit could be enfranchised. He said that this would ensure that leaseholders could enfranchise only the property in which they had an immediate financial interest. He did not say that only the smallest viable unit could be enfranchised. I do not find it at all easy to understand exactly what he meant by saying that allowing the enfranchisement of the smallest viable unit would ensure that leaseholders could enfranchise only the property in which they had an immediate financial interest. The requirement that there should be a certain proportion of tenants participating in the purchase must be enough to ensure that they have an immediate financial interest, whether the unit under consideration is the smallest viable unit or two or more self-contained viable units that, taken together, are also a viable self-contained unit. Accordingly, I do not think that the passages relied on by Mr Munro would resolve any ambiguity, if there were one.

[29] For those reasons, I would dismiss this appeal and uphold the declaration made by Judge Madge.

Black LJ said:

[30] I agree.

Sir Andrew Morrit CVO said:

[31] I also agree.

Appeal dismissed.

MORTGAGE

Helden *v* Strathmore Ltd

COURT OF APPEAL

[2011] EWCA Civ 452

11 May 2011

LORD NEUBERGER MR AND SMITH AND ELIAS LJJ

Estates Gazette 20 August 2011

[2011] 33 EG 64

Mortgage — Enforcement — Terms of loan Legal charge over property to secure loan — Parties agreeing loan amount and interest rate — Charge as executed omitting those details — Whether charge enforceable notwithstanding defect where parties reaching clear oral agreement on loan amount and interest rate — Whether breach of section 2 of Law of Property (Miscellaneous Provisions) Act 1989 or section 53 of Law of Property Act 1925

The appellant part-owned a property development company that had taken out loans with the respondent's sister company to finance the acquisition of properties. In 2005, the respondent lent £1m to the appellant, in his personal capacity, to enable him to purchase a property as his home. An agreement dated December 2005 specified the loan amount and an initial interest rate of 7.5% pa for a loan period of one year from the date of completion of the purchase, with the rate to be renegotiated in the event that the loan was not repaid within that time. The appellant executed a legal charge over the property to secure the loan. The charge purported to identify the debt and the interest rate by reference to an "offer letter", although no such letter had been prepared, and stipulated that the appellant would repay the debt at the end of March 2007. The charge was duly registered.

The appellant took a further loan of £25,000 from the respondent at an interest rate of 15% and also assumed personal responsibility for the repayment of a £90,000 loan to his company, with both sums to be secured by the 2006 charge. After it became apparent that the charge would not be redeemed by the end of March 2007, the parties agreed interest rate increases on the main loan of 10% from that date, rising to 12.5% from October 2007.

In proceedings against the respondent, the appellant disputed the enforceability of the charge on the ground that it was defective for not setting out all the relevant terms in writing, contrary to section 2 of the Law of Property (Miscellaneous Provisions) Act 1989; or, alternatively, that it contravened section 53 of the Law of Property Act 1925. He further contended that the main £1m loan was unenforceable under the Financial Services and Markets Act 2000.

Dismissing the claim, the judge held that the appellant was bound by the oral agreement on the loan amount and interest rate notwithstanding the defects in the 2006 charge. In respect of the 2000 Act, he held that although the respondent had breached section 19 in providing credit on the main loan, it was just and equitable, within section 28, to permit its enforcement. The appellant appealed.

***Held*: The appeal was dismissed. (1) The parties had agreed that the main loan would be the £1m advanced to the appellant to enable him to purchase his property and that the rate of interest would be 7.5% pa. The error in the drafting of the 2006 charge had arisen as a result of a mistaken assumption by the solicitor acting for both parties that an offer letter had been prepared, and F an oversight on the part of the parties when they came to sign the charge. The appellant was bound, on the ground of estoppel by convention, by the oral agreement as to the size of the loan and the interest payable thereon. The same result could also have been reached by ordering the rectification of the 2006 charge. Further, the judge had been entitled to find that the parties had agreed the subsequent interest rate increases. (2) Section 2 of the 1989 Act was not a bar to the enforceability of the charge. Section 2 is directed at tightening up the formalities required for contracts for the creation or sale of legal estates or interests in land. It is not concerned with documents that actually create or transfer such estates or interests. G Although a contract to transfer a freehold or a lease in the future, or a contract for a mortgage in the future, falls within the reach of the section, an actual transfer, conveyance or assignment or a lease or an actual mortgage does not. (3) Although section 53 of the 1925 Act does apply to mortgages, it is less prescriptive than section 2 of the 1989 Act and does not give rise to problems so far as estoppel or rectification are concerned. (4) The judge had been entitled to find on the facts before him that section 19 of the 2000 Act had been contravened but that it was just and equitable to allow the agreement to be enforced.**

The following cases are referred to in this report. H

Amalgamated Investment & Property Co Ltd (in liquidation) v *Texas Commerce International Bank Ltd* [1982] QB 84; [1981] 3 WLR 565; [1981] 3 All ER 577; [1982] 1 Lloyd's Rep 27, CA
Helden v *Strathmore Ltd* [2010] EWHC 2012 (Ch); [2011] Bus LR 59
Khodari v *Tamimi* [2009] EWCA Civ 1109
McLaughlin v *Duffill* [2008] EWCA Civ 1627; [2010] Ch 1; [2009] 3 WLR 1139; [2009] 2 EGLR 71; [2009] 34 EG 80
Oun v *Ahmad* [2008] EWHC 545 (Ch); [2008] 13 EG 149 (CS)
R v *Johnstone*; *R* v *Harrison*; *R* v *Mayron Multimedia Ltd*; *R* v *Eley*; *R* v *Ho*; *R* v *Croxson* [2003] UKHL 28; [2003] 1 WLR 1736; [2003] 3 All ER 884
Target Holdings Ltd v *Priestley* [1999] Lloyd's Rep Bank 175; (2000) 79 P&CR 305, Ch J
Torbay District Council v *Singh* [1999] 2 Cr App R 451

This was an appeal by the appellant, Charles Helden, from a decision of Newey J, sitting in the Chancery Division of the High Court, dismissing a claim against the respondent, Strathmore Ltd, challenging the enforceability of a legal charge over property under the Law of Property (Miscellaneous Provisions) Act 1989 and the Law of Property Act 1925.

David Christie (instructed by Morgan Hall) appeared for the appellant; Fred Philpott (instructed by Neumans LLP) represented the respondent. K

Giving judgment, **Lord Neuberger MR** said:

[1] This is an appeal brought by Mr Charles Helden against a decision of Newey J (see [2010] EWHC 2012 (Ch)*), who held that a charge that he had granted to Strathmore Ltd was enforceable against him, notwithstanding the facts that it was defectively drafted and it was entered into in breach of the Financial Services and Markets

---

* Editor's note: Reported at [2011] Bus LR 59

A Act 2000 (the FSMA). Mr Helden also appeals against the judge's award of indemnity costs against him.

Facts as agreed or found by the judge

[2] Mr Helden initially worked as a commercial conveyancing clerk, and then as a mortgage broker until 1999, when he gave up work to look after his dying first wife. In 2002, he joined forces with Mr John Jordan, an experienced house builder and property developer and they embarked on a business of developing property, mainly in London.

[3] In 2004, through one of their companies, C&J, Mr Helden and Mr Jordan were negotiating the purchase of 23 Catherine Place
B (Catherine Place). Mr Helden asked Mr Neil Wright, a solicitor at Coldham Shield & Mace, whether he knew of a source of funding and Mr Wright suggested Mr and Mrs Ashton.

[4] Mr Ashton initially worked in finance for 10 years and, in 1986, he started working for himself as a property investor and developer. He set up Strathmore Ltd (owned at all material times by Mr and Mrs Ashton) as a vehicle through which to build up a portfolio of commercial properties, which were then mostly sold off. He also had another company called Sandworth Ltd. The judge said that "Mr Ashton has very considerable financial acumen. However, he has never attended any courses on financial matters, nor acquired any
C post-school qualifications": see [10].

[5] As a result of Mr Wright's introduction, C&J was subsequently lent £300,000 by Sandworth to assist with the purchase of Catherine Place. Later, in 2005, Sandworth made a loan of some £5.4m to enable another company owned by Mr Helden and Mr Jordan to buy a property in Belgravia (Belgravia). Mr Jordan also borrowed money from Sandworth to acquire a house at Tewin Wood in Hertfordshire in the name of a third company.

[6] By late 2005, Mr Helden wished to buy 58 Chelsea Crescent (the property) as his home. Mr Helden originally intended to buy the property through a company, but he abandoned that idea when he realised that
D it would make remortgaging difficult. He approached Mr Ashton for funding, and the Ashtons agreed to make him a loan of £1m (the main loan). Fifty thousand pounds were initially advanced in December 2005 for the deposit, and a further £995,462.83 was advanced on 31 March 2006, when the purchase was completed.

[7] A contemporaneous note made by Mr Ashton recorded that the main loan was to bear interest at 7.5% pa and would be repaid in a year and Mr Helden accepted that in his evidence.

[8] An agreement (the December 2005 agreement) was entered into by Strathmore and Mr Ashton on 19 December 2005. One of the recitals to this agreement stated that "Ashton intends to make a
E loan of approximately one million pounds… through Strathmore to [Mr Helden]" and that the purpose of this loan was "to enable" him "to purchase… the Property at an initial interest rate of 7.5% pa". The recital also stated that the period of the main loan was to be "for one year from the date of completion of the purchase of the Property", and that "in the event that the Loan is not repaid within that time at a rate to be negotiated". The December 2005 agreement also stated that the main loan would be secured by a charge over the property and that Strathmore would hold the charge and any moneys received from Mr Helden on trust for Mr Ashton.

[9] On 31 March 2006, Mr Helden executed a legal charge (the 2006
F charge) in favour of Strathmore. This provided for the property to be charged "by way of a legal mortgage with the payment to [Strathmore] of the debt to be paid by [Mr Helden]". The 2006 charge, stipulated that Mr Helden would pay to Strathmore "on [31 March 2007] the amount of the Debt". Against "The Debt" and "Interest Rate" there was written "In accordance with the offer letter". No such letter had in fact been prepared. The 2006 charge included a provision for Mr Helden's fiancée, now his second wife, to countersign in the presence of an independent solicitor to confirm her consent, which she duly did. The 2006 charge was duly registered at HM Land Registry.

[10] Mr Wright acted both for Mr Helden and for Mr Ashton and Strathmore; in evidence, he accepted that he should not have done so. Mr Wright did not appreciate that there was any question of the FSMA applying to the transaction, as he was not familiar with its requirements and so made no mention of it to Mr Helden or Mr Ashton. Neither did Mr Helden, Mr Ashton or Mrs Ashton realise that the FSMA could be of any relevance.

[11] Around this time, Sandworth lent C&J around £2.3m to enable it to repay sums that it had borrowed from United Trust Bank for the purchase of Catherine Place and a site in East Sussex. Also around this time, Mr Helden borrowed a further sum of £30,000 from Strathmore on a personal basis, and he signed a promissory note, in which he promised: "to pay to Strathmore… the sum of [£30,000] for value received on or before 30th June 2006 with interest at the rate of 15% per annum on the principal sum". This loan was not secured on the property.

[12] Within a short period, £50,000 was repaid when Belgravia was sold in May 2006. By that time, Mr Helden and the Ashtons had become friends. The Ashtons attended the wedding when Mr Helden remarried in Italy on 19 May 2006. Mr Wright and his wife were also guests.

[13] Shortly afterwards, Mr Helden asked Mr Ashton for a further loan of £25,000 (the £25,000 loan), saying: "as before I accept the flat interest rate of 15%". The judge accepted that the £25,000 loan was made on the agreed basis that it should be secured by the 2006 charge.

[14] When it became clear that the 2006 charge would not be redeemed at the end of March 2007, as had been agreed, Mr Helden had discussions with Mr Ashton, as a result of which it was agreed that the interest rate on the main loan would rise to 10% pa with effect from 31 March 2007. Subsequently, on 14 September 2007, Mr Helden agreed in a conversation with Mr Ashton that the interest rate would increase to 12.5% pa from 1 October 2007.

[15] In reaching his conclusions as to the facts summarised above (many of which were in dispute), the judge was understandably influenced by contemporaneous notes of discussions made by Mr Ashton. In a number of those notes, it appears to have been assumed that interest on Mr Helden's indebtedness fell to be compounded on, broadly, a quarterly basis. It is also relevant to mention that Mr Ashton referred in an e-mail to Mr Helden of 15 January 2008 to interest being "calculated every three months and added to the principal in the normal way".

[16] Around the end of 2007, a dispute arose over a loan of £35,000 that had been made by Sandworth to C&J in January 2005. As a result of C&J's failure to repay this loan and the unusual terms of the loan, C&J now owed more than £90,000 in respect of the initial loan. The judge concluded that Mr Ashton and Mr Helden agreed that repayment of this amount (the £90,000 loan) should be the personal liability of Mr Helden and should be secured by the charge.

[17] In early 2008, while negotiating with Bank of Scotland plc for a remortgage of the property, Mr Helden began to investigate whether the 2006 charge was open to attack under the FSMA. This eventually resulted in his issuing the instant proceedings in March 2009.

[18] In late 2009, three experienced firms of estate agents, Savills, Strutt & Parker and Winkworth, each suggested that the property should be marketed at £1.8m.

[19] In addition to their loans referred to above, the Ashtons made loans through Strathmore and Sandworth to members of their family and to a Mr Michael Harrison. As the judge explained, the loans made to Mr Harrison were not affected by the FSMA for two reasons: they were made before the relevant law came into force and the relevant properties were not used as or in connection with a dwelling by Mr Harrison.

[20] On the basis of these facts, the judge had to consider a number of issues. The first concerned the terms of the main loan and the 2006 charge, the £25,000 loan, and the £90,000 loan (the three loans). The second issue was whether, quite apart from the FSMA, those terms were enforceable. The third issue was whether the main loan was unenforceable by virtue of the FSMA. The final issue was the question of costs. I shall consider those issues in the same order.

Terms of the three loans

[21] Before the judge, Mr Helden disputed that Strathmore was the creditor in respect of the three loans, a contention that the judge rejected, a finding against which, very sensibly, there is no appeal.

[22] More relevantly for present purposes, the judge held that, although the 2006 charge was defective in its provisions, because it failed to identify the loan that was thereby secured and the rate of interest it was to carry, as it sought to do so by reference to an offer letter that did not exist, it was clear that the parties had agreed what the loan was and the interest that was to be payable in respect of it. As the judge said, the point was hardly pleaded by Mr Helden, and his statement of case sought no relief in this connection. However, the point was taken rather more forcefully before us.

[23] In my view, there is nothing in the point. On their respective pleaded cases, both Mr Helden and Strathmore contend that they agreed and intended that the main loan would be the £1m advanced to Mr Helden to enable him to purchase the property, and that they also agreed that the rate of interest would be 7.5% pa. Those terms are also apparent from the 2005 agreement. It seems clear that the error in the drafting of the 2006 charge arose as a result of a mistaken assumption by Mr Wright that there had been an offer letter and an oversight by Mr Ashton and Mr Helden when they came to sign the 2006 charge.

[24] The judge held that Mr Helden was bound by the oral agreement as to the size and identity of the loan secured by the 2006 charge and the rate of interest payable thereon, as a result of estoppel by convention, and he cited *Amalgamated Investment & Property Co Ltd (in liquidation)* v *Texas Commerce International Bank Ltd* [1982] QB 84 in support. I agree. I also consider that the same result could have been arrived at through ordering rectification of the 2006 charge.

[25] The judge also found that the parties had agreed that: (a) interest was to be compounded with quarterly rests on the three loans; and (b) there would be an increase in the rate of interest payable on the main loan from 7.5% pa to 10% from 31 March 2007 and to 12.5% from 1 October 2007. In the light of the facts summarised in [14] above, I see no basis for challenging finding (b). Finding (a) is a little more controversial in the light of the reference to the "flat interest rate" in the request described in [13] above, but, given the facts summarised in [15] above, I am firmly of the view that the judge was entitled to reach the conclusion that he did.

Section 2 of the 1989 Act and section 53 of the 1925 Act

[26] Mr David Christie, on behalf of Mr Helden, contended that it was not open to the judge to fill the gap, as it were, in the 2006 charge, in respect of the amount of the main loan or the rate of interest, through estoppel or rectification in the light of section 2 of the Law of Property (Miscellaneous Provisions) Act 1989 (section 2), or, alternatively, section 53 of the Law of Property Act 1925 (section 53).

[27] Mr Helden's case on section 2 is hopeless. It proceeds on a fundamental misunderstanding of the reach and purpose of that section, a misunderstanding, it is fair to say, that appears to be not uncommon. Section 2 is concerned with contracts for the creation or sale of legal estates or interests in land, not with documents that actually create or transfer such estates or interests. So a contract to transfer a freehold or a lease in the future, a contract to grant a lease in the future or a contract for a mortgage in the future are all within the reach of the section, provided of course that the ultimate subject matter is land. However, an actual transfer, conveyance or assignment, an actual lease, or an actual mortgage are not within the scope of section 2 at all.

[28] As is spelt out in its opening words, section 2 is concerned with "a contract for the sale or other disposition of an interest in land". Its purpose is also clear from the fact that it replaced section 40 of the Law of Property Act 1925, and from the contents (and indeed the title) of the interesting and full Law Commission report that initiated it Transfer of Land: Formalities for Contracts For Sale etc of Land (Law Com no 364). The section was directed to tightening up the formalities required for contracts for the creation or transfer of interests or estates in land and it was not concerned with documents that actually create or transfer legal estates or interests in land. This conclusion is consistent with the view
expressed by the Chancellor of the High Court in *McLaughlin* v *Duffill* G
[2008] EWCA Civ 1627; [2010] Ch 1*, in [20] to [21], approving the reasoning of HH Judge Hicks QC in *Target Holdings Ltd* v *Priestley* (2000) 79 P&CR 305, in [51].

[29] Mr Helden's case on section 53 is only marginally less weak. The section does indeed apply to mortgages, as, unlike section 2, it is concerned with the "creat[ion] or disposi[tion]" of any "interest in land". However, it is far less prescriptive than section 2, which requires every term of the arrangement to be included in a document or identical documents signed by both parties. Section 53 merely requires the arrangement to be in a document signed by the person
creating or disposing of the interest. Section 2 therefore may give rise H
to problems when it comes to estoppel or rectification (as discussed in the thoughtful judgment of Morgan J in *Oun* v *Ahmad* [2008] EWHC 545 (Ch)†, in [41] to [55]), but no such problems arise in connection with section 53.

Effect of FSMA on the loans: background

[30] As the judge explained, in [64]:

> Section 19 of FSMA bars anyone but an "authorised person" or an "exempt person" from carrying on a "regulated activity" in the United Kingdom (the "general prohibition"). Section 22(1) provides that an activity is a "regulated
> activity" if, among other things, it is "an activity of a specified kind which is J
> carried on by way of business" and either (under section 22(1)(a)) "relates to an investment of a specified kind" or (under section 22(1)(b)) "in the case of an activity of a kind which is also specified for the purposes of this paragraph, is carried on in relation to property of any kind". The Financial Services and Markets Act 2000 (Regulated Activities) Order 2001… specifies kinds of activity for the purposes of section 22 of FSMA (see article 4). The activities specified include certain activities relating to "regulated mortgage contracts". The expression is defined in article 61.

[31] Mr Helden's argument, which was accepted by the judge, is that, in so far as Strathmore agreed to provide credit to Mr Helden on the security of the 2006 charge, it entered into a regulated mortgage contract, because the underlying main loan was secured on land to be K
used, "in connection with a dwelling by the borrower": see para (1) of article 61 of the Financial Services and Markets Act 2000 (Regulated Activities) Order 2001 (article 61).

[32] Section 23(1) provides that a breach of the general prohibition is an offence and conviction can lead to a fine, imprisonment, or both. Section 23(3) states that it is "a defence for an accused to show that he took all reasonable precautions and exercised all due diligence to avoid committing the offence".

[33] So far as the civil consequences of a breach of the general prohibition are concerned, the relevant provisions for present purposes
are sections 26 and 28. Section 26 states: L

> (1) An agreement made by a person in the course of carrying on a regulated activity in contravention of the general prohibition is unenforceable against the other party.
>
> (2) The other party is entitled to recover
>
> (a) any money or other property paid or transferred by him under the agreement; and
>
> (b) compensation for any loss sustained by him as a result of having parted with it.

As the judge said in [69], accepting a point made by Mr Fred Philpott,
who appeared for Strathmore, the way in which section 26(2) is framed M
is more obviously suited to a case in which the "other party" has made an investment than one where, as here, it has been lent money.

[34] Section 28 is in the following terms:

> (1) This section applies to an agreement which is unenforceable because of section 26…
>
> (2) [Deals with the amount of compensation].
>
> (3) If the court is satisfied that it is just and equitable in the circumstances of the case, it may allow

---

* Editor's note: Also reported at [2009] 2 EGLR 71

† Editor's note: Reported at [2008] 13 EG 149 (CS)

(a) the agreement to be enforced; or

(b) money and property paid or transferred under the agreement to be retained.

(4) In considering whether to allow the agreement to be enforced or (as the case may be) the money or property paid or transferred under the agreement to be retained the court must

(a) if the case arises as a result of section 26, have regard to the issue mentioned in subsection (5); or

…

(5) The issue is whether the person carrying on the regulated activity concerned reasonably believed that he was not contravening the general prohibition by making the agreement.

…

(7) If the person against whom the agreement is unenforceable

(a) elects not to perform the agreement, or

(b) as a result of this section, recovers money paid or other property transferred by him under the agreement,

he must repay any money and return any other property received by him under the agreement.

[35] There is, rightly, no challenge to the judge's conclusion that the provision of £1m secured by the 2006 charge amounted to "an activity of a specified kind". However, by way of respondent's notice, Strathmore challenges the judge's conclusion that it made the main loan pursuant to the 2006 charge as "an activity of a specified kind which is carried on by way of business". Moreover, Mr Helden, by way of appeal, challenges the judge's decision, pursuant to section 28(3) that it was "just and equitable in the circumstances of the case" to allow Strathmore to enforce the main loan and the 2006 charge.

[36] Similar issues arise in connection with the £25,000 loan. There are no issues in respect of the £90,000 loan as the judge concluded that he should not grant relief to Strathmore in relation to that loan under section 28(3), because it was not just and equitable to do so, and Strathmore does not challenge that conclusion.

Effect of FSMA on the loans: "carried on by way of business"

[37] Article 3A of the Financial Services and Markets Act 2000 (Carrying on Regulated Activities by Way of Business) Order 2001 (article 3A), provides that a "person is not to be regarded as carrying on by way of business an activity… unless he carries on the business of engaging in that activity.". As the judge explained, in [68]:

> In this context, as in some others within the financial services legislation, a distinction is drawn between (a) an activity "carried on by way of business" and (b) carrying on "the business of engaging in that activity". Whereas arranging or advising on regulated mortgage contracts will be a specified activity only if the person in question "carries on the business of engaging in that activity" (within the meaning of article 3A …), "entering into a regulated mortgage contract as lender" or "administering a regulated mortgage contract" is a specified activity if the activity is merely "carried on by way of business" (within section 22 of [the FSMA]) (provided that, in the case of "administering a regulated mortgage contract", the relevant contract was "entered into by way of business" see article 61(2)…).

[38] In the court below, Mr Philpott relied on a number of cases dealing with the now repealed section 6 of the Moneylenders Act 1900, but I agree with the judge that they are not of much assistance because they were concerned with a differently worded provision in a somewhat different statutory context.

[39] Mr Philpott also relied below, as he did before us, on one case on section 189(2) of the Consumer Credit Act 1974, *Khodari* v *Tamimi* [2009] EWCA Civ 1109, which the judge thought was of some assistance. In my view, that case helps to the extent of emphasising the point that, whether a person carries on the business of engaging in a specific activity is a matter of secondary fact, or inference, which is essentially for the trial judge, who must determine the issue by reference to all the relevant facts of the case.

[40] The judge's conclusion on this issue is to be found in [87] to [89]. In the first two of those paragraphs, the judge set out a concession, and the principal submissions, made on behalf of Strathmore on this point and repeated before us. Mr Philpott realistically accepted that the issue had to be judged by treating the Ashtons, Strathmore and Sandworth as a composite unit. However, he pointed out that: (i) apart from the loans to Mr Helden, no other regulated activity had been carried on by them; (ii) those to whom they lent money were friends and family; (iii) all such transactions were conducted informally; (iv) there was no business "set-up"; and (v) Mr Ashton saw himself as making personal investments rather than carrying on a business.

[41] The judge none the less concluded, in [89], that the making of each of the three loans to Mr Helden involved the "carr[ying] on by way of business" within the meaning of the FSMA, for the following reasons:

> i) Even excluding loans to members of the Ashton family, Strathmore/Sandworth made a sizeable number of loans. … Mr Harrison explained that he was assisted with the purchase of around 100 properties, albeit that many of the properties were acquired in portfolios of up to 25 properties;
>
> ii) The loans were made over a period of years and with some regularity. The first loan to Mr Harrison dates from 2002. Thereafter, as Mr Ashton said in his witness statement, he "made a series of very short term loans to [Mr Harrison] to assist him to build up a buy to let portfolio". Money was advanced to Mr Helden, Mr Jordan and companies associated with them between 2004 and 2006;
>
> iii) Substantial amounts of money were advanced. Sandworth lent as much as £5.4 million in connection with the purchase of [Belgravia]. C&J came to owe upwards of £3 million on the security of … Catherine Place. Mr Helden was lent more than £1 million. Mr Harrison reckoned that he never owed more than about £1 million at any time, but £1 million is a considerable amount;
>
> iv) The loans were made with a view to profit. During his oral evidence, Mr Ashton referred more than once to having tried to get the best possible return on his money. … ;
>
> v) Mr Ashton came to meet Mr Helden and Mr Jordan because they were seeking funding. The Ashtons' friendship with Mr Helden and Mr Jordan grew out of their financial relationship, not the other way round. Likewise, the relationship between Mr Harrison and the Ashtons began as a "purely professional" one… ;
>
> vi) It is possible to exaggerate the informality with which matters were conducted. Solicitors were often instructed. The loans were generally secured. Mr Ashton kept a record, even if in manuscript, of the transactions. In any case, informality is not necessarily inconsistent with business;
>
> vii) The loans to Mr Helden formed part of a chain of not dissimilar transactions, albeit that they were the only ones involving "activity of a specified kind"; and
>
> viii) Strathmore is a limited company with, I understand, commercial objects.

[42] No attack has been, or I think could be, made on the accuracy of the eight factors that the judge identified in that passage. It is, in my opinion, impossible, in those circumstances, to conclude that the judge was wrong, or at least not entitled as a matter of law, to conclude that, taking Strathmore together with Sandworth and the Ashtons, the making of the main loan and the 2006 charge were "an activity carried on by way of business", on the ground that Strathmore "carrie[d] on the business of engaging in that activity."

Effect of FSMA on the loans: section 28(3)

[43] The judge decided that it would be just and equitable to permit Strathmore to enforce the 2006 charge (and the obligation to repay the £25,000 loan), together with the two increases in the rate of interest agreed in 2007. In reaching that conclusion, he took into account one specific factor, which he set out in [97], and some more general factors, which he identified three paragraphs later.

[44] The specific factor was:

> 97. … [I]t was reasonable for the Ashtons (and Strathmore) to fail to realise that FSMA was in point. After all:
>
> i) … Strathmore employed solicitors to represent them in connection with the loan for the purchase of [the property], and those solicitors did not inform them that FSMA was (or could be) applicable;
>
> ii) The financial services legislation had not until quite recently [October 2004] extended to any mortgages…;
>
> iii) Neither the Ashtons nor their companies usually entered into transactions to which FSMA applied. It did not apply, even after 2004, to loans such as those granted to C&J, [the other companies in which Mr Helden was interested] or Mr Harrison; and

iv) Neither Mrs Ashton nor even Mr Ashton had ever attended any training courses on financial matters, let alone a course concerned with FSMA.

[45] The more general factors that the judge took into account were explained thus:

> 100. The case for allowing enforcement of the agreement relating to the [main] loan… is, as it seems to me, particularly compelling. The reasons include these:
>
> i) Mr Helden has had the use of the property which Strathmore's loan enabled him to buy… since 2006 without making any rent or interest payments;
>
> ii) The property has increased substantially in value. Whereas it was bought for £1 million, agents last year suggested that it should be marketed at £1.8m… . The loan from Strathmore has thus enabled Mr Helden to achieve a large profit;
>
> iii) … I accept, that… Strathmore would not have been willing to make the loan on an unsecured basis;
>
> iv) … Strathmore could be expected to have generated a return on the £1 million by investing it elsewhere had it not been lent to Mr Helden. They have lost that potential profit as a result of lending the money to Mr Helden;
>
> v) There is no question of Mr Helden having been taken advantage of. He had considerable experience in property matters, including as a mortgage broker. Further, the rates of interest charged were agreed with Mr Helden and were not exorbitant;
>
> vi) Mr Helden preferred not to pursue alternative funding because of his concern that he should be able to make lump sum repayments without penalty…;
>
> vii) Mr Helden has not identified respects in which he would have been better placed if Strathmore had been an "authorised person" for FSMA purposes; and
>
> viii) The Ashtons… did not realise that FSMA could apply, and it was reasonable for them not to do so.

[46] In my view, the only possible objection that could be taken to the analysis in [97] and [100] of the judgment and the resulting conclusion, is the implicit interpretation of section 28(5) that the judge appears to have adopted. He seems to have thought that that subsection assisted Strathmore, in the sense that it applied in a case such as this, where the person in breach of the general prohibition was unaware of the FSMA or its possible prohibition of the transaction in question, and where it was, in all the circumstances, reasonable for that person to be so unaware. However, there is a powerful argument for saying that a person cannot contend that he "reasonably believed that he was not contravening the general prohibition by making [an] agreement", if he was wholly unaware of the existence of the prohibition at the time of the agreement.

[47] It seems to me that there is considerable force in the simple linguistic point that a person cannot "believe that he [is] not contravening" a rule, if he is wholly unaware of the rule. Believing that one is not doing something is simply not the same thing as not believing one is doing something: to believe wrongly that one is not committing an act requires a degree of knowledge as to what that act is or entails, whereas wrongly not believing one is committing an act requires a degree of absence of knowledge, which renders it easier to contend that it would apply where one is ignorant of the existence of the act.

[48] Against that, there is some force in the point that it is unlikely that parliament could have intended that a person who wrongly, but reasonably, believes that he is not contravening a statute should be better off than a person who was, reasonably, unaware that the statute applied. Having said that, the answer to that point may be that people who carry on regulated activity and are ignorant of the law, even if reasonably so, should be more at risk, because they are more of a danger to the public, than those who carry on such activity, and are aware of the law, and reasonably, albeit wrongly, conclude that it does not apply.

[49] In support of his contention that the judge was right in his apparent interpretation of section 28(5), Mr Philpott referred to the decision of the House of Lords in *R* v *Johnstone* [2003] UKHL 28; [2003] 1 WLR 1736. That case involved consideration of the criminal liability for unauthorised use of a registered trade mark under section 92(1) of the Trade Marks Act 1994 (the 1994 Act). Section 92(5) provides that it is a defence for an accused person "to show that he believed on reasonable grounds that the use of the sign in the manner in which it was used, or was to be used, was not an infringement of the registered trade mark".

[50] In [42], Lord Nicholls of Birkenhead referred to a decision of the Divisional Court, *Torbay District Council* v *Singh* [1999] 2 Cr App R 451, where it had been held that section 92(5) was not available as a defence "where the defendant does not know of the existence of the registered trade mark in question". He then referred to later cases where this holding had been doubted and said, in the following paragraph:

> 43. I share these doubts. The interpretation adopted in the *Singh* case draws a distinction Parliament cannot have intended. The language of the subsection gives no support to this distinction. Section 92(5) is concerned to provide a defence where the person charged has a reasonable belief in the lawfulness of what he did. Those who act honestly and reasonably are not to be visited with criminal sanctions. It makes no sense to confine this defence to cases where the defendant is aware of the existence of the registered trade mark and exclude altogether those cases where the defendant is not. Section 92(5) provides a defence where the defendant believes on reasonable grounds his use of the sign does not infringe a registered trade mark of whose existence he is aware. It would be extraordinary if the subsection does not equally furnish a defence in the stronger case where the reason why the defendant believes his use of the sign does not infringe a registered trade mark is that he reasonably believes no relevant trade mark is registered. Section 92(5) is to be interpreted as including the latter case as well as the former.

[51] Lord Nicholls was concerned with a case where there were reasonable grounds to believe that a trademark had not been registered. This case is more akin to a defendant who was unaware that there was a system of registration at all. For reasons I have given in [47] above, it is more difficult to bring ignorance of the law within the scope of section 28(5). None the less, there is some support for the judge's approach in these observations of Lord Nicholls, in particular where he states that the intention is to provide a defence where the person charged has a reasonable belief in the lawfulness of what he did. Furthermore, there are points of distinction between that case and this. First, no criminal sanction is involved under section 28 of the FSMA: section 28 is concerned with enforceability and has nothing to do with criminality, which is dealt with by section 23, which has its own, not altogether dissimilar, defence in section 23(3). Second, section 28(5) identifies one factor that has to be taken into account when carrying out a much more wide-ranging exercise under section 28(3), whereas section 92(5) of the 1994 Act sets out a single, binary, potential defence. Third, ignorance of the existence of the legal requirements of the FSMA, which is what is being relied on by Strathmore in its argument, is rather different from ignorance of a simple fact, such as the existence of a trade mark.

[52] It is unnecessary to resolve this difficult issue, and I would therefore prefer not to do so, not least because I am not convinced that we have had as full submissions on the point as it may merit. Even accepting in Mr Helden's favour that the judge did assume that Strathmore could rely on section 28(5) (which he may well have done) and that he was wrong in so doing (which he may well have been), I consider that he reached the right decision on the question of whether Strathmore could rely on section 28(3). In my view, the reasons he gave in [97] and [100], when taken together, make out a strong case for Strathmore being entitled to enforce the 2006 charge, and the agreement in respect of the £25,000 loan, notwithstanding the breach of the FSMA, even if section 28(5) does not assist Strathmore's case.

[53] The judge considered the issue in a careful and thorough way, and, if he did go wrong in his interpretation of section 28(5), it is fanciful to think that that would have affected his ultimate conclusion. His well-modulated approach is demonstrated by the contrasting approach that he took to the £90,000 loan, in respect of which he said, in [103]:

> In contrast, I have not been persuaded that I should grant relief under section 28(3)… as regards the [£90,000 loan]. The following points in particular militate to my mind against doing so:
>
> i) The £91,509 derived from a £35,000 loan made less than three years before… . While I understand how the debt came to grow as it did, the increase is still equivalent to an extremely high rate of interest; and
>
> ii) The £35,000 was lent to C&J rather than to Mr Helden himself.

Effect of the FSMA on the loans: conclusion

[54] Accordingly, I would uphold the judge's decision that: (i) Strathmore contravened the FSMA by entering into the 2006 charge; and (ii) it was none the less just and equitable to permit Strathmore to enforce the 2006 charge.

Appeal on costs

[55] After he handed down his judgment dealing with the issues discussed above, the judge heard argument on various matters, including costs. He delivered an *ex tempore* judgment on 10 September 2010, in which he awarded Strathmore its costs of the proceedings on an indemnity basis.

[56] His reason for reaching this conclusion may be summarised as follows. A number of cases show that "a mortgagee is generally entitled contractually to recover all costs… in enforcing and preserving his security, and, moreover, to do so on the indemnity basis", and he "s[aw] no reason why the costs order should not reflect the contractual position". In particular, the fact that Mr Helden had succeeded in defeating the claim in respect of the £90,000 loan did not alter the judge's view, as that issue involved "no significant extra costs".

[57] In my view, that analysis was fundamentally flawed. It is perfectly true that the majority, probably the great majority, of mortgages contractually provide that the mortgagee should be entitled to recover its costs of enforcing the mortgage on an indemnity basis, and that, in the absence of a good reason to the contrary, the court will give effect to that provision when exercising its discretionary power in relation to costs.

[58] However, in this case, the 2006 charge did not contain a provision entitling Strathmore to recover the costs of enforcing its rights, on an indemnity basis or at all. In those circumstances, with due respect to the judge, who gave an excellent judgment on the substantive issues, he went wrong in dealing with costs effectively on the assumption that the 2006 charge did so provide. Where an agreement contains no provision stating how the costs of proceedings will be dealt with, it seems to me that it would be wrong in principle for the court to proceed on the basis that there is such a provision, simply because the majority of agreements of the type in question do contain such a provision. If anything, one would presume that the parties intentionally departed from the norm.

[59] It seems to me that the judge should have proceeded on the basis that he had the normal discretion enjoyed by the court when it comes to costs, namely a discretion to be judicially exercised by reference to the facts of the case, with no special presumption as there would often be if there had been a specific term dealing with costs agreed between the parties.

[60] On that basis, it seems to me that this court must consider for itself what order for costs to make. Strathmore substantially succeeded in the result: in so far as it failed on the £90,000 loan, we should respect the judge's view that it was a minor matter. Further, there were many issues of fact, on almost all of which Strathmore succeeded.

[61] However, there are two important points the other way. First, Strathmore failed on the question of whether the FSMA applied, and that involved some investigation of the nature and extent of its business (as well as that of the Ashtons and Sandworth). Second, in order to succeed in its claim to enforce the 2006 charge, Strathmore needed to seek the indulgence of the court under section 28(3) of the FSMA: while that is not quite the same as seeking relief from forfeiture or another sanction, it is a significant reason for depriving it of costs.

[62] On behalf of Mr Helden, Mr Christie submitted that the correct order for costs was that Mr Helden should pay 60% of Strathmore's costs on the standard basis. In my view, that was an entirely reasonable suggestion and I would adopt it.

Conclusion

[63] In these circumstances, Mr Helden's appeal is dismissed, save in respect of the issue of costs and Strathmore's respondent's notice is dismissed. I would uphold the judge's order, save that I would substitute for the provision for indemnity costs against Mr Helden, a provision that he pays 60% of Strathmore's costs, to be assessed on the standard basis if not agreed.

[64] No doubt, counsel can agree a form of order and, if they cannot agree on the costs of this appeal and cross-appeal, we will deal with them on written submissions.

Smith LJ said:

[65] I agree.

Elias LJ said:

[66] I also agree.

Appeal dismissed.

NEGLIGENCE

Edehomo *v* Bowling & Co Solicitors

CHANCERY DIVISION

[2011] EWHC 393 (Ch)

2 March 2011

Roth J

Estates Gazette 28 May 2011

[2011] 21 EG 96

Negligence — Solicitor — Limitation of action — Limitation Act 1980 — Husband and wife jointly owning property — Husband forging wife's signature on sale of property — Wife claiming damages against solicitor in negligence — Whether damage suffered giving rise to cause of action at date of exchange of contracts or on completion of fraudulent sale — Whether claim time-barred under section 2 of 1980 Act

In 2002, the respondent's husband instructed the appellant solicitor to act for him and the respondent on the sale of an investment property that the couple owned as registered joint tenants. Contracts were exchanged in November 2002 and completion took place the following month. The appellant sent a cheque, made out to the respondent and her husband, for the net proceeds of sale of £86,143. However, the respondent was unaware of the sale; her signature had been forged on the contract and on the transfer document and she did not receive any of the sale moneys.

In December 2008, the respondent issued a claim against her estranged husband and the appellant. Against the latter, she claimed damages in negligence, alleging various breaches of a duty of care that it owed to her in respect of the sale. The appellant applied to strike out the claim or for summary judgment in its favour on the ground that the claim was time-barred. It contended that the cause of action had accrued on exchange of contracts, such that the six-year time limit within section 2 of the Limitation Act 1980 had expired. In the county court, the judge found for the respondent, holding that she had not suffered damage and the cause of action had not arisen until completion of the sale. The appellant appealed.

***Held*: The appeal was allowed. The respondent had suffered loss on exchange such that she could then have begun proceedings against the appellant. Her claim was therefore time-barred. Had she sought to sell her share in the property after the date of exchange of contracts on the fraudulent sale, and assuming for that purpose that she was presumed to have been aware of the fraud, she would have had to disclose to a purchaser the existence of the forged sale by her husband. The value of her interest would thereby have been significantly diminished. No potential purchaser would choose to enter into a binding obligation to buy an interest in property at its open market value, where it has been informed that the property has already been sold to someone else, simply on the assurance that the vendor's signature had been forged on the earlier contract so that it could be disregarded. The potential purchaser would not know whether that was correct, and no one would want to buy into a lawsuit. In such circumstances, damage can arise not only on completion but on exchange of contracts if that is on an earlier date.**

***Per curiam*: A contract in which one co-owner's signature has been forged is not a nullity but remains valid in respect of the fraudulent co-owner. Although ownership in law will not pass to the purchaser because a joint tenancy in law cannot be severed, and although the purchaser can choose to have the transaction set aside for fraud, that transaction will none the less be effective to pass the fraudulent co-owner's beneficial interest in the property and so sever the beneficial joint tenancy. Whether such severance will cause damage to the other co-owner will depend on the facts.** F

The following cases are referred to in this report.

Ahmed v *Kendrick* (1988) 56 P&CR 120, CA G
Bell v *Peter Browne & Co* [1990] 2 QB 495; [1990] 3 WLR 510; [1990] 3 All ER 124, CA
Edwards v *Lloyds TSB Bank plc* [2004] EWHC 1745 (Ch)
Forster v *Outred & Co* [1982] 1 WLR 86; [1982] 2 All ER 753, CA
Nouri v *Marvi* [2010] EWCA Civ 1107; [2010] 3 EGLR 79; [2010] 50 EG 64

This was an appeal by the appellant, Bowling & Co Solicitors, from a decision of HH Judge Hands QC, sitting in Central London County Court, dismissing an application to strike out on limitation grounds a claim by the respondent, Rose Edehomo, for damages for professional negligence.

James Carpenter (instructed by Bowling & Co Solicitors) appeared for the appellant; Philippa Daniels (instructed by Alomo Solicitors, of Romford) represented the respondent. H

Giving judgment, **Roth J** said:

[1] This case raises the question of when an innocent vendor, whose signature is forged on the documents for the conveyance of land, suffers damage for the purposes of limitation of an action arising from a solicitor's breach of duty. Is it on the exchange of contracts, in which case the present claim is said to be time-barred or is it only on completion, in which case the claim was just commenced in time? In a judgment delivered in Central London County Court on 3 August 2010, HH Judge Hand QC held that it is the latter. The second defendant (Bowlings) appeals, contending that it is the former. J

Facts

[2] The issue arises on Bowlings' application to strike out the claim against it or, alternatively, for summary judgment. It is accepted that the facts alleged by the claimant, Mrs Rose Edehomo, should therefore be assumed, for present purposes, to be established.

[3] The relevant facts are relatively simple. The first defendant, Mr Jerome Edehomo, is the estranged husband of the claimant. In around December 1989, the claimant and Mr Edehomo together purchased the property known as 18 Stock Street, Plaistow, London E13 (the property), as an investment. The purchase price was £75,000, of which they paid a deposit of £5,000 and the balance was funded by a mortgage. The property was registered in their names as joint tenants. K

[4] In 2002, Mr Edehomo instructed Bowlings to act for him and the claimant regarding the sale of the property to a third party. Contracts were exchanged on 21 November 2002 and completion took place on 2 December 2002. However, the claimant was at the time unaware either of those instructions or of the sale and played no part in the transaction. Her signature had been forged, it would appear by

A Mr Edehomo, on both the contract and the TR1 form of transfer. It is also suggested, although the claimant is not sure of this, that Mr Edehomo may have procured someone to impersonate her and attend a meeting at Bowlings when the original instructions were given.

[5] As a result, the property was sold and there is no suggestion that the purchasers were aware of the fraud. Bowlings sent Mr Edehomo a cheque made out to him and the claimant for the net proceeds of sale, amounting to £86,143.35. The claimant has never received any of the money.

Proceedings

B [6] On 1 December 2008, the claimant issued a claim form claiming against both Mr Edehomo and Bowlings, alleging with regard to the solicitor that it was in breach of a duty of care to her. Mr Edehomo did not serve a defence, and it may be presumed that he is unlikely to be able to satisfy a judgment against him. The claimant's claim against Bowlings, as pleaded in her particulars of claim, is framed in both contract and tort, but it has been proceeded with only in tort because her case is that she was never a client of Bowlings and, accordingly, there was no contract between them. The allegations made by her against Bowlings are in para 11 of the particulars of claim, which reads as follows:

> C It was an implied term of the aforesaid agreement [that is the agreement resulting from Mr Edehomo's instructions] or alternatively it was the duty of [Bowlings] their servants or agents:
>
> (a) to take reasonable or appropriate steps to establish the identity of [Mr Edehomo] and the claimant believes [sic] the persons unknown that were impersonating the claimant, at all material times
>
> (b) to undertake the aforesaid contractual and conveyancing work with all due care, skill, competence and diligence to be expected of solicitors experienced in the conduct of conveyancing;
>
> (c) to take or advise the claimant to take all necessary steps to protect her interests in respect of the proposed sale.

D The argument for the claimant on this appeal, as in the court below, has been largely based on (a) of the above allegations.

[7] Section 2 of the Limitation Act 1980 provides:

> An action founded on tort shall not be brought after the expiration of six years from the date on which the cause of action accrued.

[8] The extensive delay in the commencement of these proceedings appears to be due, at least in part, to the conduct of the claimant's previous firm of solicitors, which firm has since been dissolved. In any event, on her own case the claim form was issued just one day within the six-year limitation period. However, Bowlings contends that if it E was in breach of duty (which it denies), the claimant suffered damage as a result on 21 November 2002, when contracts for the sale of the property were exchanged, and therefore more than six years before this action was started.

[9] I raised with counsel at the outset of this appeal whether there was not a continuing duty on Bowlings up to the date of completion to verify the identity of its clients, such that there was a further breach when it released the transfer form (or a continuing breach up to that point). If that were the position, then even if the claimant suffered damage on exchange of contracts, she might be able to claim on the basis of a further breach at the time of completion that gave rise to F further damage. It was explained to me that the reason that this was not pursued is that, under normal conveyancing practice, solicitors are expected to verify the identity of their client in a face-to-face meeting at the outset, but are not expected to do so again thereafter unless there are circumstances that should reasonably alert the solicitor to some doubt regarding its client's true identity. Whether that is a complete answer to the point where it is alleged that the solicitor never properly verified the identity of its client is not for me to determine on this appeal.

[10] However, Ms Philippa Daniels, for the claimant, told me that she wants to contend that, on the facts here, there were matters that should have alerted Bowlings to the problem after exchange of contracts and, accordingly, she intended to seek permission to serve a reply raising those matters. No draft reply was before the court at the hearing, but, pursuant to my direction, a draft reply was produced subsequently that the claimant will seek permission to serve out of time. Although determination of that application may affect the form of order to be made on disposal of this appeal, it does not obviate the need to decide the issue raised. The effect on the limitation point of any circumstances on which the claimant now seeks to rely is presently unclear, and in any event Bowlings would be entitled to dispute the accuracy or implications of those circumstances.

Appeal

[11] In his judgment below, Judge Hand held that the claimant did not suffer damage for the purpose of accrual of her cause of action until completion on 2 December 2002. However, since that judgment was given, the Court of Appeal has decided *Nouri* v *Marvi* [2010] EWCA Civ 1107*. Mr Jamie Carpenter, for Bowlings, placed the reasoning of that judgment at the forefront of his submissions and, as a result, the argument on this appeal has assumed a rather different shape from that in the court below.

[12] Before turning to that case, it is appropriate to consider what is meant by "damage" for this purpose. In *Forster* v *Outred & Co* [1982] 1 WLR 86, the Court of Appeal held that the plaintiff suffered damage caused by her solicitor's alleged negligence at the time she executed a mortgage on her freehold property as security for a loan and not only at the later time, when she was required to pay on that security. In the course of his judgment, with which the other members of the court agreed, Stephenson LJ accepted the submission of counsel for the defendant solicitor, expressed as follows, at p94C-D:

> What is meant by actual damage? Mr Stuart-Smith says that it is any detriment, liability or loss capable of assessment in money terms and it includes liabilities which may arise on a contingency, particularly a contingency over which the plaintiff has no control; things like loss of earning capacity, loss of a chance or bargain, loss of profit, losses incurred from onerous provisions or covenants in leases. They are all illustrations of a kind of loss which is meant by "actual" damage.

And Dunn LJ said, at p100A-C:

> In this case, as soon as she executed the mortgage the plaintiff not only became liable under its express terms but also and more importantly the value of the equity of redemption of her property was reduced. Before she executed the mortgage deed she owned the property free from encumbrances; thereafter she became the owner of a property subject to a mortgage. That, in my view, was a quantifiable loss and as from that date her cause of action against her solicitor was complete, because at that date she had suffered damage. The actual quantum of damages would, of course, depend on events between that date and the date when the damages had finally to be assessed, but the cause of action was complete when she executed the mortgage, without proof of special damage.

[13] In *Nouri*, the claimant was the registered owner of a leasehold flat. He allowed the defendant to live in the flat while he was out of the country. While in occupation of the flat, the defendant remortgaged the property in the claimant's name, forging the claimant's signature on the mortgage deed. Then, on 2 April 2001, he sold the flat to himself, posing as the claimant in instructing a solicitor, as vendor, and instructing another solicitor to act in his own name, as the purchaser. In that case, exchange of contracts and completion were simultaneous and the defendant forged the claimant's signature on the transfer. On 4 July 2001, using the transfer, the defendant obtained registration of the flat in his own name and he subsequently sold it to an unconnected third party.

[14] On 2 July 2007, the claimant commenced proceedings against the defendant and also the solicitor that had acted for the defendant when he had impersonated the claimant. Therefore, that action was commenced more than six years after exchange and completion but less than six years after registration. The allegation against the solicitor was similar to those in the present case and the issue raised on limitation was whether the claimant suffered damage only on registration or already

---

* Editor's note: Reported at [2010] 3 EGLR 79

on 2 April 2001. The Court of Appeal upheld the decision of the trial judge that the claim against the solicitor was statute-barred because the claimant had suffered actual damage on completion of the sale and was not simply exposed to the contingent loss of his title, which would depend on whether Mr Marvi applied for registration as proprietor based on the forged transfer.

[15] Since, in that case, exchange and completion were simultaneous, it was unnecessary for the court to make any distinction between them. I accept Mr Carpenter's submission that where Patten LJ, in the leading judgment, refers to damage arising on completion, that was by distinction with registration and should not be taken to be a determination that it would not have arisen already on exchange of contracts had that occurred still earlier.

[16] Critical to the reasoning of the Court of Appeal was the finding that, as from 2 April 2001, there was a blot on Mr Nouri's title that significantly decreased the marketable value of the property. The essential reasoning of Patten LJ (with whom Sir Mark Waller and Rix LJ agreed) is contained in the following paragraphs:

> 29. There was no valuation evidence before the judge directed to this issue but Mr Jones [Counsel for Mr Nouri] accepts that had a potential purchaser been told of the forged transfer in the period prior to registration of Mr Marvi's title, this would undoubtedly have led to a diminution in the price he would have been willing to pay. Although the contract and the transfer to Mr Marvi were of no legal effect, they did expose any purchaser to the risk of possible litigation even if that risk was more apparent than real. This is the point taken in the respondent's notice.
>
> 30. On this basis there will have been a diminution in the open market value of the Flat following completion on 2nd April 2001. Mr Jones contests this conclusion because it proceeds, he says, on a false assumption: ie that a potential purchaser would have been made aware of the difficulties caused by the forged transfer. He submits that there was no realistic possibility of Mr Nouri having discovered the fraud before registration on 4th July and therefore no question of his being under any obligation to disclose those matters to a potential purchaser. The facts were that he remained in ignorance of the fraud until much later.
>
> 31. I do not accept this analysis. It is well established that a cause of action in tort can accrue for the purposes of the Limitation Act without the claimant being aware of it. The decisions of the House of Lords in *Cartledge* v *E Jopling & Sons Ltd* [1963] AC 758 and *Pirelli General Cable Works Ltd* v *Oscar Faber & Partners* [1983] 2 AC 1 led to limited changes to the Act to provide alternative time limits of three years running from the date of knowledge. These are now contained in ss11 and 14A of the 1980 Act.
>
> 32. In this case Mr Nouri was not assisted by s14A because he had the requisite knowledge of the fraudulent sale to Mr Marvi by 2002. But the provision of these alternative time periods to address the problem of lack of knowledge has left intact the principle that time will run under s2 even if the claimant is unaware of the circumstances which have led to it. The existence of actual damage for these purposes does not therefore depend on the claimant's state of knowledge in relation to the breach of duty or its consequences but on whether the breach has in fact caused actual loss.
>
> 33. For this purpose the correct hypothesis must be to ask whether Mr Nouri could have maintained an action for damages against [the solicitors] following completion of the sale to Mr Marvi. That requires one to assume that he was aware of the breach and to assess the impact of that on his property or other assets as at that date. Consistently with this hypothesis, any diminution in the value of the Flat attributable to the breach has to be determined on the basis that Mr Nouri was aware of the forged transfer and was under a corresponding duty to alert a purchaser to that problem. The Flat was unsaleable without such disclosure. It must therefore follow that its open market value was what a purchaser would pay for it with knowledge of the forged transfer. There was therefore actual damage suffered by Mr Nouri as of 2nd April 2001.

[17] Mr Carpenter submitted that same analysis would apply here. Had the claimant sought to sell her share in the property after 21 November 2002, and for this purpose is presumed to be aware of the fraud, she would have had to disclose to a potential purchaser the existence of the forged sale by her husband. Although she would of course have been able at any time previously to sever the joint tenancy and thus seek to dispose of her beneficial interest in her tenancy in common, after 21 November 2002 no purchaser would have been prepared to enter into an agreement with her until the situation regarding the contract made by Mr Edehomo had been sorted out. Thus, it was submitted, her interest in the property had become unsaleable, or at the very least could be sold only at a significant discount or, perhaps more **G** likely, on a conditional contract whereby a purchaser was obliged to proceed only on condition that the earlier fraudulent contract had been set aside within a specified period. Whichever of these alternatives is the more likely, the value of the claimant's interest in the property was significantly diminished.

[18] In response, Ms Daniels said that the claimant had no wish to sell her share in the property at the relevant time. However, that is no answer to the point because if the value of the property is diminished that clearly constitutes a loss even if the loss would be realised only if and when the property were to be sold. Second, Ms Daniels submitted that whether the existence of the earlier contract would in fact make **H** it impossible for the claimant to sell her interest in the property, or significantly reduce its value, was a question for expert evidence and none had been produced. There is a difference, she argued, between a conveyance, as in *Nouri*, which is good against all the world, and a contract, which is binding only on the contracting parties. However, it seems to me clear beyond argument that no potential purchaser would choose to enter into a binding obligation to buy an interest in property at its open market value when he or she was told that it had already been sold to someone else, simply on the basis of being informed that the vendor's signature on the earlier contract had been forged so that it could be disregarded. The potential purchaser could not know whether **J** that was correct, and no one wants to buy themselves into a lawsuit. Despite Ms Daniels' valiant efforts to persuade me otherwise, I do not think that this is a matter that calls for expert evidence.

[19] Accordingly, I hold that the claimant did indeed suffer loss on exchange, such that she could then have started proceedings against Bowlings. It follows that she had a cause of action against it in tort that accrued more than six years before the commencement of these proceedings.

[20] This ground, which was not considered in the judgment below, is therefore sufficient to dispose of the only issue to be determined on the appeal. However, since they were fully argued, I address briefly the **K** other points raised on behalf of Bowlings.

[21] First, it is said that the judge was wrong to find that the contract here did not have the effect to sever the joint tenancy, and that such severance of the claimant's interest in the whole constituted a loss.

[22] As to severance, the position is as follows. On behalf of the claimant, it is now conceded that the above analysis is correct and that severance took place. I think that concession is rightly made. It is well established that a contract in which one co-owner's signature has been forged by the other is not a nullity but remains valid in respect of the fraudulent co-owner: see *Edwards* v *Lloyds TSB Bank plc* [2004] EWHC 1745 (Ch), in [17] to [18]. Thus, although ownership in law **L** did not pass to the purchasers, because a joint tenancy in law cannot be severed, the transaction had the effect of passing Mr Edehomo's beneficial interest in the property to the purchasers, although the purchasers could undoubtless have set aside the entire transaction for fraud if they so wished. That the effect of the passing of the beneficial interest of the fraudster to a third party is to sever the beneficial joint tenancy is made clear by *Ahmed* v *Kendrick* (1988) 56 P&CR 120. Although the judge below distinguished *Ahmed* on the basis that it concerned the effect of transfer on completion and not exchange, I do not, with respect, consider that this distinction is relevant. In that case, nothing turned on whether severance occurred at exchange or transfer, **M** but as a matter of first principle an exchange of contracts constitutes a sufficient alienation to have the effect of severing a joint tenancy; severance is not postponed until completion: see *Megarry & Wade: The Law of Real Property* (7th ed) 2008, in paras 13-037, 13-039.

[23] Accordingly, on 21 November 2002, the claimant's joint tenancy in the property was severed. However, did that in itself cause the claimant loss? Since the judge below did not find that any severance took place, he did not address this question. Mr Carpenter submitted that severance did cause damage because the claimant lost the benefit of survivorship. However, as he recognised in the circumstances of this case survivorship was rather double-edged. Had the joint tenancy continued, it would have meant that if the claimant had died the entire

A property would have passed to her estranged husband, which one may presume would have been the last thing she would have wanted. The question of whether loss is suffered is fact-specific, and although in some circumstances loss of the joint tenancy might constitute damage, I do not think that it did so in this case. Accordingly, although by a different process of reasoning from that of the judge, I would uphold his conclusion that no loss is established on this basis.

[24] Second, it was submitted for Bowlings that if the claimant had become aware of the facts, she would sensibly have taken steps to alert Bowlings to the reality of what was going on, and incurred expense in doing so, possibly in consulting a solicitor. Reliance was placed on the
B much-quoted observations of Nicholls LJ in *Bell* v *Peter Browne & Co* [1990] 2 QB 495. There, on the breakdown of a marriage, the plaintiff husband executed a transfer of the matrimonial home into his wife's sole name, but the defendant solicitor that he consulted took no steps, by way of arranging a declaration of trust or mortgage, to protect the one-sixth share in the proceeds of sale that he agreed with his wife he would receive whenever such sale occurred. The issue was whether he suffered damage when the transfer to his wife was executed without any formal agreement to protect his interest, or only on the subsequent sale of the house when his (by then) former wife failed to account to him for his agreed share. Nicholls LJ stated, at p503F-G:

C

> In considering whether damage was suffered in 1978 one can test the matter by considering what would have happened if in, say, 1980 the plaintiff had learned of his solicitors' default and brought an action for damages. Of course, he would have taken steps to remedy the default. But he would have been entitled at least to recover from the defendants the cost incurred in going to other solicitors for advice on what should be done and for their assistance in lodging the appropriate caution. The cost would have been modest, but not negligible.

[25] However, I agree with the judge below that this cannot be taken out of context or to extremes. In the first place, the claimant could have simply gone into Bowlings' offices to identify herself and told it that she had never signed any contract that it was purporting to transact on her behalf. Unsurprisingly, Bowlings did not suggest to this court that that would have been insufficient for it to investigate what was going on. That may have cost the claimant a telephone call and a bus fare, but I do not regard that as the kind of expense that Nicholls LJ had in mind. Further, as Judge Hand pointed out, the cost to which Nicholls LJ referred was that of remedying an alteration to the plaintiff's legal position. The analysis may be different where no such alteration had yet occurred and the hypothetical cost was that which might be incurred in declaring the existing legal position so that no such alteration could take place in the future. However, it is not necessary to reach a concluded view on this point since this aspect has become academic in the light of my determination that exchange of contracts caused actionable damage to the claimant on other grounds.

Appeal allowed.

PREMISES LICENCE

R (on the application of Albert Court Residents' Association and others) *v* Westminster City Council

COURT OF APPEAL

[2011] EWCA Civ 430

13 April 2011

LLOYD, WILSON AND STANLEY BURNTON LJJ

Estates Gazette 18 June 2011

[2011] 24 EG 110

Premises licence — Variation — Late representations — Licensing Act 2003 — Application to vary licence to increase permitted activities — Licensing authority having practice of notifying residents in immediate vicinity of application premises — Residents of block of flats close to application premises excluded notwithstanding some flats within usual radius for notification — Section 35 of 2003 Act — Residents claiming decision to grant variation of licence unlawful for breach of legitimate expectation — Whether such unlawfulness justifying quashing of licence variation

The respondents were the residents' association, the management company, the freeholder and four residents of flats (Albert Court) situated in the vicinity of the Royal Albert Hall, London SW7. The second appellant council were the local licensing authority. They received an application from the first appellant, under section 34 of the Licensing Act 2003, to vary the premises licence for the hall to add boxing and wrestling to the list of permitted activities and to extend the opening hours and the hours for serving late-night refreshments. The first appellant advertised its application, as required by the Act. In addition, the second appellants sent approximately 100 letters to residents of nearby properties, pursuant to their policy of notifying those in the "immediate vicinity" of premises for which licensing applications were made. No such letters were sent to residents of Albert Court, notwithstanding that some of the flats fell within the radius of 30yd from the application premises that the second appellants judged to constitute the "immediate vicinity".

The second appellants received no representations concerning the likely effect of the licence variation within the statutory deadline for such representations. A number of Albert Court residents made late representations after that date. The second appellants disregarded the late representations as not being "relevant representations" within section 35(6) and granted the licence variation accordingly.

The respondents brought judicial review proceedings to quash the second appellants' decision to grant the variation. Allowing the claim, the judge held that although the refusal to consider late representations was lawful, the decision to grant the licence variation was unlawful because the second appellants had failed to fulfil a legitimate expectation of the respondents that they would be notified of licensing applications relating to the Royal Albert Hall: see [2010] EWHC 393 (Admin); [2010] 2 EGLR 65; [2010] 20 EG 146. The appellants appealed.

***Held*: The appeal was allowed. Even assuming that the second appellants' decision not to notify the respondents of the application was irrational and in breach of their public law duties, whether as a breach of a legitimate expectation or otherwise, that could not justify the quashing of the licence variation. Neither the 2003 Act nor the regulations made thereunder oblige a licensing authority to advertise an application for the variation of a premises licence. The second appellants had been reasonably satisfied that the first appellant had complied with its duty to advertise the application. Having received no "relevant representations" within section 35(5) and (6) of the 2003 Act, the second appellants had been under a duty to allow the application and the first appellant had a right, enforceable at public law, to the grant of the variation for which it had made a valid application: see section 35(2). The court could not grant any relief that would have the effect of preventing the second appellants from complying with those duties or depriving the first appellant of its public law right. It is an incident of the principle of legislative supremacy that an otherwise legitimate expectation cannot require a public authority to act contrary to statute. The same limitation applies to judicial review sought on the ground of irrationality or any other common law ground. Consequently, the respondents were not entitled to have the licence variation quashed. If their concerns were realised, the appropriate course would be to apply for a review of the licence under section 51 of the 2003 Act.**

***Per curiam*: A decision by an authority to refrain from notifying persons affected by a licensing application could give rise to relief if challenged sufficiently promptly. An order might be sought requiring the authority to reconsider their decision and the court might require the authority to notify residents if that notification could lead to representations being made within the statutory time limit. However, that would be the limit of the available relief.**

The following cases are referred to in this report.

Miskovic v *Secretary of State for Work and Pensions* [2011] EWCA Civ 16; [2011] 2 CMLR 20

R (on the application of Albert Court Residents' Association) v *Westminster City Council* [2010] EWHC 393 (Admin); [2010] 2 EGLR 65; [2010] 20 EG 146

R (on the application of Nadarajah) v *Secretary of State for the Home Department*; *R (on the application of Abdi)* v *Secretary of State for the Home Department* [2005] EWCA Civ 1363

R (on the application of Niazi) v *Secretary of State for the Home Department*; *R (on the application of Bhatt Murphy (a firm))* v *Independent Assessor* [2008] EWCA Civ 755

R v *North and East Devon Health Authority, ex parte Coughlan* [2001] QB 213; [2000] 2 WLR 622; [2000] 3 All ER 850, CA

R v *Secretary of State for Education and Employment, ex parte Begbie* [2000] 1 WLR 1115, CA

Walumba Lumba (previously referred to as WL) (Congo) 1 & 2 v *Secretary of State for the Home Department* [2011] UKSC 12

This was an appeal by the appellants, the Corporation of the Hall of Arts and Sciences and Westminster City Council, from a decision of McCombe J, sitting in the Administrative Court, allowing a claim by the respondents, Albert Court Residents' Association, Albert Court (Westminster) Management Co Ltd, Albert Court (Westminster) Freeholds Co Ltd and Mr Kerry Rubie (four residents), for judicial review of a decision to grant a variation of a premises licence under the Licensing Act 2003.

Philip Coppel QC and Saima Hanif (instructed by Jeffrey Green Russell) appeared for the first appellant; David Matthias QC (instructed by the legal department of Westminster City Council) appeared for the second appellants; John Steel QC and Andrew Sharland (instructed by Russell-Cooke LLP) represented the respondents.

Giving judgment, **Stanley Burnton LJ** said:

Introduction

[1] This is an appeal against the order made by McCombe J dated 2 March 2010 (see [2010] EWHC 393 (Admin)*) quashing the decision of Westminster City Council (to which I shall refer as Westminster) dated 25 May 2009 granting the application of the Corporation of the Hall of Arts and Sciences, which manages the Royal Albert Hall, for the variation of its premises licence under the Licensing Act 2003 (the Act) and similarly quashing the variation of that licence. For convenience, I shall refer to the Corporation of the Hall of Arts and Sciences as "the Albert Hall" and to the respondents, who were the claimants in the judicial review proceedings heard by the judge, as "the residents".

Facts in outline

[2] I can take the facts from the judgment of McCombe J.

[3] On 22 December 2008, the Albert Hall applied to Westminster for a variation of its premises licence under the Act. It sought to:

(a) vary the hours when late-night refreshment may be provided from 11pm-1am to 11pm-1.30am;

(b) add boxing and wrestling to the permitted licensable activities;

(c) vary the plans;

(d) vary the start time for licensable events from 11am to 9am;

(e) vary the opening time from 11am to 8am; and

(f) vary the closing time from 1am to 1.30am.

[4] The Albert Hall advertised its application, as required by the Act, in a local newspaper, the *Paddington, Marylebone & Pimlico Mercury*, and by placing notices outside the hall.

[5] On 6 January 2009, Westminster sent around 100 letters to residents of neighbouring premises. Albert Court is very close to the Albert Hall and parts of it are within 30m of it, but no such letters were sent to any of the residents of Albert Court, whereas a substantial number were sent to residents of Albert Hall Mansions, which are to the north-east of the hall and a similar distance from it.

[6] The letters that were sent were dispatched pursuant to a practice of Westminster to notify businesses and residents, in the "immediate vicinity" of subject premises, of licensing applications in respect of such premises. This practice is referred to in a Westminster leaflet (*Licensing in Westminster*) and on Westminster's website. Westminster have an internal guideline for the assessment of "immediate vicinity" for these purposes, which at the relevant time was 30m from the application site.

[7] As at 19 January 2009, the required cut-off date under the Act (as to which see below), Westminster had received only one representation about the likely effect of the variation of the licence; that single representation was from their own environmental health officer, but it was withdrawn in March 2009 without a hearing having been held in respect of it. A number of residents of Albert Court made representations to Westminster after 19 January 2009. Those representations addressed many matters, including apprehended problems of anti-social behaviour, public safety, noise and disturbance and the degradation of the surrounding area. Westminster declined to consider those late representations or to act on them in any way, taking the view that the Act required them to take that stance.

[8] On 25 May 2009, Westminster granted the Albert Hall's application.

[9] Before the judge, the residents contended that the decision to grant the licence was unlawful for two reasons: first, because Westminster were wrong in law to conclude that they were prohibited from considering late representations against the application; and, second, because having promised to notify residents in the immediate vicinity, they failed adequately to do so, so frustrating a "legitimate expectation".

[10] The judge held that Westminster's decision to refuse to consider late representations was lawful, but that their decision to grant the variation of the licence was unlawful because they had failed to fulfil the legitimate expectation of the residents of Albert Court that they would send them notification of the application. He therefore made the order referred to above.

Contentions of the parties before the judge

[11] In their claim, the residents sought relief on the grounds that Westminster had:

(1) "failed to comply with their policy/practice to consult residents within 30m of the licence premises for which the variation is being sought"; and

(2) misdirected itself as to its discretion to consider and take into account representations made out of time or to hold a hearing to consider the licence variation sought.

[12] The Albert Hall had been served with the proceedings as an interested party. It had filed and served detailed grounds for contesting the residents' claim. It summarised its grounds as follows:

> (a) The 2003 Act requires that applications *must be granted* if there are no representations made within the statutory period (or, if made, have been withdrawn). It follows by necessary implication, and also under the Licensing Act (Hearings) Regulations 2005, that there is no power to hold a hearing to consider an application if there are no extant representations in relation to it.
>
> (b) A comprehensive process of consultation is prescribed by the 2003 Act. This was adhered to… Westminster's notification by additional means, of some selected residents (who were already included within the wider statutory consultation) does not create a parallel consultation which itself needs to be comprehensive.

However, the Albert Hall did not appear at the hearing of the claim for judicial review.

[13] Westminster similarly served detailed grounds of opposition. They contended:

(1) They had no discretion to consider objections outside the statutory period.

(2) They had no power to consider or to hold a hearing to consider the grant of the variation to a licence where no representations had been received during the statutory period.

(3) They were bound to grant the variation applied for.

(4) Although they had an informal practice to send out letters notifying residents in the immediate vicinity of licensed premises of a licensing application, there was no legitimate expectation that any particular residents would receive such notification.

(5) They had not acted irrationally in deciding to whom to give notification of the Albert Hall's application.

[14] Westminster's detailed grounds in respect of (3) included the contention:

> Section 35(2) of the Act requires a licensing authority to grant a licence where there are no relevant representations. The grant must be in the terms sought. …

McCombe J's judgment

[15] The judge accepted Westminster's submission that they could not take late representations into account. However, he held that the residents had a legitimate expectation that they would be notified of a licensing application relating to the Albert Hall, and that in failing to comply with that expectation, and in deciding on a process of notification of the licensing application that did not include Albert Court, Westminster had acted irrationally and unlawfully, and he therefore made the order under appeal.

[16] Despite the reference to the duty of a licensing authority under section 35(2) of the Act in Westminster's grounds, no submission was made to the judge on their behalf that the provisions of the Act precluded the grant of the relief claimed by the residents on the grounds asserted by them.

---

* Editor's note: Reported at [2010] 2 EGLR 65

Contentions of the parties on the appeal

[17] The contentions of the parties on this appeal have differed substantially from those advanced before the judge.

[18] For the Albert Hall, Mr Philip Coppel QC's principal submission was that the judge had been wrong to find that the residents had any legitimate expectation that they would be notified of the application for a variation of the licence. He also contended that the relief proposed by the residents was inconsistent with the provisions of the Act and the regulations made under it.

[19] For Westminster, Mr David Matthias QC, who did not appear before the judge, did not dispute the judge's finding as to legitimate expectation and its breach; but he submitted that any failure of Westminster to satisfy any such legitimate expectation could not affect Westminster's duty to grant the application for the variation of the licence.

[20] It was because the order made by the judge arguably prevented Westminster from complying with their statutory duty, and arguably deprived the Albert Hall of the variation of its licence to which it was entitled, and because this is a pure point of law that does not depend on any evidence, let alone any disputed evidence, that the court allowed Westminster to argue this point, even though it had not clearly been argued before the judge. Moreover, in my view, the point was in Westminster's detailed grounds, and is obvious (as to which, see the discussion in *Miskovic* v *Secretary of State for Works and Pensions* [2011] EWCA Civ 16*).

[21] The residents also sought to raise an issue that had not been before the judge as to the Albert Hall's advertising of its application, namely whether the Albert Hall had duly advertised its application and whether Westminster had reasonably been satisfied (as required by section 35) that it had done so. The court refused to allow this issue to be argued. The residents had known of this issue at the hearing before the judge and had decided not to advance it. Furthermore, its determination would involve additional evidence from all three parties. It followed that, in this respect, the appeal proceeded on the same basis as before the judge, namely that Westminster had been properly satisfied that the Albert Hall had complied with the statutory requirements, as required by section 35(1)(b).

[22] On this basis, Mr John Steel QC, for the residents, pointed out that they alleged both that Westminster had failed to comply with the legitimate expectation found by the judge but also that, irrespective of any legitimate expectation, Westminster's decision as to whom to notify of the licensing application was irrational. It was therefore unlawful and should be quashed. Westminster had failed to take into account the considerations to which they were required to have regard by section 4 of the Act. The judge had been right to grant the relief he had. Furthermore, the court should make an order substituting a later date on which the application for variation of the licence should be taken to have been given to Westminster, thereby enabling the late representations, or further representations, to be effective. Westminster would then be required to hold a hearing to consider the representations and to decide whether to grant any variation of the premises licence.

Statutory framework

[23] Sections 4, 34 and 35 of the Act are, so far as relevant, as follows:

> 4 General duties of licensing authorities
>
> (1) A licensing authority must carry out its functions under this Act ("licensing functions") with a view to promoting the licensing objectives.
>
> (2) The licensing objectives are
>
> (a) the prevention of crime and disorder;
>
> (b) public safety;
>
> (c) the prevention of public nuisance; and
>
> (d) the protection of children from harm.
>
> (3) In carrying out its licensing functions, a licensing authority must also have regard to
>
> (a) its licensing statement published under section 5, and
>
> (b) any guidance issued by the Secretary of State under section 182.
>
> …
>
> 34 Application to vary premises licence
>
> (1) The holder of a premises licence may apply to the relevant licensing authority for variation of the licence.
>
> (2) Subsection (1) is subject to regulations under
>
> (a) section 54 (form etc of applications etc);
>
> (b) section 55 (fees to accompany applications etc).
>
> …
>
> (5) The duty to make regulations imposed on the Secretary of State by subsection (5) of section 17 (advertisement etc of application) applies in relation to applications under this section as it applies in relation to applications under that section.
>
> 35 Determination of application under section 34
>
> (1) This section applies where the relevant licensing authority
>
> (a) receives an application, made in accordance with section 34, to vary a premises licence, and
>
> (b) is satisfied that the applicant has complied with any requirement imposed on him by virtue of subsection (5) of that section.
>
> (2) Subject to subsection (3) and section 36(6), the authority must grant the application.
>
> (3) Where relevant representations are made, the authority must
>
> (a) hold a hearing to consider them, unless the authority, the applicant and each person who has made such representations agree that a hearing is unnecessary, and
>
> (b) having regard to the representations, take such of the steps mentioned in subsection (4) (if any) as it considers necessary for the promotion of the licensing objectives.
>
> (4) The steps are
>
> (a) to modify the conditions of the licence;
>
> (b) to reject the whole or part of the application;
>
> and for this purpose the conditions of the licence are modified if any of them is altered or omitted or any new condition is added.
>
> (5) In this section "relevant representations" means representations which
>
> (a) are about the likely effect of the grant of the application on the promotion of the licensing objectives, and
>
> (b) meet the requirements of subsection (6).
>
> (6) The requirements are
>
> (a) that the representations are made by an interested party or responsible authority within the period prescribed under section 17(5)(c) by virtue of section 34(5),
>
> (b) that they have not been withdrawn, and
>
> (c) in the case of representations made by an interested party (who is not also a responsible authority), that they are not, in the opinion of the relevant licensing authority, frivolous or vexatious.

[24] Section 17(5) is as follows:

> (5) The Secretary of State must by regulations
>
> (a) require an applicant to advertise his application within the prescribed period
>
> (i) in the prescribed form, and
>
> (ii) in a manner which is prescribed and is likely to bring the application to the attention of the interested parties likely to be affected by it;
>
> (b) require an applicant to give notice of his application to each responsible authority, and such other persons as may be prescribed, within the prescribed period;
>
> (c) prescribe the period during which interested parties and responsible authorities may make representations to the relevant licensing authority about the application.

[25] Sections 51 to 53 make provision for review by a licensing authority of a licence. If they receive an effective application for the review of a licence, section 52(2) requires them to hold a hearing to consider any relevant representations they receive, and section 52(3) imposes on the authority a duty:

> having regard to the application and any relevant representations, [to] take such of the steps mentioned in subsection (4) (if any) as it considers necessary for the promotion of the licensing objectives.

[26] Section 52(4) specifies:

> (4) The steps are
>
> (a) to modify the conditions of the licence;
>
> (b) to exclude a licensable activity from the scope of the licence;
>
> (c) to remove the designated premises supervisor;

* Editor's note: Reported at [2011] 2 CMLR 20

A (d) to suspend the licence for a period not exceeding three months;
(e) to revoke the licence;

[27] In the exercise of the powers conferred on him by the Act, the Secretary of State has made the Licensing Act 2003 (Premises Licences and Club Premises Certificates) Regulations 2005. They include the following provisions:

> 3. These Regulations apply to applications, notices, representations and reviews.
>
> 4. A person applying for a premises licence, a provisional statement, a variation of a premises licence, a review of a premises licence or a transfer of
> B a premises licence or giving an interim authority notice shall comply with the appropriate provisions of Parts 2 and 4.
>
> 5. A club applying for a club premises certificate or a variation of a club premises certificate or a person applying for a review of a club premises certificate shall comply with the appropriate provisions of Parts 3 and 4.
>
> 6. The relevant licensing authority in relation to an application, notice, representations or a review shall comply with the appropriate provisions of Parts 4 and 5.

[28] Parts 4 and 5 of the regulations do not impose any obligation on a licensing authority to advertise or to give notification of an application to vary a premises licence.

C [29] Regulation 22 provides:

> (1) An interested party or a responsible authority making representations to a relevant licensing authority, may make those representations
>
> (a) in the case of a review of a premises licence following a closure order, at any time up to and including seven days starting on the day after the day on which the authority received the notice under section 165(4) in relation to the closure order and any extension to it;
>
> (b) in any other case, at any time during a period of 28 consecutive days starting on the day after the day on which the application to which it relates was given to the authority by the applicant.

[30] The advertisement of applications is the subject of
D regulations 25 and 26:

> 25. In the case of an application for a premises licence under section 17,… to vary a premises licence under section 34 [(except where the only variation sought is the inclusion of the alternative licence condition)]… , the person making the application shall advertise the application, in both cases containing the appropriate information set out in regulation 26
>
> (a) for a period of no less than 28 consecutive days starting on the day after the day on which the application was given to the relevant licensing authority, by displaying a notice,
>
> (i) which is
>
> (aa) of a size equal or larger than A4,
>
> E (bb) of a pale blue colour,
>
> (cc) printed legibly in black ink or typed in black in a font of a size equal to or larger than 16;
>
> (ii) in all cases, prominently at or on the premises to which the application relates where it can be conveniently read from the exterior of the premises and in the case of a premises covering an area of more than 50 metres square, a further notice in the same form and subject to the same requirements every fifty metres along the external perimeter of the premises abutting any highway; and
>
> (b) by publishing a notice
>
> (i) in a local newspaper or, if there is none, in a local newsletter, circular or
> F similar document, circulating in the vicinity of the premises;
>
> (ii) on at least one occasion during the period of 10 working days starting on the day after the day on which the application was given to the relevant licensing authority.
>
> 26. …
>
> (3) In the case of an application to vary a premises licence or a club premises certificate, the notices referred to in regulation 25 shall briefly describe the proposed variation.
>
> (4) In all cases, the notices referred to in regulation 25 shall state
>
> (a) the name of the applicant or club;
>
> (b) the postal address of the premises or club premises, if any, or if there is no postal address for the premises a description of those premises sufficient to enable the location and extent of the premises or club premises to be identified;
>
> (c) the postal address and, where applicable, the worldwide web address where the register of the relevant licensing authority is kept and where and when the record of the application may be inspected;
>
> (d) the date by which an interested party or responsible authority may make representations to the relevant licensing authority;
>
> (e) that representations shall be made in writing;…

[31] Although, as mentioned above, Part 4 of the regulations does not impose any obligation on a licensing authority to advertise or to notify any application they receive for the variation of a premises licence, regulation 38 in Part 5 does impose a duty on them to advertise an application for the review of a premises licence under section 51.

[32] The Secretary of State has issued guidance under section 182 of the Act. In large part, it is a paraphrase of the statutory provisions. For present purposes, only the following paragraph is relevant:

> It is open to licensing authorities to notify residents living in the vicinity of premises by circular of premises making an application, but this is not a statutory requirement.

Discussion

[33] I have real doubts as to whether the residents did have any legitimate expectation as to their being notified of any application for the variation of the premises licence relating to the Albert Hall. However, for the purposes of this appeal, I am prepared to assume that, either for that reason or otherwise, as argued by Mr Steel (see [22] above), Westminster's decision to send circulars concerning the Albert Hall's application to residents in the neighbourhood but not to send such a circular to any resident of Albert Court was irrational and therefore in breach of Westminster's public law duties.

[34] Neither the Act nor the regulations impose any duty on a licensing authority to advertise such an application or to take any steps to notify anyone affected by them that it has been made. As has been seen, the sole duty to advertise and to give notice of the application is placed on the party making the application, in this case the Albert Hall. Before the judge, it was accepted that the Albert Hall had complied with that duty and, more relevantly, it was not contended that Westminster had not reasonably been "satisfied that the applicant has complied with any requirement imposed on him by virtue of" section 34(5): see section 35(1)(b). It is now accepted that Westminster did not receive any "relevant representations" as defined in section 35(5), because the residents' representations were received after the deadline imposed by the regulations and therefore did not fulfil the requirements of subsection (6). The duty of the authority under subsection (3) to hold a hearing arises only if a relevant representation is received: this, too, is common ground. It follows that Westminster were under the duty imposed by section 35(2) to grant the application, and the Albert Hall had a right, enforceable at public law, to the grant of the application. This court cannot grant any relief that would have the effect of preventing Westminster from complying with their statutory duty or deprive the Albert Hall of its public law right to the grant of the variation of its licence, for which it had made a valid application. Any failure by an authority to act in respect of their extra-statutory notifications cannot give rise to any right to interfere with the performance of their statutory duties.

[35] If authority is required for the proposition that an otherwise legitimate expectation cannot require a public authority to act contrary to statute, it is to be found in the seminal judgment of this court, given by Lord Woolf MR as he then was, in *R* v *North and East Devon Health Authority, ex parte Coughlan* [2001] QB 213, in [86] and the judgment in *R (on the application of Bhatt Murphy (a firm))* v *Independent Assessor* [2008] EWCA Civ 755, in [46], stating that the claim in *R* v *Secretary of State for Education and Employment, ex parte Begbie* [2000] 1 WLR 1115 failed "principally because to enforce the expectation would have required the Secretary of State to act contrary to statute". The judgment of Laws LJ in *R (on the application of Nadarajah)* v *Secretary of State for the Home Department* [2005] EWCA Civ 1363 includes an illuminating exposition of the principle of legitimate expectation that was approved by the Supreme Court in *Walumba Lumba (previously*

referred to as WL)(Congo) 1 and 2 v *Secretary of State for the Home Department* [2011] UKSC 12. Laws LJ said, in [69]:

> … Of course there will be cases where the public body in question justifiably concludes that its statutory duty (it will be statutory in nearly every case) requires it to override an expectation of substantive benefit which it has itself generated. So also there will be cases where a procedural benefit may justifiably be overridden. The difference between the two is not a difference of principle. Statutory duty may perhaps more often dictate the frustration of a substantive expectation.

[36] Precisely the same limitation must apply to judicial review sought on the ground of irrationality or any other common law ground. It is no more than an incident of the principle of legislative supremacy.

[37] The residents' case is not assisted by reference to section 4 of the Act. When exercising any discretion or power of decision under the Act, a licensing authority must do so "with a view to promoting the licensing objectives" as there defined. However, once the authority are under an unqualified duty to carry out an act specified by the statute, there is no room for section 4 to apply. The authority simply have no choice but to perform their statutory duty.

[38] If relevant representations (by definition, representations submitted in time) have been received by the authority and have not been withdrawn, section 35(3)(b) expressly requires it to "take such of the steps mentioned in subsection (4) (if any) *as it considers necessary for the promotion of the licensing objectives*". It is significant that the italicised words do not appear in the immediately preceding subsection (2). In effect, the residents seek to reword that subsection by substituting "may grant the application" for "must grant the application" and in addition by inserting the words to the effect of those I have italicised. That is an impossible exercise in interpretation.

[39] My conclusion does not mean that a decision by an authority to refrain from notifying persons affected by a licensing application cannot be challenged. In theory, if it were thought that an authority were acting irrationally or otherwise unlawfully, an order could be sought requiring them to reconsider their decision, and if made sufficiently promptly the court might grant relief if it would have any practical value. Thus, in an appropriate case the court might require the authority to notify residents in the immediate vicinity if the notification could lead to representations being made within the statutory time limit. However, that is the limit of the relief the court could grant. The court cannot subvert the statutory scheme.

[40] Mr Steel submitted that, in the present case, the court could make orders having the effect of postponing the date of the Albert Hall's application, thereby enabling timely representations to be made by the residents. Neither the Act nor the regulations confer any statutory power on the court to grant such relief, and it has no common law power to grant it. Furthermore, as was pointed out by Mr Coppel, such relief would conflict with the provisions of the Act and the regulations and create practical problems, which it is unnecessary to set out.

[41] In [43] of his judgment, McCombe J summarised his reasons for his now unchallenged decision that a licensing authority have no power to take into account late representations:

> 43. All these considerations, [counsel for Westminster] submitted, militated against the existence of the implicit residual obligation/discretion of the Council to consider late representations, initially directed to it as licensing authority. I agree with that submission. In my judgment, for reasons already advanced, this Act contains a clear statutory procedure, with clear time limits for the making of an application, objections to it and for the resolution of such objections. It provides in clear terms that, subject to such representations, the grant of the application is to follow. Given this clear machinery, I consider that in exercising its functions under this Act, the Council is entitled to deal with representations made to it as licensing authority strictly as such. Subject to such representations, as Richards J said [in *The British Beer and Pub Association* v *Canterbury City Council* [2005] EWHC 1318 (Admin)], the applicant is entitled to the grant of his application. It would be contrary to that entitlement for the machinery to be undermined by letting in late applications through a back door not provided for by the Act itself.

[42] Unfortunately, doubtless because the point had not clearly been made to him, the judge did not draw the, to my mind, inevitable conclusion from his analysis. Once the conditions set out in section 35(1) G
were satisfied, in the absence of any relevant representation the Albert Hall was, as the judge said, entitled to the grant of its application and, to paraphrase his words, it would be contrary to that entitlement for it to be undermined by a failure of the licensing authority to carry out a notification process (and therefore a "back door") not provided for by the Act.

[43] It does not follow that the residents have no possible remedy. If their fears are realised, and the result of the variation in the licence is public disorder and public nuisance, they may apply for a review of the licence under section 51, to which I have referred. If they do so, then, subject to the provisions of section 51, Westminster will be H
under a duty to hold a hearing to consider their representations, and they will then be required by section 52(3) to take such of the steps specified in subsection (4), if any, as they consider necessary to promote the licensing objectives listed in subsection 4. Those steps include the modification of the conditions of the licence (subsection (4)(a)) or even the revocation of the licence: subsection (4)(e).

[44] For these reasons, I would allow this appeal and set aside the order made by the judge quashing the variation of the premises licence.

Wilson LJ said: J

[45] I agree.

Lloyd LJ said:

[46] I also agree.

Appeal allowed.

R (on the application of Hope and Glory Public House Ltd) *v* City of Westminster Magistrates' Court K

COURT OF APPEAL

[2011] EWCA Civ 31

26 January 2011

SIR RATHBONE WALL, PRESIDENT OF THE FAMILY DIVISION, AND LAWS AND TOULSON LJJ

Estates Gazette 7 May 2011

[2011] 18 EG 110 L

Premises licence — Variation of licence — Appeals — Licensing Act 2003 — Local licensing authority attaching conditions to premises licence for public house after receiving complaints from local residents over noise — Appeal by owner of public house to magistrates' court under section 181 of 2003 Act — Correct approach on such appeal — Whether burden lying with appellant to persuade court that decision of licensing authority wrong

The appellant owned a public house with a licence for the sale and supply of alcohol and the provision of entertainment and late-night refreshment. The local council, as the licensing authority, reviewed M
the licence following complaints from residents concerning the level of noise in the evenings as a result of customers congregating outside the premises. The licensing subcommittee attached conditions to the licence to stop the public nuisance caused by the noise, the main one being that, after 6pm, customers should not be allowed to take drink from the premises in an open container.

The appellant appealed to the magistrates' court under section 181 of, and Schedule 5 to, the Licensing Act 2003. The district judge directed himself that the decision of the licensing subcommittee should not be lightly reversed but should be overturned only if he were satisfied that it was wrong. Dismissing the appeal, he concluded that there was clear evidence of public nuisance and that the conditions imposed

A by the subcommittee were necessary and proportionate to ensure the promotion of the licensing objectives.

The appellant applied for a judicial review of that decision; permission was refused and it appealed, contending that the district judge had erred in law in placing on it the burden of disproving that the noise caused by customers amounted to a nuisance. It submitted that: (i) the appeal before the district judge was a fresh hearing, where the burden lay with the licensing authority to prove their allegations of public nuisance and that the modifications were necessary and proportionate; and (ii) to start with a presumption in favour of the licensing authority undermined that process and contravened the appellant's right, under Article 6 of
B the European Convention on Human Rights, to a fair hearing in the determination of its civil rights.

***Held*: The appeal was dismissed. The decision of the licensing authority is a relevant matter for the magistrates' court to take into consideration on an appeal from that decision. The licensing authority are performing an administrative function in discharging their licensing functions in the exercise of a power delegated by the public to decide what the public interest requires. The magistrates' court must pay careful attention to the licensing authority's reasons for reaching the decision under appeal, bearing in mind that parliament has chosen to place responsibility for making**
C **such decisions on those bodies. The weight that the magistrates should attach to those reasons is a matter for their judgment, taking into account the fullness and clarity of the reasons, the nature of the issues and the nature and quality of the evidence submitted on appeal. In order to overturn a decision of a licensing authority, the magistrates' court on appeal needs to be satisfied, on the evidence, including any fresh evidence, that the judgment below is wrong, even if it was wrong when it was made. It is for the appellant to persuade the court that the decision under appeal should be reversed; that is the usual position on appeals and is envisaged by the Magistrates' Courts Rules 1981 on statutory**
D **appeals to magistrates' courts from decisions of local authorities. That approach is not limited to cases where the decision under appeal was based on considerations of general policy. Further, the form of appeal provided by section 182 of, and Schedule 5 to, the 2003 Act satisfies the requirements of Article 6. There is no general principle in domestic or European case law that it is incompatible with Article 6 for a party aggrieved by an administrative decision to bear the responsibility of establishing its complaint.**

The following cases are referred to in this report.

Ali v *Birmingham City Council*; *Tomlinson* v *Birmingham City Council*;
E *Ibrahim* v *Birmingham City Council*; *Moran* v *Manchester City Council*; *R (on the application of Aweys)* v *Birmingham City Council* [2010] UKSC 8; [2010] 2 AC 39; [2010] 2 WLR 471; [2010] 2 All ER 175; [2010] HLR 22

Kingsley v *United Kingdom* 35605/97 (2002) 35 EHRR 10, ECtHR

R (on the application of Alconbury Developments Ltd) v *Secretary of State for the Environment, Transport and the Regions*; *R (on the application of Holding & Barnes plc)* v *Secretary of State for the Environment, Transport and the Regions* [2001] UKHL 23; [2001] 2 AC 295; [2001] 2 WLR 1389; [2001] 2 All ER 929; (2001) 3 LGR 38; (2001) 82 P&CR 40; [2001] 2 PLR 76; [2001] JPL 920, HL

R (on the application of Chief Constable of Lancashire) v *Preston Crown Court*; *R (on the application of Smith)* v *Lincoln Crown Court* [2001] EWHC 928
F (Admin); [2002] 1 WLR 1332

R (on the application of Hope and Glory Public House Ltd) v *City of Westminster Magistrates' Court* [2009] EWHC 1996 (Admin); [2010] ACD 12

Sagnata Investments Ltd v *Norwich Corporation*; *sub nom Norwich Corporation* v *Sagnata Investments Ltd* [1971] 2 QB 614; [1971] 3 WLR 133; [1971] 2 All ER 1441, CA

Stepney Borough Council v *Joffe*; *Stepney Borough Council* v *White*; *Stepney Borough Council* v *Diamond* [1949] 1 KB 599; [1949] 1 All ER 256, DC

Tower Hamlets London Borough Council v *Runa Begum* [2003] UKHL 5; [2003] 2 AC 430; [2003] 2 WLR 388; [2003] 1 All ER 731; [2003] HLR 32, HL

Tsfayo v *United Kingdom* 60860/00 [2007] HLR 19; (2009) 48 EHRR 18; [2007] LGR 1, ECtHR

This was an appeal by the appellant, Hope and Glory Public House Ltd, from a decision of Burton J, sitting in the Administrative Court, refusing permission for judicial review of a decision of the respondent, City of Westminster Magistrates' Court, dismissing an appeal under the Licensing Act 2003 against a decision of the interested party, Westminster City Council, to modify the terms of a premises licence.

Ian Glen QC and Gordon Bishop (instructed by Jeffrey Green Russell Solicitors) appeared for the appellant; the respondent did not appear and was not represented; David Matthias QC and Emma Dring (instructed by the legal department of Westminster City Council) represented the interested party.

Giving judgment, **Toulson LJ** said:

Introduction

[1] This appeal raises a question about how a magistrates' court hearing an appeal from a decision of a licensing authority under the Licensing Act 2003 (the Act) should approach the decision.

Background

[2] The appellant owns the Endurance public house in Berwick Street, Soho. The premises are licensed for the sale and supply of alcohol and for the provision of entertainment and late-night refreshment. The licence was granted on 12 March 2007 by Westminster City Council (the council) as the local licensing authority.

[3] On 15 April 2008, the council's environmental health consultation service (EHCS) applied, under section 51(1) of the Act, for a review of the licence after complaints were made by residents about the level of noise caused by customers taking their drinks out of the pub and congregating on the street during the evenings.

[4] The hearing of the review took place before the council's licensing subcommittee on 26 and 27 June 2008. The subcommittee heard submissions and evidence lasting around five hours. It decided to attach a number of conditions to the licence, the main condition being that no customer should be permitted to take drink from the premises in an open container after 6pm. The decision and the subcommittee's reasons were notified to the appellant's solicitor by a letter dated 4 July 2008. The subcommittee stated:

> We have no policy to ban outside drinking, and we have accordingly not approached the case on that basis. We were not referred to the Council's statement of licensing policy by any party. We have had regard, as we must, to the policy,… but we have reached our decision based on the evidence that has been put before us in relation to these premises, and not on any policy ground.
>
> The application was made on the grounds of public nuisance, and we first consider whether it was established that a public nuisance for the purposes of the Act exists. The evidence we heard was that large numbers of customers of the Endurance congregate on a daily basis outside the public house in Kemps Court in the evening, the numbers involved ranging from very few (five-10) to very many (180 or more). Those customers drinking and talking outside the premises make a noise. The noise is amplified by the configuration of buildings in the area. The noise causes public nuisance to surrounding residents, including, in particular residents directly opposite the public house.
>
> The licensee argued that the noise was not so bad as to constitute a nuisance and that the complaints… were exaggerated. He called expert evidence in support of that proposition. We are completely satisfied that the noise is indeed a serious nuisance…
>
> A number of local residents and other customers of the premises gave evidence about the way in which the premises were run, and we accept that the premises are valued by its customers and that a number of people enjoy being able to drink outside. We reject however the argument that a licensee has a fundamental right to, in effect, appropriate a part of the public realm for his own commercial purposes, if the effect of doing so is to cause serious public nuisance to his neighbours. Accordingly, we are persuaded that it is appropriate to take steps to prevent that public nuisance from continuing.
>
> We recognise that steps should only be taken where they are necessary and that it cannot be necessary to take disproportionate steps…

[5] The subcommittee then considered the conditions proposed by the EHCS and additional conditions proposed by the police. It concluded that most of the proposed conditions were required.

[6] The appellant appealed against the decision to the City of Westminster Magistrates' Court under section 181 of, and Schedule 5 to, the Act.

[7] At a preliminary hearing on 7 May 2009, District Judge Snow heard argument about how he should approach the decision of the subcommittee on the hearing of the appeal. He held that he was bound by the decision of the Court of Appeal in *Sagnata Investments Ltd* v *Norwich Corporation* [1971] 2 QB 614, in the light of which he ruled:

> I will therefore
> (1) Note the decision of the licensing sub-committee.
> (2) Not lightly reverse their decision.
> (3) Only reverse the decision if I am satisfied it is wrong.
> (4) I will hear evidence.
> (5) The correct approach is to consider the promotion of the Licensing Objectives. To look at the Licensing Act 2003, the Guidance made under section 182 LA03, Westminster's Statement of Licensing Policy and any legal authorities.
> (6) I am not concerned with the way in which the Licensing Sub-Committee approached their decision or the process by which it was made. The correct appeal against such issues lies by way of Judicial Review.

[8] The district judge heard the appeal over five days between 11 and 25 June 2009, during which he heard four days of evidence, considered 1,797 pages of statements and exhibits and visited the site. On 30 June 2009, he delivered a 22-page written judgment. His conclusions in summary were:

> I find, on the balance of probabilities, that given the number of Residents, Students and Teachers affected, and given the geographical spread, that the nuisance clearly is a public nuisance.
>
> …
>
> The evidence is clear, that the public nuisance arises between 6pm and 11pm. The conditions imposed by the Licensing Sub-Committee are necessary and proportionate to ensure the promotion of the licensing objectives.
>
> …
>
> On 7 May 2009 I set out that I would only interfere with the decision of the sub-committee if I was satisfied that it was wrong. In fact I am satisfied that it was right. This appeal is dismissed.

[9] The appellant applied for judicial review of the district judge's decision on various grounds. The primary argument was that the district judge's ruling about how he should approach the decision of the subcommittee was wrong in law.

[10] The appellant's application for permission to apply for judicial review was dismissed by Burton J in a judgment dated 21 July 2009: see [2009] EWHC 1996 (Admin)*.

[11] Permission to appeal was refused by Moses LJ on paper but was granted by Sir Mark Waller after an oral hearing on 19 May 2010. The permission was limited to the single question of whether the district judge's self-direction was correct. As to that, Sir Mark Waller observed:

> So far as the direction is concerned, the position may well be covered by the authority *Sagnata Investments Limited* v *Norwich Corporation* [1971] 2 QB 614, but it seems to me that the question of whether it is an appropriate direction and the question of whether that is the right way in which a magistrate should approach an appeal in which he is hearing all the evidence *de novo* is a matter of some importance. We can spend a great deal of time arguing about the arguability of the point and it is better to have a decision which clarifies the position, which at present there is not.

Fresh evidence

[12] In addition to the ground on which leave to appeal was granted, Mr Ian Glen QC sought leave on behalf of the appellant to introduce fresh evidence. The purpose of the fresh evidence was to rebut evidence given by a witness, Ms Bailey, at the hearing before the district judge to the effect that noise from the Endurance disturbed lecturers and students at the nearby Westminster Kingsway College. Ms Bailey had provided a witness statement on 15 January 2009, which had been disclosed to the appellant's representatives soon afterwards, that is, several weeks prior to the hearing before the district judge. The fresh evidence came from others at the college and was obtained in October 2010, that is, several months after Waller LJ granted limited permission to appeal. We can see no basis on which the late discovery of this evidence could provide a proper ground for judicial review of the district judge's decision and we refuse the application for permission to introduce it. G

Licensing Act 2003

[13] The short title of the Act is:

> An Act to make provision about the regulation of the sale and supply of alcohol, the provision of entertainment and the provision of late night refreshment, about offences relating to alcohol and for connected purposes. H

[14] The Act brought about major changes to the licensing system in England and Wales. The background, nature and purpose of its provisions are summarised in the explanatory notes to the Act.

[15] Essentially, the Act integrated alcohol, public entertainment, theatre, cinema, night café and late-night refreshment licensing. Previously, there was a patchwork system under which liquor licences were granted by licensing magistrates but other licensing functions, such as public entertainment licensing, were the responsibility of local authorities. The Act followed the publication, in April 2000, of a White Paper (Cm 4696) entitled *Time for reform: Proposals for the modernisation of our licensing laws*. J

[16] The Act created a unified system of regulation of the activities of the sale and supply of alcohol, the provision of regulated entertainment and the provision of late-night refreshment, referred to in the Act as the "licensable activities". The White Paper proposed that the licensing authority under the new scheme should be the local authority; and the Act follows that proposal. The government explained its reasons in the White Paper as follows:

> 117. The current responsibility of magistrates for liquor licensing reflects their traditional role in maintaining the peace and the association of alcohol with crime. Entertainment licensing came on the scene at a time when the magistrates' role had moved a long way from law enforcement towards the administration of justice. With an integrated system of licensing it is necessary to decide if the responsibilities should fall to the magistrates or the local authorises or some third body which might involve both. K
>
> …
>
> "123. There are three compelling reasons in favour of giving the local authority (at district level) the responsibilities we have described in this White Paper. They are:
> "Accountability: we strongly believe that the licensing authority should be accountable to local residents whose lives are fundamentally affected by the decisions taken L
> "Accessibility: many local residents may be inhibited by court processes, and would be more willing to seek to influence decisions if in the hands of local councillors
> "Crime and disorder: Local authorities now have a leading statutory role in preventing local crime and disorder, and the link between alcohol and crime persuasively argues for them to have a similar lead on licensing.
> 124. In reaching our conclusion, we do not in any way seek to devalue the importance of the wider contribution the local licensing justices have made for so many years. While in our proposals they would be relieved of administrative licensing responsibilities, they would retain, in their capacity as magistrates, the responsibility for dealing with people charged with offences under licensing law and for the imposition of sanctions and penalties in respect of personal licence holders. M

[17] Magistrates also have an appellate function, which lies at the heart of this appeal.

[18] Section 4 sets out general duties of licensing authorities. It identifies "licensing objectives" that licensing authorities are to promote. These include the prevention of public nuisance. Section 5 requires licensing authorities to produce statements of licensing policy for three-year periods. In carrying out their licensing functions, a licensing authority must have regard to their licensing statement and to any guidance issued by the Secretary of State for Culture, Media and Sport under section 182. Before determining their policy for a

* Editor's note: Reported at [2010] ACD 12

three-year period, a licensing authority must go through a process of public consultation: section 5(3). Section 6 provides for licensing authorities to conduct their licensing functions through licensing committees. Section 9 deals with proceedings before licensing committees and empowers the Secretary of State to make regulations about them.

[19] There are various types of "personal licence" and "premises licence" that a licensing authority may grant. The present case concerns a premises licence granted under section 18. It is open to a licensing authority to attach such conditions to a licence under section 18 as they consider necessary for the promotion of the licensing objectives identified in section 4.

[20] Under section 51, an "interested party" or a "responsible authority" may apply to the licensing authority for a review of a premises licence. An interested party includes anyone living or involved in a business in the vicinity: see section 13(3). A responsible authority includes the local authority that have statutory responsibilities in respect of the protection of the environment and human health: see section 13(4)(e). In the present case, the applicant for the review was the council, acting through the EHCS. Section 53 expressly permits a local authority to make an application under section 51 for a review of a premises licence in their capacity as a responsible authority and to determine the application in their capacity as the licensing authority.

[21] Section 52 provides that a licensing authority that receive an application under section 51 may, after holding a hearing to consider it and any relevant representations:

> take such of the steps mentioned in subsection (4) (if any) as it considers necessary for the promotion of the licensing objectives.

The steps mentioned in subsection (4) include modifying the conditions of the licence.

[22] Section 52(10) requires the licensing authority to notify their determination, and their reasons for making it, to the holder of the licence, the applicant, any person who made relevant representations and the local chief officer of police.

[23] Section 181 and Schedule 5 provide a system for appeals from decisions of a licensing authority to a magistrates' court. Paragraph 8 of Schedule 5 deals with appeals against decisions made under section 52. It provides:

> (1) This paragraph applies where an application for review of a premises licence is decided under section 52.
> (2) An appeal may be made against that decision by
> (a) the applicant for the review,
> (b) the holder of the premises licence, or
> (c) any other person who made relevant representations in relation to the application.

[24] The powers of a magistrates' court on an appeal from a decision of the licensing authority are to dismiss the appeal, to substitute any other decision that could have been made by the licensing authority or to remit the case to the licensing authority to dispose of it in accordance with the direction of the court: see section 181(2).

[25] The Magistrates' Courts Rules 1981 (made under the Magistrates' Court Act 1980) provide that where a statutory appeal lies to a magistrates' court against a decision or order of a local authority or other authority, the appeal shall be by way of complaint for an order: see r 34. The rules also provide that on the hearing of a complaint, it is for the complainant to go first in calling evidence: see r 14.

Appellant's submissions

[26] Mr Glen submitted that the district judge wrongly placed the burden on the appellant to disprove that the noise caused by customers of the Endurance was such as to amount to a public nuisance and that the conditions imposed by the licensing authority were necessary and proportionate. He submitted that it was for the EHCS to prove their allegation of public nuisance and to establish that the modifications to the licence were necessary and proportionate. The hearing before the district judge was a hearing *de novo*, at which evidence was given and tested by cross-examination. Mr Glen pointed out that the licensing subcommittee itself stated that its decision was not based on any policy ground. Rather, it turned on the subcommittee's assessment of the facts. On factual issues of that kind, it undermined the nature of an appeal process by way of rehearing if the court started with a presumption in favour of the licensing authority. Moreover, such an approach did not comply with the requirement of Article 6 of the European Convention on Human Rights that, in the determination of his civil rights, everyone is entitled to a fair and public hearing within a reasonable time by an independent and impartial tribunal established by law. In support of this submission, he relied on the following passage from *Paterson's Licensing Acts* 2009, in para 5.4:

> Assuming we are correct in saying that the hearing in the magistrates' court needs to be article 6 compliant, then the magistrates would not be an "independent and impartial" tribunal if the court starts off from a position favouring the decision of the licensing authority. The licensing authority will be a party to any appeal and the success or failure of the appeal should depend on the evidence which is given and the arguments which are put forward.

[27] Mr Glen also cited the decision of the Divisional Court in *R (on the application of Chief Constable of Lancashire)* v *Preston Crown Court* [2001] EWHC 928 (Admin)*. That case concerned an appeal from licensing justices to the crown court under the Licensing Act 1964. It was argued that there was a breach of Article 6 because the composition of the court included two members who belonged to the same licensing committee as the magistrates whose decision was under appeal. The argument was rejected, but Mr Glen relied on a passage (in [18]) where Laws LJ, who delivered the main judgment, referred to the crown court conducting "a rehearing in the full and proper sense". If it was to be a rehearing in that sense, Mr Glen submitted that it must follow that the burden of proof on the appeal was the same as on the original hearing.

[28] Mr Glen cited a number of other authorities for the proposition that an appeal against a licensing decision have long been recognised to be a rehearing. It is not necessary to refer to them because it is not in dispute that the appeal is a rehearing at which the affected parties are all entitled to call evidence and that the court must make its decision on the full material before it. The issue is what is the proper approach to the original decision and, in particular, the reasons given for it. Mr Glen did not submit that they should be disregarded. He accepted that the court hearing the appeal could properly take into consideration the reasons given by the licensing authority, but not to the point of placing a legal burden on the appellant.

[29] Mr Glen submitted that the district judge went wrong in attaching too much significance to a sentence from a judgment of Lord Goddard CJ in *Stepney Borough Council* v *Joffe* [1949] 1 KB 599 cited by Edmund Davies LJ in *Sagnata Investments*. In *Sagnata Investments*, an application was made under the Betting Gaming and Lotteries Act 1967 for a permit to open an amusement arcade in Norwich. The application was refused by the local authority and the applicant appealed to quarter sessions. The recorder who heard the appeal had written reasons for the refusal furnished by the town clerk and evidence of witnesses on both sides as to the merits of the application. He did not have any information about what had happened before the licensing committee. He allowed the appeal. The local authority appealed to the Divisional Court (whose judgment is not reported) and then to the Court of Appeal (Lord Denning MR, Edmund Davies and Phillimore LJJ). Its appeal was dismissed by the majority, Lord Denning dissenting. Lord Denning considered that the local authority were entitled to their opinion that it was socially undesirable to have such arcades in Norwich and that the recorder was wrong to substitute his view for those of the elected body responsible for making such decisions.

[30] The majority considered that the recorder had been entitled to conclude that the local authority had effectively decided that they would not grant any permit under the Act for an amusement place in Norwich and that there was no error of law in his decision to allow the appeal.

---

* Editor's note: Reported at [2002] 1 WLR 1332

Edmund Davies LJ, at p633F, quoted Lord Denning in the course of argument as summarising the issue in this way:

Is the hearing to be treated as a new trial to be determined on evidence *de novo*, without being influenced by what the local authority has done? Or is the hearing to be treated as an appeal proper, in which the local authority's decision is to be regarded as of considerable weight, and is not to be reversed unless their decision is shown to be wrong?

[31] Edmund Davies LJ considered that this was a false antithesis. From the reasons that he gave for preferring an intermediate position, he must have understood the second of Lord Denning's alternatives ("an appeal proper") as being confined to deciding whether the local authority's decision was wrong in law on the material before it. He went on to say, at p636D-E:

The provision for an appeal to quarter sessions seems to me largely, if not entirely, "illusory" if the contention of the appellant council is right. If it is, I am at a loss to follow how the recorder would set about discharging his appellate function. Lacking all information as to what had happened before the local authority, save the bare knowledge that they had refused the application and their written grounds for refusal, he would be powerless, as I think, to make any effective examination of the validity of those reasons.

[32] Edmund Davies LJ expressed his conclusion as follows, at p636G:

I hold that the proceedings before this recorder were by way of a complete rehearing.

But, contrary to what has been contended, this conclusion does *not* involve that the views earlier formed by the local authority have to be entirely disregarded by quarter sessions. It is true that in *Godfrey* v *Bournemouth Corporation* [1969] 1 WLR 47, after observing that an appeal to quarter sessions under Schedule 6 to this same Act was by way of a complete rehearing, Lord Parker CJ said, at p52, "the discretion is a discretion which the recorder in the present case had to arrive at himself uninfluenced by what the local authority had done." But with respect, I do not accept this. It went much too far, it was in direct conflict with the view which Lord Parker had earlier expressed in *Reg* v *Essex Quarter Sessions, Ex parte Thomas* [1966] 1 WLR 359, 363, it was contrary to the approach adopted both by the recorder and by Lord Parker CJ himself in the instant case, and it was, with deference, an uncalled-for observation. Here again, *Stepney Borough Council* v *Joffe* [1949] 1 KB 599 establishes what I regard as the proper approach, for, having made the point that there was in that case an unrestricted appeal, Lord Goddard CJ continued at pp602, 603:

"That does not mean to say that the court of appeal, in this case the metropolitan magistrate, ought not to pay great attention to the fact that the duly constituted and elected local authority have come to an opinion on the matter, and ought not lightly, of course, to reverse their opinion. It is constantly said (although I am not sure that it always sufficiently remembered) that the function of a court of appeal is to exercise its powers when it is satisfied that the judgment below is wrong, not merely because it is not satisfied that the judgment was right."

Phillimore LJ's judgment was to similar effect.

[33] Mr Glen observed that that case was one in which the local authority's decision had been based on a general policy and that it was therefore right for the recorder to attach weight to the local authority's policy, although he still had to form his own judgment on the evidence as to whether a permit should be granted. The decision, he submitted, provided no support for taking a similar approach where (as the licensing subcommittee recognised in the present case) no question of licensing policy was involved. The core question in this case was whether the noise caused by the customers of the Endurance amounted to a public nuisance, and this was a matter for the EHCS to establish on the evidence called before the district judge.

Council's submissions

[34] Mr David Matthias QC submitted that Burton J was right in his approach to *Stepney Borough Council* and *Sagnata Investments* and his dismissal of the appellant's claim. Burton J said in his judgment:

43. I conclude that the words of Lord Goddard approved by Edmund Davies LJ are very carefully chosen. What the appellate court will have to do is to be satisfied that the judgment below "*is wrong*", that is to reach its conclusion on the basis of the evidence put before it and then to conclude **G** that the judgment below *is wrong*, even if it *was not wrong* at the time. That is what this district judge was prepared to do by allowing fresh evidence in, on both sides.

44. The onus still remains on the claimant, hence the correct decision that the claimant should start, one that cannot be challenged as I have indicated.

45. At the end of the day, the decision before the district judge is whether the decision of the Licensing Committee is wrong. Mr Glen has submitted that the word "wrong" is difficult to understand, or, at any rate, insufficiently clarified. What does it mean? It is plainly not "*Wednesbury* unreasonable" because this is not a question of judicial review. It means that the task for the District Judge having heard the evidence which is now before him, and specifically addressing the decision of the court below is to give a decision whether, because **H** he disagrees with the decision below in the light of the evidence before him, it is therefore wrong.

[35] Mr Matthias submitted that, as a matter of principle as well as precedent, there are good reasons why the magistrates' court should pay great attention to the decision of the licensing authority and should allow an appeal only if satisfied on the evidence before it that the decision was wrong. He pointed out that parliament had chosen to make the local authority central to the promotion in their area of the licensing objectives set out in the Act because local councillors are accountable to the local electorate and are expected to be sensitive to the needs and concerns of the local populace. In licensing matters, there **J** is often no single "right answer". Mr Matthias pointed to the conditions that the licensing authority attached to the licence on the review in the present case as an example. The ban imposed on customers taking drink from the premises in an open container after 6pm might equally have been imposed somewhat earlier or somewhat later. It is normal for an appellant to have to show that the order challenged was wrong. The only unusual feature about this type of appeal is that all parties have carte blanche to call evidence. It does not, however, follow that the respondent to the appeal should bear the responsibility of showing that the order should be upheld and so should be required to present its case first. **K**

[36] On the Article 6 issue, Mr Matthias's propositions may be paraphrased as follows:

(1) The decision of the licensing authority was an administrative decision, which admittedly involved a determination of the appellant's "civil rights" within the meaning of Article 6, as it has been interpreted in the European case law.

(2) The extent to which Article 6 requires such a decision to be subject to review by an independent and impartial tribunal depends greatly on the nature of the decision. Article 6 is an important expression of the rule of law, but the rule of law itself allows proper scope for democratic process in administrative decision-making. **L**

(3) Administrative decisions often involve making judgments and assessing priorities on matters of social and economic policy. It accords with democratic principles for such decisions to be taken primarily by democratically accountable bodies. The power of the High Court in judicial review proceedings to review the legality of such decisions and the procedures followed is sufficient to ensure compatibility with Article 6.

(4) Some administrative decisions, although not necessarily involving wide issues of policy, call for particular knowledge or experience on the part of the decision maker. Often such decisions will involve an **M** evaluative judgment and the exercise of discretion. In such cases, too, the availability of judicial review in the High Court is sufficient to meet the requirements of Article 6. It would be perverse if Article 6 were to require a full fact-finding appeal to a tribunal that lacked the degree of knowledge and expertise of the original decision maker.

(5) There may be cases where an administrative decision does not depend on what may be described as democratic questions (questions of local or national policy, such as belong to the political forum), but that depends essentially on a question of fact requiring no special knowledge or experience on the part of the decision maker. In such a case, Article 6 may require that an aggrieved person whose civil rights are determined by the decision should be entitled to have it reviewed by

a tribunal whose power includes whatever factual review is necessary for justice to be done.

(6) There is nothing in domestic or Strasbourg case law to suggest that there is a general principle that it is incompatible with Article 6 for a person aggrieved by an administrative decision to bear the responsibility of establishing his complaint.

[37] Mr Matthias's concession that Article 6 is engaged in the present case followed from the decision in *Kingsley* v *United Kingdom* 35605/97 (2002) 35 EHRR 10, in [34], where it was held that Article 6 is engaged in proceedings that determine whether an individual is entitled to undertake licensable activities. For his other submissions, he cited a number of authorities, including particularly *R (on the application of Alconbury Developments Ltd)* v *Secretary of State for the Environment, Transport and the Regions* [2001] UKHL 23; [2001] 2 AC 295, *Tower Hamlets London Borough Council* v *Runa Begum* [2003] UKHL 5, [2003] 2 AC 430, *Tsfayo* v *United Kingdom* 60860/00 (2009) 48 EHRR 47; [2007] LGR1 and *Ali* v *Birmingham City Council* [2010] UKSC 8; [2010] 2 AC 39.

[38] Mr Matthias submitted that, in this case, the appellant's right of appeal to the district judge amply satisfied the requirements of Article 6.

Conclusion

[39] Since Mr Glen accepted (in our view rightly) that the decision of the licensing authority was a relevant matter for the district judge to take into consideration, whether or not the decision is classified as "policy based", the issues are quite narrow. They are:

(1) How much weight was the district judge entitled to give to the decision of the licensing authority?

(2) More particularly, was he right to hold that he should allow the appeal only if satisfied that the decision of the licensing authority was wrong?

(3) Was the district judge's ruling compliant with Article 6?

[40] We do not consider that it is possible to give a formulaic answer to the first question because it may depend on a variety of factors: the nature of the issue, the nature and quality of the reasons given by the licensing authority and the nature and quality of the evidence on the appeal.

[41] As Mr Matthias rightly submitted, the licensing function of a licensing authority is an administrative function. By contrast, the function of the district judge is a judicial function. The licensing authority have a duty, in accordance with the rule of law, to behave fairly in the decision-making procedure, but the decision itself is not a judicial or quasi-judicial act. It is the exercise of a power delegated by the people as a whole to decide what the public interest requires: see the judgment of Lord Hoffmann in *Alconbury*, in [74].

[42] Licensing decisions often involve weighing a variety of competing considerations: the demand for licensed establishments, the economic benefit to the proprietor and to the locality by drawing in visitors and stimulating the demand, the effect on law and order, the effect on the lives of those who live and work in the vicinity and so on. Sometimes, a licensing decision may involve narrower questions, such as whether noise, noxious smells or litter coming from premises amount to a public nuisance. Although such questions are in a sense questions of fact, they are not questions of the "heads or tails" variety. They involve an evaluation of what is to be regarded as reasonably acceptable in the particular location. In any case, deciding what (if any) conditions should be attached to a licence as necessary and proportionate to the promotion of the statutory licensing objectives is essentially a matter of judgment rather than a matter of pure fact.

[43] The statutory duty of the licensing authority to give reasons for their decision serves a number of purposes. It informs the public, who can make their views known to their elected representatives if they do not like the licensing subcommittee's approach. It enables a party aggrieved by the decision to know why it has lost and to consider the prospects of a successful appeal. If an appeal is brought, it enables the magistrates' court to know the reasons that led to the decision. The fuller and clearer the reasons, the more force they are likely to carry.

[44] The evidence called on the appeal may or may not throw a very different light on matters. Someone whose representations were accepted by the licensing authority may be totally discredited as a result of cross-examination. By contrast, in the present case, the district judge heard a mass of evidence over four days, as a result of which he reached essentially the same factual conclusions as the licensing authority had reached after five hours.

[45] Given all the variables, the proper conclusion to the first question can be stated only in very general terms. It is right in all cases that the magistrates' court should pay careful attention to the reasons given by the licensing authority for arriving at the decision under appeal, bearing in mind that parliament has chosen to place responsibility for making such decisions on local authorities. The weight that the magistrates should ultimately attach to those reasons must be a matter for their judgment in all the circumstances, taking into account the fullness and clarity of the reasons, the nature of the issues and the evidence given on the appeal.

[46] As to the second question, we agree with the way in which Burton J dealt with the matter in [43] to [45] of his judgment.

[47] We do not accept Mr Glen's submission that the statement of Lord Goddard in *Stepney Borough Council*, applied by Edmund Davies LJ in *Sagnata Investments* is applicable only in a case where the original decision was based on "policy considerations". We doubt whether such a distinction would be practicable because it involves the unreal assumption that all decisions can be put in one of two boxes, one marked policy and the other not. Furthermore, *Stepney Borough Council* was not itself a case where the original decision was based on "policy considerations". In that case, three street traders had their licences revoked by London County Council after they were convicted of selling goods at prices exceeding the maximum fixed by statutory regulations. On appeal, the magistrate decided that they were still fit to hold the licences. The county council unsuccessfully argued before the Divisional Court that the magistrate's jurisdiction was limited to considering whether there was any material on which the council could reasonably have arrived at their decisions to revoke the licences. The court held that the magistrate's power was not limited to reviewing the decision on the ground of an error of law, but that he was entitled also to review the merits. It was in that context that Lord Goddard went on to say that the magistrate should, however, pay great attention to the decision of the elected local authority and should reverse it only if he was satisfied that it was wrong.

[48] It is normal for an appellant to have the responsibility of persuading the court that it should reverse the order under appeal, and the Magistrates Courts Rules envisage that this is so in the case of statutory appeals to magistrates' courts from decisions of local authorities. We see no indication that parliament intended to create an exception in the case of appeals under the Act.

[49] We are also impressed by Mr Matthias's point that in a case such as this, where the licensing subcommittee has exercised what amounts to a statutory discretion to attach conditions to the licence, it makes good sense that the licensee should have to persuade the magistrates' court that the subcommittee should not have exercised its discretion in the way that it did rather than that the magistrates' court should be required to exercise the discretion afresh on the hearing of the appeal.

[50] As to Article 6, we accept the propositions advanced by Mr Matthias and we agree that the form of appeal provided by section 182 of, and Schedule 5 to, the Act amply satisfies the requirements of Article 6.

[51] Although the point is academic in the present case, we doubt the correctness of part of the district judge's ruling where he said:

> I am not concerned with the way in which the licensing sub-committee approached their decision or the process by which it was made. The correct appeal against such issues lies by way of judicial review.

[52] Judicial review may be a proper way of mounting a challenge to a decision of the licensing authority on a point of law, but it does not follow that it is the only way. There is no such express limitation in the

Act, and the power given to the magistrates' court under section 181(2) to "remit the case to the licensing authority to dispose of it in accordance with the direction of the court" is a natural remedy in the case of an error of law by the authority. We note also that the guidance issued by the government under section 182 and laid before parliament on 28 June 2007 states, in para 12.6:

> The court, on hearing any appeal, may review the merits of the decision on the facts and consider points of law or address both. B

However, this point was not the subject of any argument before us.

[53] For the reasons that we have given, the appeal is dismissed.

Appeal dismissed.

REAL PROPERTY AND CONVEYANCING

Irwin *v* Wilson and others

CHANCERY DIVISION

[2011] EWHC 326 (Ch)

23 February 2011

SIR WILLIAM BLACKBURNE

Estates Gazette 11 June 2011

[2011] 23 EG 88

Sale of land — Contract — Waiver of term — Vendor required to provide purchasers with specified documentation regarding correction of defect in title plans — Either party entitled to terminate contract by notice if vendor unable to provide documentation by given date — Vendor giving such notice — Purchasers thereafter purporting to waive requirement as to provision of documents — Whether term capable of waiver — Whether waiver possible once notice to terminate contract given

In October 2009, the claimant contracted to sell to the defendants a leasehold interest in a flat for £580,000. The contract incorporated the standard conditions of sale (4th ed) so far as they applied to a sale by private treaty. By clause 25.1 of the contract, completion was to take place 14 days after the claimant served the defendants with certain information documenting the correction of an error in the lease plan and registered title plans for the flat. Clause 25.2 provided for either side to terminate the agreement on five days' notice in the event that the claimant, having used all reasonable endeavours, was unable to provide all the clause 25.1 documentation by a date in February 2010.

The claimant proved unable to provide the defendants with all the requisite information by the specified date. He accordingly served five days' notice on the defendants to terminate the contract under clause 25.2. On the final day of the notice period, the defendants served a notice on the claimant, purporting to waive the requirements of clause 25.1 and demanding that the sale proceed to completion.

The claimant sought a declaration that he had lawfully terminated the contract. The defendants applied for specific performance of the contract and sought summary judgment on that relief. That application was dismissed in the county court on the ground that clause 25.1 was not for the defendants' sole benefit and, accordingly, it was not open to them to waive it. The defendants appealed.

On the appeal, the claimant further argued that: (i) clause 25.1 was not severable from the remainder of that clause and therefore could not be waived in isolation from clause 25.2; and (ii) once the claimant had given notice to terminate the contract under clause 25.2, it was too late for the defendants to exercise a waiver.

***Held*: The appeal was dismissed.**

(1) A party to a contract can waive a contractual condition or provision provided that it is solely for that party's benefit and is severable. Whether the stipulation is, in terms or by necessary implication, for the exclusive benefit of the party in question must be considered by reference to the contract as a whole in the light of the surrounding circumstances. The term in clause 25.1 as to the service of documents was for the exclusive benefit of the defendants as purchasers. It concerned the rectification of a defect in the vendor's title and was therefore akin to the term in a contract for the sale of land stating the nature of the title that the vendor had to give. It is open to a purchaser to waive a term F **dealing with the title to the purchased property. The existence of clause 25.2 did not compel a different conclusion: it was a distinct term of the contract conferring on both parties separately the right to terminate the contract in the event that the vendor had not secured performance of the documents service term by February 2010. The existence of a right for either party to terminate the contract if a particular condition is not performed by the due date is not inconsistent with the condition in question being for the exclusive benefit of one party and with that party having the right to waive it. (2) Clause 25.1 could be severed from** G **the remainder of clause 25 notwithstanding that it formed part of a larger provision that determined the completion date by reference to a date 14 days after compliance with clause 25.1. Waiver of clause 25.1 did not render the date for completion uncertain. The consequence of waiver was that the contract would proceed to completion as though either clause 25.1 had been performed or the contract contained no such requirement. On the latter view, although the contract would fail to make provision for fixing the completion date, the court could supply a date by reference to what had to be done to complete the contract. On any view, a reasonable period for completion had passed by the time the defendants had** H **issued their application for specific performance. (3) Although it had been open to the defendants to waive clause 25.1, it was too late to do so once clause 25.2 had been exercised by the service of a notice to terminate the contract. The notice took effect when it was given such that no obligation to complete could arise thereafter. Accordingly, once the claimant had served a valid notice under clause 25.2, the time for completion had passed and the defendants could not then seek to waive the claimant's clause 25.1 obligations. The claimant had validly terminated the contract by his notice.**

The following cases are referred to in this report. J

Akzo Nobel UK Ltd v *Arista Tubes Ltd* [2010] EWCA Civ 28; [2010] 2 All ER (Comm) 1198; [2010] 05 EG 112 (CS)
Bennett v *Fowler* (1840) 2 Beav 302
Chaitlal v *Ramlal* [2003] UKPC 12
Globe Holdings Ltd v *Floratos* [1998] 3 NZLR 331, CA, NZ
Hawker v *Vickers* [1991] 1 NZLR 399, CA, NZ
Heron Garage Properties Ltd v *Moss* [1974] 1 WLR 148; [1974] 1 All ER 421; (1973) 28 P&CR 54; 229 EG 439
Moreton v *Montrose Ltd* [1986] 2 NZLR 496, CA, NZ
Spiro v *Glencrown Properties Ltd* [1991] Ch 537; [1991] 2 WLR 931; [1991] 1 All ER 600; [1991] 1 EGLR 185; [1991] 02 EG 167
Sudbrook Trading Estate Ltd v *Eggleton* [1983] 1 AC 444; [1982] 3 WLR 315; K [1982] 3 All ER 1; (1982) 44 P&CR 153; [1983] 1 EGLR 47; (1982) 265 EG 215, HL
Yewbelle Ltd v *London Green Developments Ltd* [2007] EWCA Civ 475; [2007] 2 EGLR 152

This was an appeal by the defendants, Celia Wilson, Florence Wilson and Alice Wilson, from a decision of HH Judge Madge, sitting in Central London County Court, dismissing an application for summary judgment on their counterclaim for specific performance of a contract for the sale of land, made in the course of proceedings by the claimant, Guy Irwin, for declaratory relief as to the valid rescission of that contract.

Timothy Carlisle (instructed by Colemans Solicitors LLP, of Maidenhead) represented the claimant; Gregory Hill (instructed by Hutchinson Mainprice) appeared for the defendants.

Giving judgment, **Sir William Blackburne** said:

Introduction

[1] This is an appeal against a decision of HH Judge Madge sitting in Central London County Court on 16 July 2010. It arises in proceedings started in March of last year in which the claimant seeks a declaration that a contract dated 28 October 2009 for the sale by him to the three defendants, who are sisters, of a leasehold flat at 1 Cranworth Gardens, London SW9, was lawfully terminated by a notice given on his behalf by his solicitor on 1 February 2010. In addition to the declaration, the claimant seeks possession of the flat and associated monetary orders.

[2] The defendants countered by issuing an application in the proceedings for summary relief by way of specific performance of the contract. Their application is supported by a document headed "Particulars of Different Remedy Sought", which is in effect a defence and counterclaim to the claimant's particulars of claim.

[3] The leasehold interest in the flat, for which the price to be paid is £580,000, is comprised in a lease dated 21 April 1983 and in another dated 16 April 1999. Title to the property so to be sold is registered with absolute title under title number SGL 372337. By clause 5 of the contract, the title to be deduced is expressed to consist of official copies of the entries and of the title plan appearing on the register of that title number as at 16.46 on 26 August 2009. The contract incorporates the fourth edition of the standard conditions of sale so far as applicable to a sale by private treaty (as this is) and in so far as they are not varied by or inconsistent with the terms of the contract.

[4] The dispute between the parties centres on clause 25. Headed "Completion Date", it provides that:

> 25.1 The Completion Date shall be 14 days after the Seller's solicitors shall have served (by fax or by email) the Buyer's solicitors with copies of:
>
> (a) The Office Copy entries relating to Title SGL32337, as amended by the Land Registry to show that [sic] the correct location of the Property on the Filed Plan;
>
> (b) A Deed of Variation to the Lease dated 21st April 1983, executed by the Freeholder and Seller to replace the existing Lease Plan with a plan showing the correct location and floor layout of the floor plan, drawn to scale and showing a north point and being fully Land Registry compliant.
>
> (c) The consent of the Seller's mortgagees to the proposed variation to the Lease.
>
> (d) Same as (a) but relating to the adjacent flat.
>
> (e) Same as (b) but relating to the adjacent flat.
>
> (f) Same as (c) but relating to the adjacent flat.
>
> 25.2 If the Seller, having used all reasonable endeavours, shall not be able to provide all of the information referred to in 25.1 to the Buyer's solicitors by February 1st 2010, then either side may give notice of five working days to terminate this Agreement, whereupon the Buyer shall vacate the Property and the Seller's solicitors shall return that part of the deposit which they were holding as stakeholders to the Buyer's solicitors and the Seller shall return that part of the deposit which was released to him on exchange of contracts, to the Buyer's solicitors.
>
> 25.3 For the avoidance of doubt, the Buyer shall not be required to vacate the Property in accordance with clause 25.2 unless and until the Seller's solicitors confirm that they are in funds to reimburse the whole of the deposit including the £6,000 which they held as agents for the seller on exchange of contracts.

The title number referred in clause 25.1(a) accidentally omitted a figure "7". There is no dispute that it is intended to refer to the title number mentioned in clause 5, namely SGL 372337.

[5] Clause 26 permits the buyer to go into occupation of the property from exchange and sets out the terms, including a monthly payment, on which that is to occur. I understand that the defendants did take up and have since remained in occupation.

[6] It appears that the reason for the amended plans referred to in clause 25.1(a) was that, presumably by accident when the leases of the claimant's flat and the adjoining flat were granted, the two lease plans were transposed. That error, I understand, was carried into the filed plans recorded on the respective Land Registry titles.

[7] It is common ground between the parties that the claimant used all reasonable endeavours but was unable to provide the defendants' solicitor with all the information referred to in clause 25.1 by 1 February 2010. At 16.02 on Monday 1 February, the claimant's solicitor faxed a letter to the defendants' solicitor, in which the following was stated:

> In accordance with Clause 25.2 of the agreement dated 28th October 2009 between our clients, we hereby give five working days' notice to terminate that Agreement when our client requires your clients to vacate the property and we shall in turn return the deposit paid on exchange of contracts.

The defendants had paid a 10% deposit, equal to £58,000, of which £6,000 was held by the claimant's solicitor as his agent and the remaining £52,000 held as stakeholder. This explains the reference to the £6,000 in clause 25.3.

[8] By the standard conditions applicable to the contract, a notice that is received after 4pm on a working day is to be treated as having been received the next working day. It follows therefore that, having been faxed after 4pm, the notice contained in the letter of 1 February is to be treated as having been received by the defendants on the following day, Tuesday 2 February. It is common ground that the day of service is to be excluded in calculating the five working days. The five days given by the notice therefore expired on Tuesday 9 February.

[9] On Monday 8 February 2010, the defendants' solicitor sent the following letter to the claimant's solicitor:

> We write with further reference to your letter of 1st February.
>
> That letter purported to give notice terminating the Agreement of 28th October 2009 with effect from close of business today. Your letter was sent following non-compliance with the conditions set out in Special Condition 25.1 of the Contract. Special Condition 25.1 is a unilateral condition for the benefit of the Buyer. As such the Condition is capable of being waived by the Buyer and the Buyer hereby waives that Condition.
>
> The Contract in consequence still being live, the Buyer is now making arrangements to complete, having made appropriate financial arrangements to avoid using mortgagees' monies. Please let us have an immediate completion statement as at tomorrow's date.
>
> If notwithstanding this letter you do not confirm by midday tomorrow that your client will now complete, we shall be serving a completion notice, followed if necessary by an application to the Court for specific performance on the expiry of that notice.
>
> The Contract no longer being conditional, and our clients being entitled forthwith to complete the purchase, their right to possession of the property is incidental to the right to complete and they will therefore not be vacating the property in the interim.

[10] That letter was also faxed. It was faxed at 15.37. By the standard conditions applicable to the contract, a notice is given when it is received, and if sent by fax is to be treated as received, subject to proof to the contrary, one hour after despatch. It follows, therefore, there being no proof to the contrary, that the notice of purported waiver of clause 25.1 was given at 16.37 on that Monday and, since that was after 4pm, is to be taken as given on the following day, Tuesday 9 February. That was on the fifth and last of the five working days' notice given by the claimant's notice of the previous week.

[11] When the defendants' summary judgment application came before him on 16 July 2010, Judge Madge directed that the question of whether, as the defendants' solicitor's letter of 8 February had contended, clause 25.1 was capable of waiver by them as being for the buyer's (that is, the defendants') benefit should be argued first. In an admirably short and clear judgment, he held that the clause was not simply for the benefit of the purchaser, with the result that it was not open to the defendants to waive it. He therefore dismissed their application. In so doing, he referred to and placed reliance on what was said by Brightman J (as he then was) in *Heron Garage Properties Ltd* v *Moss* [1974] 1 All ER 421*, at p426. He also referred to and placed reliance on the presence in clause 25.2 of a right in the seller (that is, the claimant) to give notice of termination if all the information

* Editor's note: Also reported at (1973) 229 EG 439

referred to in clause 25.1 had not been given to the buyer's solicitor by 1 February 2010.

This appeal: Issues

[12] It is against that decision, and the dismissal of the summary judgment application that thereby resulted, that the defendants appeal. Norris J refused the defendants permission to appeal on paper, but, on their renewed application, Mann J gave them permission. I shall continue to refer to them as the defendants rather than as the appellants. I shall also continue to refer to the claimant as such. For completeness, I should mention that the claimant has served a respondent's notice.

[13] Three broad issues were argued before me, namely whether: (i) clause 25.1 is for the sole benefit of the buyer (that is, the defendants); (ii) even if it is, clause 25.1 is severable from the remainder of clause 25, in particular the right of termination by notice give by clause 25.2, and how, if it is capable of waiver and was otherwise waived, the date for completion of the contract is to be determined; and (iii) even if the first and second issues are resolved in the defendants' favour, it was open to the defendants to seek to waive the condition and call for completion after the claimant had given notice, under clause 25.2, to terminate the contract, albeit that the defendants' purported waiver was notified before the expiry of the five working days provided by that notice. The second and third issues were not argued before Judge Madge, but I was invited by counsel to determine them if they arose. I indicated that I was willing to do so: they do not involve contested issues of fact; the matters were fully argued; costs and time will be saved.

Defendants' case

[14] For the buyers, Mr Gregory Hill submitted that, properly understood, the term in clause 25.1 for the service on the buyers' solicitor of copies of the various Land Registry entries, deeds of variation and mortgage consents ("the documents service term", as I shall call it) was for the exclusive benefit of the buyers, that it was capable of waiver by them and that by the letter of 8 February they did waive it. Benefit, he submitted, is to be assessed by reference to the substance of the relevant provision and by reference to the position that results if the contract is performed. The presence of a right given to either side to terminate the transaction if the term is not performed does not necessarily show that the performance of that term is intended to benefit both of them. The purchaser needs the termination right to ensure that he is not kept waiting indefinitely for the term to be performed: the right gives him a let-out from the contract if the term is not fulfilled. Correspondingly, the seller needs the termination right to ensure that he cannot be made to keep trying indefinitely to procure the performance of the term, especially since his obligation is merely to use "all reasonable endeavours" and performance of the term is dependent on the co-operation of others, which is not something he can compel. This is so whether performance of the term is for the exclusive benefit of the purchaser or is advantageous to both of them. The contractual purpose of the seller's right to terminate is therefore to give him the certainty that he is free of obligation when his termination notice takes effect. *Heron Garage* does not assist the claimant, and the court below was wrong to think that it did.

[15] On the second issue the severability of the documents service term from the remainder of clause 25 Mr Hill submitted that the only question that arises is how the date for completion is to be fixed. This, he submitted, is a matter of mechanics only. The court should construe the contract in a way that enables effect to be given to a waiver of that part of clause 25.1 constituted by the documents service term and also provides a completion date. He submitted that although the effect of a waiver is ordinarily to treat the relevant term as omitted from the contract, there is no intrinsic reason why, in appropriate cases, the term in question should not be treated as performed. He submitted that the court can and should adopt the analysis that is most appropriate to what makes the transaction effective and leads to its performance; it should avoid an approach that leads to the transaction going off on what he referred to as a "technicality". He directed my attention, by way of an analogy, to the approach of Hoffmann J (as he then was) *Spiro* v *Glencrown Properties Ltd* [1991] Ch 537*, at p544G-H, where what was in issue was the characterisation of an option to buy land. If, therefore, waiver is equated with performance, clause 25.1 itself provides that completion is to follow 14 days later. In the events that happened, that would have been 14 days after the giving of the notice by the defendants' solicitor in its letter dated 8 February 2010. This analysis enables the contract to be completed substantially as the parties intended. The only alteration to their obligations as a result of the waiver is that the claimant does not actually have to produce the corrected plans. This releases him of a burden and causes him no detriment whatever.

[16] Even if the result of an effective waiver is to treat the waived provision as though it had been omitted so that, in the instant case, the terms of clause 25.1 can no longer be applied as the completion date is fixed by reference to the performance of a term that no longer exists, there nevertheless remains an obligation to complete, in that if and in so far as a contract for the sale of land does not specify a time for completion, jurisprudence in the field of vendor and purchaser allows an implication that completion is to take place after a time that is reasonable in all the circumstances: see, for example, *Chaitlal* v *Ramlal* [2003] UKPC 12, in [22]. Once the relevant part of clause 25.1 was waived, therefore, completion could have taken place as soon as the necessary transfer had been agreed and a completion statement prepared. Whatever a reasonable time for those purposes might have been, it must have elapsed by 8 April 2010, when the defendants issued their application notice for specific performance and, *a fortiori*, by the time of the hearing before Judge Madge on 16 July. Mr Hill referred also to *Sudbrook Trading Estate Ltd* v *Eggleton* [1983] 1 AC 444† as an example of the willingness of the court to substitute its own machinery where the machinery fixed by the contract, being a subsidiary and non-essential part of the contract, has broken down for any reason. In that case, the machinery in question was for the ascertainment of the sale price of certain leasehold reversions that, by the contract, was to be at a fair and reasonable price by the application of objective standards.

[17] On the third issue whether it was open to the defendants to waive the provision of the corrected plans after the claimant had given notice under clause 25.2 to terminate the contract Mr Hill submitted that, on the correct construction of that subclause, termination occurs only on the expiry of the five working days. The reference in the clause is to "notice of 5 working days to terminate" the contract. That is, moreover, a perfectly sensible result for the parties to have intended to produce, in that it is appropriate to give the other party an opportunity to do whatever is necessary to ensure that the contract completes, rather than going off: if the seller gives the notice, the buyer can waive the documents service term (as the defendants have done) and if it is the buyer who gives the notice the seller has a last chance to bring to fruition his efforts to fulfil that term. The fact that the contract remains alive while the notice is running, and that the right to terminate will cease on the corrected plans being obtained, even after 1 February, support this conclusion. There is no significant prejudice to whichever party gives the notice if the position is crystallised when the notice expires rather than when it is given: either way, the party's interest in knowing where he stands is protected. In short, Mr Hill submitted, the right of the party giving the notice is not an unqualified right to cry off: it is a right to cry off unless, when the notice expires, the party receiving the notice can complete in a way that ensures that the giver of the notice gets what he contracted for and is not prejudiced by completing.

Claimant's case

[18] For the claimant and in support of the decision by Judge Madge in the court below, Mr Timothy Carlisle submitted that the documents service term was not one the performance of which was exclusively for the benefit of the defendants as buyers. He submitted that the provision of corrected plans and Land Registry entries benefits the seller in that his position as the owner of the leasehold interest is correctly defined

* Editor's note: Also reported at [1991] 1 EGLR 185
† Editor's note: Also reported at [1983] 1 EGLR 47

A in respect of: first, the superior (freehold) title; second, his mortgagee; and, third, the Land Registry title. It also minimises any risk of liability or costs as against either his freeholder or his mortgagee if the incorrect documentation should remain in place. Those benefits, he said, would continue after completion of the sale to the defendants.

[19] At the forefront of Mr Carlisle's submissions was the provision in clause 25.2 conferring the right to terminate the contract in the event that, despite the use of all reasonable endeavours, the seller is not able to provide all the information referred to in clause 25.1 by 1 February 2010. The existence of that right is a real benefit to the seller. Moreover, had the parties thought that the provision of the information was for the
B benefit of the buyer and not the seller, one might have expected the buyer alone to be permitted to give notice terminating the contract in the event that the term could not be performed. The contract could easily have so provided. However, the contract is not so worded. It empowers either side to terminate the contract in the event of non-performance of the documents service term. The fact, therefore, that it is open to the seller, having used all reasonable endeavours but having failed to perform that term, to take advantage of that failure by giving notice of termination of the contract is, Mr Carlisle submitted, a clear indication that the term cannot have been simply for the buyer's benefit. He submitted that Judge Madge was right so to have concluded. He referred me to
C *Heron Garage* (on which the judge below had also placed reliance) and certain observations in *Moreton* v *Montrose Ltd* [1986] 2 NZLR 496, at pp505-505 [sic]. He submitted that the right is or may be of benefit to the seller in so far as it allows what Mr Carlisle described as a "change of mind" on the seller's part enabling him to "re-assess the market". I took this to mean that it would enable the seller to attempt to achieve a higher price for the flat by selling it to someone else.

[20] On the second issue severability and the relationship of the completion date to the fulfilment of the documents service term Mr Carlisle submitted that the term was closely interlinked with clause 25.2, which modified what might otherwise appear to be an
D absolute obligation on the seller's part (in clause 25.1) to serve the buyer's solicitor with copies of the various documents in question by stating that the obligation in that subclause is no more than one to use all reasonable endeavours. It was therefore not open to the buyer to waive clause 25.1 (or any part of it) in isolation from clause 25.2. He went on to submit that, in any event, the date for completing the contract is fixed by reference to the service by the seller's solicitor on the buyer's solicitor of the various documents referred to in the documents service term. If that part of clause 25.1 is waived, there is no ascertainable date for completion. The standard conditions do not assist because clause 12.3 of the contract has disapplied standard condition 6.1, which
E sets out, in the absence of anything to the contrary in the contract itself, when the completion date is to occur.

[21] On the third issue the availability of a power to waive after notice of termination is given pursuant to clause 25.2 Mr Carlisle submitted that the proper construction of clause 25.2 is that, on giving the notice, all primary obligations under the contract cease, leaving as secondary obligations only the vacation of the property by the buyer and repayment of the deposit by the seller. He submitted that the purpose of the notice is not to give to the parties a last chance to obtain and serve copies of the relevant documents. If it were, the contract could and would have said so. The aim of the five days was simply to give the
F buyer (if in possession) time to make arrangements to vacate and to give the seller time to make arrangements to repay the deposit.

Conclusions

[22] Beguilingly straightforward as the matter appeared to Judge Madge, I consider that the issues raised are far from simple. They break down into four separate points.

(a) Benefit of condition

[23] The test for determining whether a contract term is for the exclusive benefit of one party, failing which and in the absence of any express power of waiver the term is not capable of unilateral waiver by the party to the contract that claims to have the benefit of it, is that stated by Brightman J (as he then was) in *Heron Garage*, at p426e-h:

> Without seeking to define the precise limits within which a contracting party seeking specific performance may waive a condition on the ground that it is intended only for his benefit, it seems to me that in general the proposition only applies where the stipulation is in terms for the exclusive benefit of the plaintiff because it is a power or right vested by the contract in him alone…, or where the stipulation is by inevitable implication for the benefit of him alone… If it is not obvious on the face of the contract that the stipulation is for the exclusive benefit of the party seeking to eliminate it, then in my opinion it cannot be struck out unilaterally. I do not think that the court should conduct an enquiry outside the terms of the contract to ascertain where in all the circumstances the benefit lies if the parties have not concluded the matter on the face of the agreement they have signed.

[24] In a decision of the New Zealand Court of Appeal, to which Mr Carlisle referred me, namely *Globe Holdings Ltd* v *Floratos* [1998] 3 NZLR 331 (and to which I shall return later), there is (at p334) a citation from an earlier decision of the same court (*Hawker* v *Vickers* [1991] 1 NZLR 399, at pp402-403) setting out the following statement of the approach in law:

> A party may waive a condition or provision in a contract which is solely for that party's own benefit and is severable. In such a case the other party is denied the right to treat the condition as unsatisfied and is obliged to complete notwithstanding the loss of that advantage. The question is one of construction of the contract. It turns on whether the stipulation is in terms or by necessary implication for the exclusive benefit of the party, and the answer is derived from consideration of the contract as a whole in the light of the surrounding circumstances…

That seems to me, with respect, to be an entirely accurate summary of the relevant approach.

[25] Measured by that approach and confining my gaze simply to what I have referred to as the documents service term, there can be no doubt that the term is for the exclusive benefit of the buyer. It is akin to the term in a contract for the sale of land stating what the title is that the vendor has to give. In such a case, as Lord Langdale MR stated in *Bennett* v *Fowler* (1840) 2 Beav 302, at p304:

> the obligation to which a vendor is subject to make out a title is intended for the benefit of the purchaser only, and… if he thinks fit to waive it, he has a right to do so.

[26] Mr Carlisle sought to persuade me that the term is not of that nature because title is dealt with by clause 5 of the contract. He is right that that clause states what the title is that the seller agrees to provide but it does not follow that the documents service term is not also concerned purely with a matter of title, as, in my judgment, it clearly is. The term is concerned with the rectification of a defect in the expression of the vendor's title that had come about, as I have mentioned, by a misidentification of the leasehold flat at 101 Cranworth Gardens by the plan attached to the lease creating that interest, which defect, unsurprisingly, was reflected in the filed plan on the registered title. The nub of the matter, as Mr Hill submitted, is whether the *performance* of the term is for the exclusive benefit of the buyer. If performed and the transaction completes the claimant no longer has any interest in the property of which he was once but is no longer the owner. Defects in his former title cease to be of concern to him. It might have been otherwise if there were evidence that the claimant was the owner or had some interest in the adjoining flat (and was affected therefore by the defect because of the transposition of plans referred to earlier) and in consequence (in light of that other flat) interested in having the error corrected. There was no suggestion, however, that that is this case. None of the other matters identified by Mr Carlisle as constituting "benefit" to the seller from the existence of the documents service term is properly to be so characterised: they all related to the claimant's position as owner of the flat and therefore assumed, or appeared to assume, that the contract would not be completed.

(b) Severability

[27] However, as Mr Carlisle was at pains to point out, the documents service term does not stand alone. It appears in clause 25.1 and is part of a larger provision determining the completion date of the contract. Does that mean I do not at this stage look at clause 25.2 that it is not open to

the buyer to waive performance of that term (assuming that the term is in all other respects exclusively for the buyer's benefit)?

[28] I do not consider that it does. Common sense suggests that it ought to be open to the buyer to waive a term dealing with the title to that property. However, if he does, how is the completion date to be determined? As Mr Carlisle rightly pointed out, it cannot be by reference to the standard conditions because standard condition 6.1 which is the material condition has been expressly disapplied by clause 12.3 of the contract. How then does the matter stand?

[29] There are, as Mr Hill submitted, two ways of approaching the matter. The first is to treat the waived stipulation as though it had been performed. The other is to treat the waiver as if it had deleted the stipulation from the contract altogether. (This last was the approach favoured by Brightman J in *Heron Garage* (at p427) in the passage that I have set out above, where there is reference to the party claiming the exclusive benefit of a stipulation "seeking to eliminate it" and to the stipulation in question being "struck out unilaterally".)

[30] Which of these two approaches is the appropriate one to follow must depend on the contractual context. In the instant case, each leads to the same result. If the former approach is followed and one treats the waived stipulation as though it had been performed, clause 25.1 can take effect according to its terms: completion is to take place 14 days after waiver. The latter approach treating the waived term as though it had been struck out of the contract merely means that clause 25.1 cannot take effect according to its terms and, as a consequence, the contract fails to make any provision (or any effective provision) for fixing the completion date. That is not, however, fatal to the contract since, as Mr Hill submitted, the jurisprudence establishes that if the parties do not set out in their contract when completion is to occur, the court steps in to supply the date. The decided cases show that the courts will go to some lengths to supply a completion date where, for any reason, the parties have failed to do so. In the instant case, that means, as Mr Hill went on to submit, that the date is to be measured by what needs to be done to complete the contract once any need to perform the documents service term is to be ignored. I can see no good reason why the date that falls 14 days from the date on which that term is waived (assuming that it is otherwise validly waived) should not be taken to be the completion date. Fourteen days was the period that the parties themselves agreed on the footing that the term was performed. It is difficult to see why there should be any greater period merely because the term is waived. On any view, a reasonable period had passed by 8 April 2010 when the defendants issued their application for specific performance.

[31] I should add that on this point Mr Carlisle drew my attention to a later passage in *Heron Garage* on which he placed some reliance. In that passage (at p427c-d), Brightman J said:

> There is an added difficulty in the way of Heron's case. The decision in *Hawksley* v *Outram* [1892] 3 Ch 359 suggests that a stipulation cannot be waived if it is inextricably mixed up with other parts of the transaction from which it cannot be severed. Clause 8 of the sale agreement specifies the date for completion. That date under cl 8 is dependent on the date when Heron receive planning consent without conditions or when Heron is deemed to have approved conditions attached to the planning consent. Nothing is said about the date for completion if Heron waive the condition for planning consent. So it would seem that Heron's unilateral elimination of cl 7 of the sale agreement will also eliminate cl 8 and leave the date for completion in the air.

Mr Carlisle submitted that the position is the same in the instant case if it is otherwise open to the buyer to waive the documents service term.

[32] Mr Hill submitted that this passage was not part of the *ratio* in *Heron Garage*, that the view there expressed was "tentative" and that I should not be obliged to follow it. I agree. The fact that a contract for the sale of land provides, as it well may, that completion is to take place a stated number of days after the vendor has shown that he has a particular title to the land in question does not mean that the purchaser cannot waive the vendor's obligation to show that title merely because to do so "will leave the date for completion in the air" if by that is meant either that the stipulation in question cannot be waived or, if it were otherwise to be waived, the contract will cease to be enforceable. The only consequence of the waiver by the purchaser in such a case is that the contract proceeds to completion either as though title had been shown or (which comes to the same thing) as though the contract provided that the vendor was under no obligation to show title. I cannot for one moment think that the court in such a case would have any difficulty in fixing a date for completion by reference to what remained to be done in order to carry the contract to completion. G

(c) Right to terminate given by clause 25.2

[33] What, then, of the presence in clause 25.2 of the right, conferred separately on both the buyer and the seller, to terminate the contract in the event that, despite having used all reasonable endeavours, the seller has not secured performance of the documents service term by 1 February 2010? It was the presence of this right that persuaded Judge Madge to conclude that clause 25.1 was not capable of waiver by the defendants. H

[34] In my judgment, the presence of that right is irrelevant to whether the documents service term is for the exclusive benefit of the seller. The principle is that a party may waive a contract term if that term, *if performed*, is of benefit to him but not to the other party (or parties) to the contract. By contrast, the right to terminate the contract conferred by clause 25.2 is exercisable if and only if the term *cannot be or is not performed*.

[35] This very point was discussed in *Globe Holdings*, the New Zealand decision referred to earlier. In that case, a condition of the contract for the sale of an apartment block stipulated that, within 60 days of acceptance, the purchaser would obtain planning consent from the local council for the subdivision of the block. The contract contained a general condition that in relation to any financial or other conditions either party could, at any time before the condition was fulfilled or waived, avoid the contract by giving notice. Within the 60 days the purchaser's solicitor gave notice that the special condition was waived and that, accordingly, the contract could be regarded as unconditional. The question was whether the notice was legally effective. In the course of a judgment dealing with a number of points, the New Zealand Court of Appeal (at p339) cited a passage from the earlier decision of *Hawker*, which stated that: J K

> there is nothing inconsistent in providing expressly or by necessary implication for unilateral waiver of a condition up to a certain date and thereafter for allowing either party to avoid the contract for nonfulfilment of the condition. Such a provision simply recognises the commercial reality that the nature and significance to the parties of a condition in a contract may change over time or at a point in time. If the contract [sic] is fulfilled or waived, the parties then have the certainty of an unconditional contract. If not fulfilled or waived by the nominated date, each is free to end the contract by appropriate notice to the other.

The court then pointed out that: L

> The argument against waiver rests upon the desirability of certainty for a vendor from being able immediately to bring the contract to an end, or see it immediately collapse, once the given time has elapsed. But certainty is achieved by a different rule, namely that any waiver must occur on or before the condition date, or at least before the contract is actually brought to an end (if it is not automatically void). It has to be remembered that we are at this point concerned with a situation in which it is to be accepted that there is no substantive benefit to [the vendors]. Therefore, their only legitimate interest is in knowing whether the transaction is to proceed or not. Once the time allowed for the fulfilment of the condition expires they can forthwith give notice of cancellation if they have not already been informed that the sale will go ahead. M It matters not to them whether it does so because of fulfilment or because the purchaser elects to proceed anyway. The achieving of certainty is in the vendors' own hands if there has been no action by the purchaser. If there has been a waiver the transaction proceeds as it would have done if the condition had been satisfied on the date of the waiver.
>
> ...
>
> We conclude therefore that a distinction is to be drawn between the benefit of the substance of the condition and the benefit of the time limit…

[36] The reasoning in that passage thus distinguishes between the benefit of the condition here, the documents service term contained in clause 25.1 and the benefit of the right to terminate the contract if the condition has not been fulfilled by the due date here, the right to

A terminate the contract after 1 February if the information in question has not been provided. These are two distinct terms of the contract. The existence of the right in either party to terminate the contract if a particular condition is not performed by the due date is not inconsistent with the condition in question being for the exclusive benefit of the other party to the contract and with that other party having the right, if necessary by implication of law, to waive the condition.

[37] *Heron Garage* is not authority for a contrary view. The condition in that case was that the purchaser would obtain a particular planning consent. Obtaining that consent was a condition precedent to the contract. Brightman J put the matter thus, at p426d-e:

B

> The town planning consent is expressed in cl 7 of the sale agreement as a condition fundamental to the enforceability of the sale agreement as a whole. It is not expressed as a condition which is precedent only to the liability of Heron as purchaser. Clause 7 is not a clause which is expressed only to confer rights on Heron. It is expressed to confer a right also on the vendors.

It is perhaps the presence of the last sentence in that passage that needs some elaboration. The right there referred to was a right in either party, if the stipulated planning consent should not have been obtained within six months or within such extended period as the parties might agree, to terminate the agreement by notice in writing to the other. It was part
C and parcel of the very clause stating that the contract was conditional on the particular planning consent being obtained. The purchaser, Heron, had shortly before the expiry of the six-month period given notice in writing to the vendor's solicitor purporting to waive what it described as the benefit of clause 7 of the contract. It is not surprising, therefore, that Brightman J concluded that the condition was not capable of waiver by the purchaser: it was a condition precedent to the very existence of the contract and, what is more, it contained a provision expressed to be for the benefit of both parties.

[38] The only other authority I need to refer to, and can do so very briefly, is *Yewbelle Ltd* v *London Green Developments Ltd*
D [2007] EWCA Civ 475; [2007] 2 EGLR 152, in which there was an agreement for the sale by the vendor to the purchaser of a development in south London with a view to its redevelopment in accordance with a particular planning permission. The planning permission was subject to a suitable agreement being entered into under section 106 of the Town and Country Planning Act 1990. The vendor was obliged to use all reasonable endeavours to secure the section 106 agreement and the purchaser was under no obligation to complete in the absence of such an agreement. One of the questions was whether a term should be implied into the contract under which the vendor could rescind the contract if, despite the exercise of all reasonable endeavours,
E the necessary section 106 agreement could not be achieved and the buyer did not waive that condition. Implicit in that question was that it was open to the buyer to waive the condition. The Court of Appeal, upholding Lewison J on this point, saw no inconsistency between the presence in the contract of a condition the obtaining of the necessary section 106 agreement that was for the sole benefit of the buyer and the implication into the contract of a term to the effect that if the seller complied with his obligation to use all reasonable endeavours but was unable to complete the section 106 agreement, the seller should give to the buyer an opportunity to complete the sale without that agreement (in other words, to waive the condition requiring that agreement), but if the
F buyer should choose not to complete the sale on that basis the vendor would have the right to terminate the contract.

(d) Notice to terminate

[39] The final question is whether the claimant's notice, set out in his solicitor's letter of 1 February 2010, was an effective exercise of the power to terminate the contract given by clause 25.2.

[40] It is not in dispute that, subject only to the effect of the notice contained in the letter of 8 February sent by the defendants' solicitor, the claimant's notice was effective and the contract was terminated. The defendants accept that, by the time the claimant's notice was given, the claimant had used all reasonable endeavours to provide all the information referred to in clause 25.1 but had not been able to do so.

[41] The outcome of this question turns on whether the claimant's notice was effective to terminate the contract when it was given that is, when it was received by the defendants' solicitor or whether it took effect only on the expiry of the five working days to which (as clause 25.2 required) the notice referred. This question was not a matter dealt with by Judge Madge. Given his conclusion that it was not open to the defendants to waive clause 25.1, it necessarily followed that the letter of 8 February, which alone is relied on as negating the effect of the claimant's notice, was of no effect. I should add that Mr Carlisle did not suggest that once 1 February had passed it was, irrespective of the giving of notice under clause 25.2, too late for the defendants to seek to waive the benefit of clause 25.1.

[42] Is a notice given under clause 25.2 effective to bring about a termination of the contract only on the expiry of the five working days, which is what Mr Hill submitted? Or does such a notice, as Mr Carlisle submitted, take immediate effect as a termination of the contract when given, but give five working days to the buyer to vacate the property (assuming that the buyer is in occupation) and a like period to the seller (and his solicitor) to return the deposit?

[43] On first considering the point, I am inclined to the view that it was the former. As Mr Hill pointed out, clause 25.2 refers to "notice of 5 working days to terminate this Agreement, whereupon…". This construction treats the notice, if given by the buyer, as effectively a notice to perform: its service gives to the seller a final five working days within which to provide the stipulated information, failing which the contract comes to an end without further ado. If given by the seller, it is effectively a notice putting the buyer to his election whether he wishes to waive the benefit of the documents service term in clause 25.1 and complete his purchase notwithstanding the incorrect plans or whether he still expects performance of the term but is leaving it to the seller to terminate the contract by allowing the notice to expire.

[44] Mr Carlisle was able, however, to draw my attention to a very recent decision of the Court of Appeal in *Akzo Nobel UK Ltd* v *Arista Tubes Ltd* [2010] EWCA Civ 28*, in which the same point had arisen for decision. In that case, the claimant, Akzo, had sought specific performance of an agreement made on 31 December 1998 under which the defendant, Arista, had agreed to take underleases of five units (referred to in the agreement as "Business Property") in certain factory premises. The landlord's consent was needed, both to the grant of the underleases to Arista and also to an assignment to Arista of the headlease. Akzo agreed to use all reasonable endeavours to procure those consents (referred to in the agreement as "Property Consents") and Arista was allowed in the meantime to take up occupation of the property in question under a licence arrangement. As matters turned out, Arista remained in occupation under the licence arrangement for eight years. For its part, although it made efforts to do so, Akzo was unable to obtain the property consents or an assignment to itself of the headlease until 13 December 2007. However, by then, Arista had served notice on Akzo to terminate the agreement. It had done so by a notice on Akzo dated 30 October 2008, pursuant to the following term (referred to as para 11) of the agreement:

> If by [31 December 1999] all Property Consents shall not have been obtained in respect of any Business Property then either the Seller or the Purchaser may, by three months notice in writing to the other, terminate on the date of expiry of that notice, the obligations of the parties hereto in respect of that Business Property… (but without prejudice to antecedent breach) in which event the Purchaser shall vacate the Business Property in question by the end of such notice period.

[45] After observing, in [17], that there was no dispute that Arista was entitled to serve the notice on Akzo on 30 October 2008, Mummery LJ (with whom the two other members of the court agreed) said of the notice:

> It was not a notice to complete the parties' transaction by giving Akzo a last chance and by setting a deadline for it to take a transfer of the lease and to obtain

---

* Editor's note: Reported at [2010] 2 All ER (Comm) 1198

the Property Consents. It was a notice to terminate the parties' contractual obligations, a provision which enabled either party to escape from obligations in relation to the Premises. Such a notice could be served by either party if Akzo had not obtained the Property Consents by the stipulated date. The obligations which could be terminated included Arista's obligation to pay a fee for its licence to occupy the Premises. One would not normally or reasonably expect that, in the absence of express provisions in the [agreement] or fresh agreement between the parties, a notice terminating the parties' obligations, once given, could be unilaterally revoked or reversed or that, as [counsel for Akzo] strongly contends, it has no legal effect on the parties' obligations before the notice has expired. In my view, the immediate effect of the notice was that there was no longer any obligation on Akzo to use reasonable endeavours to obtain Property Consents or an assignment of the lease, or on Arista to take underleases of the Premises.

The reference in that passage to a notice once given being "unilaterally revoked or reversed" could equally extend to an attempt, as here, to waive the benefit of the condition. Mummery LJ continued, in [19]:

It is common ground that the notice provision must be construed as a whole, in its context and in its ordinary and natural meaning. As I read paragraph 11 the parties clearly agreed that, if the Property Consents had not been obtained by 12 months after the date of the [agreement], they were entitled to serve notice in writing terminating the parties' obligations in respect of the Premises. Arista's obligations included the obligation to take the underleases, as well as its obligation to pay the fee under the Licence Agreement. On service of the notice the position of the parties was that Arista would have to vacate the Premises by the end of the notice period and it would cease to be under an obligation to pay the licence fee, or to be under an obligation to take the underleases of the Premises. The language of the paragraph does not allow Akzo to reverse or vary that position on Arista's obligations, such as by obtaining the Property Consents at any point down to the expiration of the notice. The obligation to take the underleases did not arise before the notice was given and the purpose of giving the notice was to prevent it from ever arising subsequently, given that Arista would be bound to vacate the premises by the expiry of the notice period.

[46] Although the decision in *Akzo* cannot of itself determine the construction of clause 25.2 in the instant case, as each is concerned with a differently worded clause, it is difficult to see why, given the similarity of language, the result in the instant case should be different from that in *Akzo*. The reasoning that led to the decision in *Akzo* (upholding the decision of Floyd J in the court below) is no less applicable, in my view, to the instant case.

[47] Although he sought to do so, Mr Hill was unable to provide any convincing grounds why the reasoning in *Akzo* should not apply to the notice in this case. He submitted that the factual matrix in *Akzo* was different, in that: (i) the premises there were commercial premises from which Akzo was conducting a business; (ii) the period of notice was a substantial three months to enable Akzo to wind down; and (iii) there was a need for the parties to have certainty once the notice was given. In the instant case, the premises are residential. He submitted that, by the terms of the notice provision, Arista was to vacate "*by*" the end of the notice period, so that Arista was able to go out of occupation at any time after the notice had been given, thus supporting the conclusion that once the notice was given nothing that subsequently happened could lead to the contract being completed. In the instant case, by contrast, the language used is different: the reference in clause 25.2 is to "*whereupon*", implying that there is no expectation that the buyer will vacate before the end of the notice period. Finally, he submitted, the effect of the notice in *Akzo* was to fix a final date by reference to which occupation, and payment for occupation, were to be determined. In the instant case, the position is different: clause 25.3 does not require the buyer to vacate until the seller's solicitor confirms that it is in funds to reimburse the whole of the deposit. Mr Hill told me that no such confirmation was given by the claimant's solicitor.

[48] I am not persuaded that these differences are material. The fact that the premises in the instant case are residential does not suggest that there is any less need for certainty than if there were commercial, as in *Akzo*. The use of "*whereupon*" does not mean that there is no expectation that the buyer will vacate before the notice has expired. Clause 25.3 is a protection for the buyer. It throws no light on the nature of the notice served under clause 25.2.

[49] In my judgment, the notice given under clause 25.2 enables the parties to bring an end to their relationship if one of them chooses to do so and the relevant information has not been provided by 1 February. The more natural contruction of the clause is to read it as having that effect when it is given. It is inconsistent with that purpose to allow an obligation to complete to arise (either because the documents service term is performed or the term is waived) after the notice has been given. The whole point of the notice is that the time for completion has passed.

[50] In my judgment, therefore, the claimant having, through his solicitor, served a valid notice under clause 25.2, it was too late for the defendants to seek thereafter to waive the claimant's obligation as seller, to provide all the information referred to in clause 25.1.

Result

[51] It follows that the contract was validly terminated and although for reasons different from those on which he based his decision, Judge Madge was correct to dismiss the defendants' application for specific performance.

Appeal dismissed.

Redcard Ltd and others *v* Williams and others

COURT OF APPEAL

[2011] EWCA Civ 466

20 April 2011

MUMMERY, HUGHES AND BLACK LJJ

Estates Gazette 25 June 2011

[2011] 25 EG 106

Sale of land — Contract — Companies — Execution of documents — Companies Act 2006 — Contract for sale of property — Company and its authorised signatories separately named as sellers — Individual signatories executing document without affixing common seal of company — Section 44 of 2006 Act — Whether document expressed to be executed by the company for purposes of section 44 — Whether validly executed contract binding company

The appellants were the purchasers, under a contract and supplementary agreement, for the sale of a residential building containing five flats. The term "Seller" was defined in those documents to include both the first respondent company, which was selling the freehold of the building, and several of its directors and shareholders, who were selling their long leasehold interests in individual flats. The various sales were comprised in a single composite contract for simultaneous sales at an unapportioned total price of £3.35m. Of the various signatures attached to the supplementary agreement, under the heading "SIGNED... SELLER", two belonged to parties who were not only individual sellers of their leasehold interests but were also authorised signatories of the first respondent.

The appellants subsequently refused to complete their purchase. They asserted that they were not obliged to complete because the contractual documents had not been validly executed "by or on behalf of" each party as required by section 2 of the Law of Property (Miscellaneous Provisions) Act 1989; they claimed that there had been no valid execution by the first respondent. That argument turned on section 44 of the Companies Act 2006 and its provision for companies to execute documents, with the same effect as if under common seal, by affixing the signatures of two authorised signatories on the company's behalf in a document "expressed, in whatever words, to be executed by the company". The appellants contended that a valid execution by the company required not only the signatures of the authorised signatories but also express words stating that those signatures were "by and on

A behalf of" the company; they submitted that the words used in the sale contract did not indicate that the first respondent was executing it.

In proceedings between the parties, summary judgment was given for the appellants, but that decision was reversed on appeal. Permission was granted for the appellants to bring a second appeal.

Held: The appeal was dismissed. A document signed by two authorised signatories of the company in accordance with section 44(2) will not, without more, have the same effect as if executed under the common seal of the company. The requirement in section 44(4) for a document "expressed in whatever words to be executed by the company" adds a further requirement. However,
B it does not require express words spelling out that the signatures are "by or on behalf of" the company. Where the "Seller" was defined to include both the first respondent selling its freehold and the individuals selling their leaseholds, and where the signatures to the document included those of two authorised signatories under the heading "SIGNED… SELLER", that meant that the document was expressed to be executed simultaneously by both the first respondent and by the individuals, all being included in the term "Seller". No problem would have arisen had the sales of the freehold and leaseholds been effected by two separate documents, with the freehold document signed in a similar way by the authorised
C signatories. There was no reason why the position should be different where the transactions were combined in the same document, especially since the statutory provisions are intended to expand the range of formalities that will count as execution by a company. Accordingly, the supplementary agreement had been validly executed by the first respondent.

The following cases are referred to in this report.

Hilmi & Associates Ltd v *20 Pembridge Villas Freehold Ltd* [2010] EWCA Civ 314; [2010] 1 WLR 2750; [2010] 3 All ER 391; [2010] 2 EGLR 41; [2010] 25 EG 104

Lovett v *Carson Country Homes Ltd*; *sub nom Re Carson Country Homes Ltd*
D [2009] EWHC 1143 (Ch); [2009] 2 BCLC 196

This was an appeal by the appellants, Roger Williams and Mrs Williams, from a decision of Lewison J, sitting in the Chancery Division of the High Court, reversing a summary judgment given in their favour by Master Price in proceedings against the respondents, Redcard Ltd, Ms Helen Christie, Mrs Anne Auber, Mrs Serene Ton Soo Ling, Mrs Rachel Singleton and Mrs Sarah Ferris, regarding the execution of a contract for the sale of land under the Companies Act 2006.

Timothy Dutton (instructed by Bircham Dyson Bell LLP) appeared for the appellants; Evan Ashfield (instructed by Davies Battersby)
E represented the respondents.

Giving judgment, **Mummery LJ** said:

Introduction

[1] This appeal turns on the construction of section 44(4) of the Companies Act 2006 (the 2006 Act), a statute that enjoys the dubious distinction of being the longest Act of parliament ever passed. In this judgment, reference is made to only three of the 1,300 sections. No reference is made to any of the 16 Schedules.

[2] Section 44, which came into force on 6 April 2008, relates to the execution of documents by a company. It allows the use of more
F informal methods than the affixing of its common seal. The section is in a group of sections in Part 4 of the Act (section 43 (Company contracts), section 44 (Execution of documents) and section 45 (Common seal) under the general heading "Formalities of doing business…".

[3] A company neither needs to have a common seal nor does it have to execute a document under its common seal: see section 45. Subject to detailed provisions, section 44 extended the changes made to the Companies Act 1985 by the insertion of section 36A by the Companies Act 1989 (the 1989 Act). It is enacted that a document signed on behalf of the company by two authorised signatories or by the attested signature of a director and expressed to be executed by the company has the same effect as if executed under the common seal of the company. (Since there was no attestation of signatures here, this judgment deals only with the case of signatures on behalf of a company by authorised signatories.)

Background

[4] The salient facts are that a company called Redcard Ltd was named as a party to a contract dated 28 July 2008 and a supplementary agreement dated 27 January 2009. The contracts included an agreed sale of Redcard's freehold interest in a substantial residential building at 24 Castlenau, Barnes, London SW3. The building was converted into five self-contained residential flats let on long leases to individuals who were directors and shareholders of Redcard. The contracts also included an agreed sale of the leasehold flats by the individuals.

[5] Redcard fell within the definition of "Seller" contained in those documents. The question is whether the documents, which did not bear a common seal of the company or separate signatures stated to be "for or on behalf of" Redcard, were validly executed by the company. The supplementary agreement, which varied the completion date and confirmed the earlier contract in all other respects, has been treated by the parties as the relevant document for the purposes of the submissions on the application of section 44. If it is validly executed by Redcard, that cures any defect in the earlier contract. It is agreed that the issue on execution stands or falls by whether the supplementary agreement was executed by Redcard.

[6] The supplementary agreement bore various signatures under the words "SIGNED… SELLER" including those of two individuals, who were authorised signatories of Redcard. The two individuals were named among the parties to the same document. They also fell within the defined term of "Seller", since they agreed to sell their leasehold interests in their respective flats. The contracts for the sale of the freehold and of the leasehold interests were all included in the one composite document for simultaneous sales at an unapportioned aggregate purchase price of £3.35m.

[7] The purchasers named in the supplementary agreement were the appellants, Mr and Mrs Roger Williams. Their case is that there is no contract that they can be compelled to complete because the sellers' part of the contract was not properly executed by Redcard. They refuse to complete the purchase, despite a notice to complete served on 11 February 2009. They submit that the valid execution of the documents by Redcard requires that, in the absence of a common seal, the documents should contain words expressly stating that the signatures of the authorised signatories are "by or on behalf of" Redcard. In this case, they say that no words appeared in the supplementary agreement expressly stating that the signatures of the individuals at the foot of it were "by or on behalf of" Redcard, as distinct from the signatures of individuals acting in a personal capacity on the sale of their leasehold interests in their respective flats. Accordingly, the contract of sale was not executed by Redcard and they cannot be compelled to complete it. (The appellants have taken other points as to why the sellers' claims against them should fail, but they do not arise on this appeal. It is common ground that if the appellants' point on the execution issue succeeds, that would be fatal to the sale contract on which the sellers are relying.)

[8] It is common ground that, being for the sale of land, the contract had to be signed "by or on behalf of" each party to the contract: see section 2 of the Law of Property (Miscellaneous Provisions) Act 1989. Those formalities apply to a contract made by or on behalf of a company, as well as a contract made by an individual: see section 43(2) of the 2006 Act. The main thrust of the appellants' case is that a distinction is to be drawn between the signature of an authorised signatory pursuant to the section 44 procedure for execution of a document and the signature of an agent on a company contract, as permitted by section 43(1)(b); and that the words required by section 44 are any that make it clear that the person said to have signed on behalf of the company did so under section 44 rather than as agent for the company at common law.

[9] On 18 November 2009, Master Price, on hearing cross-applications for summary judgment, held that the supplementary agreement was not executed by Redcard and that it was not valid and binding. On 23 April 2010, Lewison J held, on appeal on the execution

issue, that the supplementary agreement was executed by Redcard in accordance with the requirements of section 44 of the 2006 Act and complied with section 2 of the 1989 Act.

[10] On 29 June 2010, I refused permission for a second appeal. On 26 October 2010, the chancellor granted permission on the ground that the appeal raises a question of general public importance going to the root of all manner of transactions involving limited companies.

Judgment of Lewison J

[11] In allowing the appeal from the decision of Master Price, Lewison J concluded that the point taken by the appellants on execution was not a good one. He said:

> 21… If a contract is to be signed by a person, both on his own behalf and also on behalf of a company, I cannot see why he should not be able to sign with a single signature provided that the signature is expressed to be a signature on his own account and also on behalf of the company. After all, the general principle of English Law is that no formalities are required to make a contract and it seems to me that if statute is to impose limitations on that general principle it must do so clearly. Accordingly, I consider that provided that, on the fair interpretation of the words in a contract, the reasonable reader would understand the signatures of the natural persons are signatures both on their own account and on behalf of the company, that is sufficient to amount to proper execution for the purposes of section 44. In this respect I respectfully differ from Master Price who held that the contract must state expressly in a testimonium clause that the natural person is signing for and on behalf of the company or some words to that effect.
>
> 22. Accordingly, I apply the test which I favour to the supplemental agreement. Would a reasonable reader with the background knowledge reasonably available to the parties have understood that the contract was signed on behalf of Redcard? The signatures appear in a box headed "Seller." In order to understand what was meant by the word "seller", the reasonable reader would have looked at the front page of the supplemental agreement. On seeing that he would have seen that the seller included Redcard. He would next have looked at the signatures themselves.

[12] Lewison J then explained that, with regard to the relevant signatures, the reader would have been faced with two possibilities: either the signatures were both on their own account and also on behalf of Redcard or Redcard had failed to execute the agreement at all. In the latter case, the contract could not be enforced by or against the buyers (the appellants). The description "seller" would have been untrue since, in their personal capacities, they were part only of the seller as defined. Lewison J concluded that the supplementary agreement was validly executed in accordance with section 44 of the 2006 Act and was thus signed on behalf of Redcard for the purposes of section 2(3) of the 1989 Act. He said:

> 25. The reasonable reader would also have known from his background knowledge that all the directors of the company had committed themselves personally to performance of a contract which could only be carried out with the co-operation of the company of which they were directors and majority, if not sole, shareholders. He would also have understood that the contract was intended to be effective rather than ineffective. Faced with that choice, the reasonable reader would have concluded that the signatures of at least Ms Silberstein and Mrs Auber were appended to the contract both on their own account and also on account of Redcard.

[13] In his able argument in this court on behalf of the appellants, Mr Timothy Dutton contended that Lewison J's construction of section 44(4) was wrong. I turn to the terms of that subsection and to Mr Dutton's submissions on its construction.

Companies Act 2006

[14] Section 44 ("Execution of documents") applies to any documents, not just contracts to which section 43 applies. It provides that:

> (1) Under the law of England and Wales… a document is executed by a company
> (a) by the affixing of its common seal, or
> (b) by signature in accordance with the following provisions.
> (2) A document is validly executed by a company if it is signed on behalf of the company
> (a) by two authorised signatories, or G
> (b) by a director of the company in the presence of a witness who attests the signature.
> (3) [definition of authorised signatories]
> (4) A document signed in accordance with subsection (2) and expressed, in whatever words, to be executed by the company has the same effect as if executed under the common seal of the company.
> (5) In favour of a purchaser a document is deemed to have been duly executed by a company if it purports to be signed in accordance with subsection (2).
> A "purchaser" means a purchaser in good faith for valuable consideration and includes a lessee, mortgagee or other person who for valuable consideration acquires an interest in property. H
> (6) Where a document is to be signed by a person on behalf of more than one company, it is not duly signed by that person for the purposes of this section unless he signs it separately in each capacity.

[15] I note a number of points on the purpose, context and language of the section.

(1) Section 44, like its neighbouring sections 43 and 45, relates to the formalities of doing business, in particular the formalities for the making of contracts and the execution of documents by companies.

(2) The purpose of section 44 was to facilitate the formalities of the execution of documents by companies by allowing signatures either of two authorised signatories or of one director, if attested, to count as J execution by a company under its common seal.

(3) In order to achieve that result, section 44 uses the familiar legislative technique of "deeming." Thus, in the case of a document signed in accordance with section 44, the document has the same effect "as if" executed under the common seal.

(4) The legal effect of words in subsection (4), "and expressed in whatever words to be executed by the company", are central to Mr Dutton's contention that, in order to give rise to a deemed execution of a document by a company, the signatures relied on must be expressed in words conveying that they are by or on behalf of the company.

(5) As for subsequent purchasers for value in good faith, K subsection (5) deems, in their favour, that a document was duly executed by a company if it purports to be signed in accordance with subsection (2). We were referred to *Lovett* v *Carson Country Homes Ltd* [2009] EWHC 1143(Ch); [2009] 2 BCLC 196, in [79], for the proposition that "purports", in subsection (5), "operates to refer to the impression a document conveys", focusing on what appears to be the case rather than what actually is the case. That particular point does not arise here because the appellants are immediate not "subsequent" purchasers, but the provision is relied on by Mr Dutton as supporting his submissions on the construction of the key words in subsection (4).

(6) Similarly, subsection (6) does not apply here because only one L company is involved, but it is relied on by both sides as an aid to the construction of subsection (4). It provides that where a document is to be signed by a person on behalf of more than one company, it must, for the purposes of due execution, be signed separately in each capacity. On this subsection, we were referred to the recommendations of the Law Commission No 253, 1998, *The Execution of Deeds and Documents by or on behalf of Bodies Corporate*, in paras 3.30 to 3.35, that preceded the enactment of section 44(6).

Appellants' submissions

[16] It is accepted that the supplementary agreement was signed by M or on behalf of the individual sellers of the leasehold interests in the five flats. The question is whether it was executed "by" Redcard as the seller of the freehold interest. On this point, Mr Dutton's submissions may be summarised as follows.

[17] He said that the key to the execution requirements is in the answer to the question: was the supplementary agreement executed "by" Redcard? There is a distinction, which is recognised in section 43(1) of the 2006 Act, between the execution of a document "by" a company, which must comply with the requirements of section 44(4) and be apparent from the language of the document itself, and the execution of a document "on behalf of" a company. He cited *Hilmi & Associates Ltd* v *20 Pembridge Villas Freehold Ltd* [2010] EWCA

Civ 314; [2010] 1 WLR 2750*, in [28] to [31], as recognising, in the context of execution of documents, the distinction between a document signed by, as distinct from on behalf of, a company.

[18] For a document to be executed by a company, it must either bear the company's seal, or it must comply with section 44(4) in order to take effect as though it had been executed under seal. Subsection (4) requires that the document must not only be made on behalf of the company by complying with one of the two alternative requirements for signature in section 44(2): it must also be "expressed, in whatever words, to be executed by the company". That means that the document must purport to have been signed by persons held out as authorised signatories and held out to be signing on the company's behalf. It must be apparent from the face of the document that the people signing it are doing something more than signing it on the company's behalf. It must be apparent that they are signing it on the company's behalf in such a way that the document is to be treated as having been executed "by" the company for the purposes of subsection (4), and not merely by an agent "for" the company. The supplementary agreement does not state on what behalf each signatory was signing it, or identify the relationship between the signatories and Redcard. Mr Dutton submitted that it cannot be said that the supplementary agreement was executed "by" Redcard within section 44 because there was no expression in words in the supplementary agreement that Redcard itself was executing it, rather than by an agent. There is nothing in it to indicate who was executing it on behalf of Redcard or whether Redcard was executing it by any one of the signatures.

[19] Mr Dutton said that Lewison J's fundamental error was to equate "by" the company in subsection (4) with "on behalf of" the company in subsection (2). It was not made clear by express words in the supplementary agreement that Redcard itself was invoking the procedure supplied by section 44. It was not sufficient to identify Redcard as a party to the supplementary agreement, for that says nothing about whether Redcard was executing it itself or by an agent. It was not sufficient for the signatures to be those of authorised signatories because that was not an expression of what Redcard was doing.

[20] According to Mr Dutton, the judge was also in error in thinking that the factual context of the supplementary agreement and its availability to the reasonable reader could bring it within section 44(4). He said that Lewison J applied the wrong test. Asking whether the reasonable reader with the background knowledge reasonably available to the parties would have understood that the supplementary agreement was signed on behalf of Redcard was a test that was not warranted by the terms or statutory purpose of section 44.

[21] It is also submitted that the test applied by the judge is inconsistent with subsection (6), which applies in the case of the same signatory wanting to sign on behalf of two companies. He should have applied that provision by analogy, pointing as it does to the undesirability of situations where one person signs a document in more than one capacity. Clear words are needed before one signature is to be taken as execution by two parties. Against Mr Dutton, it was argued on behalf of the sellers that subsection (6) is otiose if separate signatures in each capacity, one personal and the other for the company, are required under subsection (4).

[22] As an aid to construction of subsection (4) Mr Dutton relied on subsection (5). He said that it indicates that a document should be regarded only as "purporting" to be signed for the purposes of that subsection and subsection (2) or as "expressed" to be signed under subsection (4), if that appears clearly from the face of the document itself.

Discussion and conclusion

[23] The appeal turns on what the words "expressed in whatever words to be executed by the company" in subsection (4) add to the presence of signatures by two authorised signatories in accordance with subsection (2) and whether that added requirement was satisfied in this case.

[24] I agree with Mr Dutton that those words must add something to the provision in subsection (2) that a document is validly executed by a company if it is signed on behalf of the company by two authorised signatures. Subsection (4) does not simply provide that a document signed in accordance with subsection (2) has the same effect as though executed under the common seal of the company.

[25] I am unable to agree with Mr Dutton that the critical words require that, in addition to the signatures of the individuals who are the authorised signatories, there must be words spelling out that those signatures are "by or on behalf of" the company. Let us suppose that, in this case, the contract for the sale of the freehold and the leasehold interests was the same and that instead of being put into one composite document there were two separate documents. Also suppose that, in the separate contract for the sale of the freehold interest, Redcard was, as here, defined as "Seller" and that the signatures of two authorised signatories appeared, as here, under the words "SIGNED… SELLER", but without stating that the signatures were "by or on behalf of" Redcard. In my judgment, it would be absurd in such circumstances to say that the contract for the sale of the freehold by Redcard was not expressed to be executed by Redcard. If Redcard is defined as "Seller", the signatures at the end of the agreement under the words "SIGNED… SELLER" could mean only that the document was expressed to be executed by Redcard.

[26] Why should the legal position regarding execution by Redcard be any different when, as here, the freehold and the leasehold transactions are combined in the same document, the authorised signatories being parties to the contract and the defined term "the seller" including both the individual leaseholders and Redcard? In my judgment, there is no conceivable sensible reason why the legal position should be any different in the case of one document rather than two, especially when the statutory provisions were intended to expand the range of formalities that would count as execution by a company.

[27] That this is a simple case within subsection (4) can be clearly demonstrated: the signatures to the supplementary agreement are under the words "SIGNED… SELLER"; "SELLER" is defined in the supplementary agreement as including both Redcard selling its freehold and the individuals selling their leaseholds; the signatures include the signatures of two authorised signatories; and the use of the defined term "SELLER" above those signatures means that the document is expressed to be simultaneously executed both by Redcard and by the individuals, all being included in the term "SELLER".

[28] As for the construction and operation of subsections (5) and (6), which do not apply to the facts of this case, I express no view. They cater for different situations and are not especially relevant to the construction of subsection (4).

Result

[29] I would dismiss the appeal. By a slightly different route, I have arrived at the same conclusion as Lewison J on the construction and application of section 44(4). The supplementary agreement was executed by Redcard.

[30] From a practical point of view, it may just be worth stating the obvious: expensive and long-drawn-out litigation concerning the execution of a document by a company can be avoided by taking greater care over compliance with the formalities at the time of execution, by, for example, adding words that expressly state the capacity in which an individual is signing a document to which a company is a party.

Black LJ said:

[31] I agree.

Hughes LJ said:

[32] I also agree.

Appeal dismissed.

* Editor's note: Also reported at [2010] 2 EGLR 41

TOWN AND COUNTRY PLANNING

Avon Estates Ltd *v* Welsh Ministers and another

COURT OF APPEAL

[2011] EWCA Civ 553

16 May 2011

RIX AND SMITH LJJ AND SIR DAVID KEENE

Estates Gazette 6 August 2011

[2011] 31 EG 54

Town and country planning — Lawful use certificate — Planning permission for limited period — Section 72 of Town and Country Planning Act 1990 — Holiday bungalows — Seasonal occupancy condition attached to permission — Development remaining in use after expiry of limited period — Bungalows acquiring immunity from enforcement action — Landowner applying for lawful use certificate in respect of use as dwellinghouses — Whether seasonal occupancy condition continuing to apply

Between 1965 and 1973, four planning permissions were granted for the erection of holiday bungalows on the appellants' land near the Welsh coast. Each permission was for a limited period and stated that it would expire and that the site was to be restored to its former use on or before a given date. Two of the permissions specified July 1985 while the others specified July 1995, but all contained a seasonal occupancy condition limiting the use to a specified part of the year only.

The bungalows remained in place and continued to be used, within the seasonal occupancy periods after the specified expiry dates. No enforcement action was taken to secure their removal and the restoration of the site, and the bungalows became immune from enforcement action by virtue of section 171B of the Town and Country Planning Act 1990.

In 2008, the appellant made two applications for certificates of lawfulness of the existing or proposed use of the bungalows as dwellinghouses, pursuant, respectively, to section 191(1) and section 192(1) of the 1990 Act. The second respondent local planning authority failed to determine those applications and the appellants appealed to the first respondents. Allowing the appeal, the first respondents found that the condition as to the temporary nature of the bungalows was spent and accordingly granted the certificates, subject, however, to the seasonal occupancy condition, which was held to remain extant.

Proceedings brought by the appellant to quash the first respondent's decision, under section 288 of the 1990 Act, were dismissed in the High Court: see [2010] EWHC 1759 (Admin). The appellant appealed. It contended that the seasonal occupancy conditions had ceased to be enforceable when the original permissions to which they were attached had expired on the specified dates.

***Held*: The appeal was allowed. (1) For a condition lawfully to endure once the development to which it relates has ceased to be authorised by the relevant permission, and has become immune from enforcement action, that condition would have to continue to relate fairly and reasonably to the permitted development. That is unlikely to be the case for a permanent condition in a permission for a temporary development. It is difficult to conceive of a condition in a temporary permission under section 72 that could sensibly relate to the development once it has ceased to be authorised by the** F **permission. The condition imposed under section 72(1)(b), requiring the discontinuance of the use and the reinstatement of the land at the end of the specified period, is an exception that does not go beyond its own scope, since the condition is expressly and precisely provided for by the statute. It does not support a general approval for conditions that bind the land once the development has ceased to be authorised and has become immune from enforcement action. (2) Although the seasonal use condition in each permission had no express time limit, it was, as a matter of objective construction, intended to be coterminous with the authorised development, with the result that the seasonal use restriction applied only during the** G **period for which the holiday bungalows were permitted. When construing a permission and its conditions, the document should be read as a whole and in such a way as to avoid internal inconsistency, if possible. It would be illogical and inconsistent to give the seasonal use condition in each permission a life after the specified date. To do so would contradict the first condition in each permission, which made it clear that the permission was being granted only for a limited and specified period.**

The following cases are referred to in this report.

Adur District Council v *Secretary of State for the Environment, Transport and the Regions* [2000] 1 PLR 1 H
Avon Estates Ltd v *Welsh Ministers* [2010] EWHC 1759 (Admin)
Essex Construction Co v *East Ham Borough Council* (1965) 16 P&CR 220
Fawcett Properties Ltd v *Buckinghamshire County Council* [1961] AC 636; [1960] 3 WLR 831; [1960] 3 All ER 503; (1960) 59 LGR 69; 12 P&CR 1; 176 EG 1115, HL
Gravesham Borough Council v *Secretary of State for the Environment* (1982) 47 P&CR 142; [1983] JPL 307, QB
Handoll v *Warner Goodman & Streat* (1994) 93 LGR 293; (1995) 70 P&CR 627; [1995] 1 EGLR 173; [1995] 25 EG 157; [1995] 1 PLR 40, CA
Mounsdon v *Weymouth and Melcombe Regis Borough Counci* [1960] 1 QB 645; [1960] 2 WLR 484; [1960] 1 All ER 538; (1960) 11 P&CR 103
Pyx Granite Co Ltd v *Ministry of Housing and Local Government* [1958] J 1 QB 554; [1958] 2 WLR 371; [1958] 1 All ER 625; (1958) 56 LGR 171; 9 P&CR 204, CA

This was an appeal by the appellant, Avon Estates Ltd, from a decision of Beatson J, sitting in the Administrative Court, dismissing a claim, under section 288 of the Town and Country Planning Act 1990, to quash a decision by the first respondents, the Welsh Ministers, on appeal from the second respondents, Ceredigion County Council, to grant lawful use certificates for the retention of holiday bungalows subject to a seasonal occupancy condition.

Rupert Warren (instructed by Clarke Willmott LLP, of Birmingham) K appeared for the appellant; Jonathan Moffett (instructed by the Treasury Solicitor) appeared for the first respondents; Paul Stinchcombe QC (instructed by the legal department of Ceredigion County Council) represented the second respondents.

Giving judgment, **Sir David Keene** said:

Introduction

[1] This appeal raises a novel point of law about the status and effect of conditions attached to a planning permission granted for a limited period, once that limited period has expired. Of course, a failure to

A remove the permitted building or to cease the permitted use after that specified period will normally be vulnerable to enforcement action under Part VII of the Town and Country Planning Act 1990 (the 1990 Act), but what is the position if the planning authority do not take such action during the years in which they could, so that the permitted development becomes immune from enforcement action by virtue of section 171B of that Act? Can other restrictive conditions on the planning permission, such as an occupancy condition, still be enforced?

[2] It must be borne in mind that planning permissions granted for a limited period, although recognised in section 72(1)(b) and section 72(2) of the 1990 Act, are unusual, if not rare. The Departmental B Circular 11/95 on the *Use of Conditions in Planning Permissions* warned (as did its predecessor, Circular 1/85) against the granting of what it called "Temporary Permissions", except in very limited circumstances. It is perhaps because such permissions are unusual creatures that the precise point of law now in issue does not seem to have been determined in previous authority.

Facts

[3] The land in question lies close to the Welsh coast and is within an area of great landscape value. Between the years 1964 and 1973, a total of four planning permissions were granted for "the erection of holiday C bungalows" or in one case the erection of "semi-detached chalets". In each case, there was a condition stating that:

> The permission hereby authorised shall expire and the site be restored to its former use on or before…

a stated date. That date varied, being 31 July 1985 in two of the permissions and 31 July 1995 in the other two. Nothing turns on the difference in the specified dates.

[4] Each permission then contained a condition requiring the proposed bungalows to be maintained to the satisfaction of the planning authority "throughout this period". Then, in each case, there D was a condition limiting occupancy to only part of the year. Thus, the 1964 permission stated as its third condition:

> The holiday bungalows shall be occupied only during the period 1st March to 31st October.

The same wording appeared in the second permission, dated 17 May 1965, while the other two permissions contained similar wording but referring to a somewhat longer part of the year. Nothing turns on those differences.

[5] In fact, the bungalows remained in existence long after the date specified for the expiry of the permissions. No enforcement action was E taken to secure their removal and the restoration of the site to its former use and, at a public inquiry in September 2009, it was agreed between the appellant and the local planning authority that the bungalows, of which there were by now 42 in number, were immune from enforcement action. It was also agreed that they had been occupied seasonally for more than four years but also that they had not been occupied outside the periods in the year referred to in the permissions.

[6] In 2008, the appellant sought a certificate of lawfulness of existing use under section 191(1) of the 1990 Act. The use described in the application was "use of buildings as 42 dwelling houses". Before any decision had been made on that application, a further application F was submitted, this time under section 192(1), for a certificate of lawfulness of the proposed use of one of the bungalows "as a dwelling house", and reference was made to Class C3 of the Use Classes Order 1987 (use as a single dwellinghouse (whether or not as a sole or main residence)). That building was, in effect, chosen as a test case for a section 192 certificate.

[7] When no decision was made on either application, the appellant (the owner of the land) appealed. The two appeals were determined by an inspector appointed by the Welsh Ministers, the first respondents. A number of matters were raised in the course of the appeals that are no longer relevant. Although the whole site was known as Gilfach holiday village, the inspector ruled that each bungalow was individually and separately occupied and was physically and functionally separate from each other and thus each was a separate planning unit. He went on to find, referring to the decision in *Gravesham Borough Council* v *Secretary of State for the Environment* (1982) 47 P&CR 142, that each bungalow constituted a dwellinghouse within Class C3 of the Use Classes Order. However, he also concluded that, although the "temporary conditions" on the planning permissions had become spent more than 14 years earlier, the seasonal use condition in each permission had not been breached and so those conditions "remain extant": see para 21. Thus, although he allowed the appeal in respect of the certificate of lawful existing use, the certificate he granted for the 42 buildings for use as dwelling houses was "subject to the conditions imposed in those permissions [that is, the four permissions between 1964 and 1974], with the exception of those conditions requiring each permission to expire and the site to be restored to its former use on or before a specified date".

[8] Taken strictly, that certificate would seem to leave in being not only the seasonal occupancy conditions but also those requiring the bungalows to be maintained to the satisfaction of the local planning authority throughout the periods specified, but no one suggests that those latter conditions still have any effect. As for the appeal in respect of the certificate of proposed lawful use, the inspector noted that the proposed use made no reference to the restriction contained in the seasonal condition and so he dismissed that appeal. In essence, that decision was based on the same proposition that guided his decision on the first appeal, namely that the seasonal conditions in the planning permissions were still "extant".

[9] The appellant challenged those decisions under section 288 of the 1990 Act, but was not successful. Beatson J, in [42], held (see [2010] EWHC 1759 (Admin)) that a permission granted for a limited period, to use the statutory words in section 72(2), did not "cease to exist" after the expiry of that period:

> What has expired is the time within which the use should have ceased and the restoration should have occurred.

Only that breach of condition had acquired immunity from enforcement action. The judge noted that the 1990 Act did not use the word "expire" and that there was no exception in section 75 in respect of a planning permission granted for a limited period. (That is the provision that states in subsection (1) that a grant of planning permission:

> shall (except in so far as the permission otherwise provides) enure for the benefit of the land and of all persons for the time being interested in it.

[10] Looking at the wording of these particular permissions, the judge noted that the occupancy conditions (as he termed the seasonal use conditions) did not refer to the period referred to in the time limit conditions and so he concluded that the seasonal restrictions were intended to apply after the end of that period. He contrasted the wording of those conditions with the wording used in the maintenance conditions. For all those reasons, he found that the seasonal use conditions were still effective and that the inspector had been right on this issue.

[11] The judge also dealt with whether the applicant was entitled to a certificate of lawful use that went beyond the use that had actually taken place "in accordance with the seasonal occupancy conditions". On this, he concluded that there was no such entitlement, relying on a passage in Circular 10/97, annex 8, dealing with such certificates. Paragraph 8.17 therein, to the extent cited by the judge, in [47], reads:

> In all cases the description must be more than simply a title or label, if future interpretational problems are to be avoided. The LDC should therefore state the characteristics of the matter so as to define it unambiguously. This is particularly important for uses which do not fall within any "use class" (that is, a "sui generis" use). So for example a LDC for a caravan site might typically include the number and type or size of caravan *found to be lawful at the application date* and, where the use is seasonal, the calendar dates on which the use then took place.

(Original emphasis.)

Statutory context

[12] Certificates of lawfulness of an existing use or a proposed use of land are dealt with by sections 191 and 192 respectively of the 1990 Act. It is unnecessary to set out those provisions in full. It is sufficient to note that "lawful", as applied to uses, operations or failures to comply with a condition on a planning permission, is so defined as to mean that it applies if the time for taking enforcement action has expired: see section 191(2) and (3). So the issue of enforceability (or lack of it) is at the heart of the concept. That, in turn, takes one to the issue of whether the seasonal use condition on each of those permissions was still extant and enforceable as a condition at the time when the certificates under sections 191 and 192 were applied for.

[13] The power to attach conditions to a planning permission derives from two separate sections of the 1990 Act. The general power is to be found in section 70(1), which, in so far as material for present purposes, provides:

> 70(1) Where an application is made to a local planning authority for planning permission
> (a) subject to sections 91 and 92, they may grant planning permission, either unconditionally or subject to such conditions as they think fit;...

Much litigation has taken place on what conditions may legitimately be imposed under this provision. Then, section 72(1) deal with two specific types of condition, one of which is relevant to this appeal:

> 72(1) Without prejudice to the generality of section 70(1), conditions may be imposed on the grant of planning permission under that section
> (a) ...
> (b) for requiring the removal of any buildings or works authorised by the permission, or the discontinuance of any use of land so authorised, at the end of a specified period, and the carrying out of any works required for the reinstatement of land at the end of that period.

That type of condition is further referred to in the next subsection, section 72(2), as follows:

> 72(2) A planning permission granted subject to such a condition as is mentioned in subsection(1) b) is in this Act referred to as "planning permission granted for a limited period".

[14] The reason why the time limit on the authorisation in such a case is secured by way of condition, rather than in the description of the development permitted, is not difficult to identify. It flows from the definition of "development" in section 55(1) as meaning the carrying out of building, engineering, mining or other operations or the making of a material change in the use of any buildings or other land. Under a temporary permission, that is a permission granted for a limited period, no "development" as such takes place at the end of the authorised period of time: the buildings (if any) are already there on site, the material change of use has already taken place, and the continued existence of the buildings and/or the continuation of the use does not constitute development. Thus, the time limit on the permitted development can be achieved only by a condition to that effect, since that may be enforced in its own right even though no development takes place at that point in time.

[15] The other type of condition given specific approval by section 72(1) is a condition relating to other land controlled by the applicant for planning permission, so long as such a condition appears to be expedient for the purposes of or in connection with the development authorised by the permission. That limitation on such a condition has some similarity to the principle well established in the case law that any condition must relate "fairly and reasonably" to the permitted development: *Pyx Granite Co Ltd* v *Ministry of Housing and Local Government* [1958] 1 QB 554 at 572; *Fawcett Properties Ltd* v *Buckinghamshire County Council* [1961] AC 636, at pp674, 678-679 and 685.

[16] Enforcement action is provided for in Part VII of the 1990 Act. Section 172(1) empowers the local planning authority to issue an enforcement notice where it appears to them that there has been a breach of planning control and that it is expedient to issue the notice, having regard to the development plan and any other material considerations. A breach of planning control is defined by section 171A(1):

> 171A(1) For the purposes of this Act
> (a) carrying out development without the required planning permission; or
> (b) failing to comply with any condition or limitation subject to which planning permission has been granted,
> constitutes a breach of planning control.

However, there are time limits on taking enforcement action, set by section 171B:

> 171B(1) Where there has been a breach of planning control consisting in the carrying out without planning permission of building, engineering, mining or other operations in, on, over or under land, no enforcement action may be taken after the end of the period of four years beginning with the date on which the operations were substantially completed.
> (2) Where there has been a breach of planning control consisting in the change of use of any building to use as a single dwellinghouse, no enforcement action may be taken after the end of the period of four years beginning with the date of the breach.
> (3) In the case of any other breach of planning control, no enforcement action may be taken after the end of the period of ten years beginning with the date of the breach.

[17] It is also to be observed that the 1990 Act makes special provision in respect of conditions that may be imposed on permissions for mineral extraction. Such permissions are limited in their duration, automatically by virtue of para 1 of Schedule 5, but they are treated as something quite separate and different from permissions "granted for a limited period" within the meaning of section 72(1)(b) and 72(2). They are expressly excluded from that definition by para (6)(1) of Schedule 5. That same schedule makes express provision for aftercare conditions to be imposed on mineral extraction permissions, to operate after the time limited for the mineral extraction has expired.

Submissions

[18] The appellant contends that the seasonal use condition in each permission was no longer extant at the dates of the applications for certificates in 2008 and 2009 and that the inspector was wrong to find otherwise. Mr Rupert Warren, on its behalf, put the construction of these permissions at the forefront of his case, but drawing support from the statutory framework. He submitted on construction that each permission and its conditions must be interpreted as a whole and in such a way as to avoid internal inconsistency. To attribute force to the seasonal use condition after the date when the buildings were no longer authorised by the permission and when the land was required to be restored to its former use without any buildings would be, it is said, illogical and internally inconsistent. If the seasonal use condition had expressly provided for its application for, say, 20 years after the time limit on the authorisation of the buildings, one would be faced with major problems of interpretation and doubt would be cast on the meaning of the first condition, the time-limit condition. Moreover, any lawful condition must relate to the permitted development, fairly and reasonably, and it is difficult to see how one could have a lawful condition with permanent effect relating to the use of buildings that by the same permission are allowed to exist only temporarily.

[19] As for the statutory context, Mr Warren submitted that the terms of section 72 mean that the permission and the conditions attached to it expire and cease to have effect at the end of the specified period. He acknowledged that the condition requiring restoration of the site to its former use may involve some work after that period and, of course, that that condition may be enforced thereafter until the time for enforcement action has passed. Such a condition relates to and acknowledges the temporary nature of the development permitted. But other than that, conditions attached to a permission that has expired are no longer enforceable. The appellant places some reliance, if only by way of analogy, on the Court of Appeal decision in *Adur District Council* v *Secretary of State for the Environment, Transport and the Regions* [2000] 1 PLR 1, a decision to which I shall return.

[20] On behalf of the Welsh Ministers, Mr Jonathan Moffett urged caution in using the word "expiry" in respect of planning permissions,

A that being an expression not found in the 1990 Act. He submitted that there is nothing in the definition of a permission granted for a limited period that precludes such a permission having other conditions that may continue after the end of the limited period. However, when pressed for an example of a condition that would do so and would still relate "fairly and reasonably" to the permitted temporary development, the only instance he could put forward was the very condition providing for the expiry of the permission and the restoration of the site found in the permissions involved in this appeal. Mr Moffett made the fair, if somewhat technical, point that that condition must remain enforceable after the specified date for the expiry of the permission. However,
B unlike the judge below, he does not regard section 75, the provision that the permission enures for the benefit of the land, as shedding any light either way on this issue.

[21] Mr Paul Stinchcombe QC, for the county council, adopted the arguments of the Welsh Ministers as to the statutory scheme and submitted that a permission within section 72(2), if it contains other conditions as well as the time limit condition, still survives. Neither it nor its conditions expire. It is simply that the time limit condition will in due course become immune from enforcement, but that does not mean that the other conditions have acquired such immunity. A local planning authority, it is argued, can properly refrain from enforcing the
C time limit on the permitted use but may still enforce other conditions restricting that use.

[22] Both respondents argue that one is concerned primarily with the construction of these permissions and their conditions. In some cases, it may well be that a condition no longer binds the land once the time limit condition can no longer be enforced. That was so in *Adur* but that is not inevitable. Mr Moffett contended that these planning permissions do not expire until the site has actually been restored to its former use, even if that post dates the date specified in the first condition (the "time limit" condition).

[23] All parties agree that, if the appeal succeeds, the matter should
D be remitted to the inspector to reconsider the issue of the certificates and their terms, in the light of this court's decision.

Discussion

[24] There is some common ground between the parties. All regard the construction of these particular permissions as being of prime importance. They differ on the issue of whether the statutory scheme in the 1990 Act allows for the possibility of conditions in a permission granted for a limited period remaining in force once the "time limit" condition has become unenforceable. That statutory context is none the less obviously relevant and I begin by considering it.

E [25] *Adur* is of limited assistance on this aspect, because it was clearly decided principally as a matter of the interpretation of the actual permission and its conditions. It was a case concerning a 1991 permission granted for a limited period, during which time a port authority was entitled to use certain land for the storage of cars. That authorisation expired on 31 December 1991. There was also a condition, condition 2, that:

> Notwithstanding the provisions of the Town and Country Planning (Use Classes) Order 1987 and the Town and Country Planning General Development Order 1988 (as amended), the site shall be used solely for storage of cars.

F By the time the case reached the Court of Appeal, there was no dispute that the combined effect of the temporary permission and the use of the land by the port authority for port-related purposes was to render the site "operational land" for the purposes of the rights granted to port operators by the General Permitted Development Order 1995, the successor in this respect to the 1988 Order referred to in condition 2. So the effect of the temporary permission was far-reaching, and it was argued by the local planning authority that condition 2 was of continuing effect and so excluded those rights.

[26] The court rejected that argument, pointing out the obvious inconsistency between a condition that granted *temporary* permission for the storage of cars and an allegedly continuing condition that not merely excluded the development order rights but expressly confined the use of the land to the storage of cars. Thus, condition 2 could not be interpreted as applying beyond the end date on the temporary permission for such storage. In the leading judgment, Pill LJ said in terms that the question of whether it would be possible to have a permanent restrictive condition on a temporary planning permission did not fall for decision on the facts of the case: see p7. None the less, it is to be observed that he went on, at p8A-B, to add:

> Similarly, a limitation imposed in condition 2 applies only while rights are being exercised under the 1991 permission. When reliance is no longer placed on the express grant, the condition does not have the effect of excluding development rights under the GPDO.

[27] That last point reflects a long-standing approach in planning law, running from *Mounsdon* v *Weymouth and Melcombe Regis Borough Council* [1960] 1 QB 645 to cases like *Essex Construction Co* v *East Ham Borough Council* (1965) 16 P&CR 220 and *Handoll* v *Warner Goodman & Streat**, whereby a condition in a permission cannot be enforced if the landowner does not have to rely on the permission to authorise its development. Those cases, however, do not (unlike *Adur*) expressly deal with the situation where the landowner did need the permission originally to sanction his development and where the permission has been implemented.

[28] Having said that, it is very difficult to conceive of a condition on a temporary permission under section 72 that could sensibly relate to a development, once that development has ceased to be authorised by the permission. The time limit and restoration condition itself does not provide an example that goes beyond its own scope, since that condition is expressly and precisely provided for by section 72(1)(b). One cannot derive a general approval from that for conditions that bind the land once the development has ceased to be authorised and has become immune from enforcement action. Such enduring conditions would, to be lawful, still have to relate fairly and reasonably to the permitted development, which was and is to be seen as a temporary development. So although I would not wish to be categoric as to the impossibility of such enduring conditions, I do regard it as very unlikely that the statutory scheme allows for what can be described as a permanent condition on a temporary permission, other than the time limit condition. The latter is a very different animal: as Mr Warren put it, such a condition circumscribes the entire authorisation of the use. It is quite unlike a condition limiting in a certain respect a use that has become an unauthorised use.

[29] It is also significant that parliament has found it necessary to make special express provision for the situation where conditions are often required to apply after the authorisation of the permitted development has expired, namely in the case of mineral extraction. Aftercare conditions have had to be given explicit sanction by the statute in such cases: see [17] (*ante*).

[30] Within that statutory context, I turn to the permissions and conditions that call for interpretation in this appeal. It is of course correct that the seasonal use condition in each permission has no express time limit on it and that it does not contain the phrase "throughout this period" as does the maintenance of the bungalows condition. The judge attached some significance to the apparent contrast between the wording of the maintenance condition and the seasonal use condition's wording in this respect, but I regard that significance as unjustified. The maintenance condition was imposing a positive obligation on the landowner to maintain the bungalows and it was necessary to word the condition in such a way as to make it clear that this was a continuing obligation throughout the period to which the temporary permission related. No such additional words were needed for the seasonal use condition, which imposed a restriction patently intended to apply throughout the life of the permitted development.

[31] There is no doubt that, when construing a permission and its conditions, the document must be read as a whole and in such a way as to avoid, if possible, internal inconsistency. In the present

* Editor's note: Reported at [1995] 1 EGLR 173

case, I accept Mr Warren's submission that it would be illogical and internally inconsistent when construing these permissions to give the seasonal use condition in each permission a life after the specified date. The inconsistency is not as glaring as in *Adur*, but it is none the less present, and it is a powerful objection to the continuing force of the seasonal use condition. The first condition in each permission makes it clear that permission is being granted at all only for a limited and specified period. To have a condition that contradicts that by assuming that development is being authorised for a longer period, albeit on a seasonal basis, would give rise to major problems of interpretation as to the duration of the authorisation. This would become obvious if the seasonal use condition had spelt out such longer duration expressly. If the permission had stated that the buildings were to be removed and their use as holiday bungalows were to cease by 31 July 1985 but the seasonal use condition was said to apply until, say, 1999, the difficulty would be readily apparent. The wording does not do that, but the interpretation advanced by the respondents is to that effect, albeit with no time limit.

[32] I am not persuaded by Mr Moffett's attempt to interpret the time limit condition as authorising the restoration of the site after the specified date. The wording of the condition, with the expression "on or before", is not ambiguous. It sets an end date, by which time the site must have been restored to its previous use, although no doubt the planning authority would not take enforcement action if at the end date restoration was clearly well under way.

[33] Planning conditions do not anticipate or provide for the consequences of the breach of other conditions in the way in which a contract might well. If a valid and effective condition is breached, the remedy is to be found in the statutory machinery for enforcement. A planning authority faced with a development that has failed to observe a time limit and restoration condition imposed under section 72 has entirely adequate enforcement powers available to it. If it sits on its hands until the time limit condition can no longer be enforced, it has only themselves to blame.

[34] In my judgment, the seasonal use conditions in these permissions applied during the period for which development was authorised by those permissions. The seasonal use conditions were, as a matter of objective construction, intended to be coterminous with the authorised development, with the result that the seasonal use restriction applied during that period for which these holiday bungalows were permitted. It follows that I would allow this appeal and, as is agreed in that event, remit the certificate appeals to the inspector to redetermine in the light of this court's decision.

Smith LJ said: I agree.

Rix LJ said: I also agree.

Appeal allowed.

R (on the application of Cala Homes (South) Ltd) *v* Secretary of State for Communities and Local Government

COURT OF APPEAL

[2011] EWCA Civ 639

27 May 2011

RIX, RIMER AND SULLIVAN LJJ

Estates Gazette 27 August 2011

[2011] 34 EG 68

Town and country planning — Development plan — Regional spatial strategies (RSSs) — Government proposing to pass legislation to abolish RSSs — Secretary of State issuing statements advising planning authorities to have regard to that matter as material consideration in planning decisions — Whether capable of being material consideration — Whether advice subverting policy and objects of existing planning legislation

The appellant hoped to carry out a large residential development in an area that was allocated for housing development in a regional spatial strategy (RSS) promulgated by the respondent's predecessor, under Part 1 of the Planning and Compulsory Purchase Act 2004.

In May 2010, following the election of a new government, the respondent issued a letter to all local planning authorities (LPAs) informing them of the government's intention to abolish RSSs and directing them to have regard to that letter as a material consideration when making planning decisions. In July 2010, the respondent made a statement of intention to revoke RSSs. In judicial review proceedings brought by the appellant, that statement was held to be unlawful on the ground that, under the existing planning legislation, including Part 5 of the Local Democracy, Economic Development and Construction Act 2009, the respondent could not revoke an RSS without replacing it: see [2010] EWHC 2866 (Admin); [2010] 46 EG 116 (CS). However, in November 2010, the respondent issued a statement, forwarded in a letter to all LPAs, advising that the government intended to abolish RSSs by legislation and indicating that LPAs and the Planning Inspectorate should still have regard to the May 2010 letter in their decisions.

The appellant brought a further judicial review claim in which it challenged the lawfulness of the respondent's communications of May and November 2010 and submitted that it would be unlawful for planning authorities to have regard to the government's intention to abolish RSSs by legislation, since to do so would subvert the policy and objects of the existing statutory planning framework. Dismissing the claim, the judge held that the intention to abolish RSSs could be a "material consideration", of which planning authorities were entitled to take account in planning decisions: see [2011] EWHC 97 (Admin); [2011] JPL 887. The appellant appealed.

***Held*: The appeal was dismissed. (1) The government's proposal to abolish RSSs could, as a matter of law, be a material consideration for the purposes of determining a planning application or appeal. The weight to be given to any prospective change to planning policy will be a matter for the decision maker's planning judgment in each case, having regard to the means by which the change is to be effected. Where the change is to be effected by new legislation, it is not necessarily an immaterial consideration prior to obtaining royal assent; instead, the stage reached in the legislative process goes to the question of weight. (2) Taking account of the government's intention to abolish RSSs would not thwart or subvert the purpose of the relevant legislation so as to preclude planning authorities from taking it into account as a material consideration when determining planning applications and appeals. The policy and objects of the legislative scheme, construed as a whole, go beyond**

A **the requirement that there should be an RSS for each region. Those responsible for determining planning applications and appeals are required to consider not only the development plan, of which RSSs form a part along with local development plan documents, but also other material considerations: see section 70(2) of the Town and Country Planning Act 1990 and section 38(6) of the Planning and Compulsory Purchase Act 2004. Those considerations can include the fact that the development plan policies have become outdated or are no longer relevant because of a change of circumstances, and they might indicate that the decision should not accord with the development plan. (3) The respondent's communications of May**
B **and November 2010 were not unlawful. Planning authorities had not been advised to ignore the policies in the RSSs, to treat them as no longer forming part of the development plan or to determine planning applications otherwise than in accordance with them, nor had they been told what weight they should give to the government's proposal. The May 2010 letter would be irrelevant in the majority of cases because regional policy would not be in issue or would be of marginal significance. Given the early stage that the proposal to abolish RSSs had reached in the legislative process, and the fact that revocation of individual RSSs would be subject to the strategic environmental assessment process, many LPAs would consider that**
C **they should give little, if any, weight to the proposed abolition of RSSs when making their decisions. However, the court could not find that, at the present stage, there were no circumstances in which a decision maker could rationally give some weight to the proposed abolition of RSSs.**

The following cases are referred to in this report.

City of Edinburgh Council v *Secretary of State for Scotland* [1997] 1 WLR 1447; [1997] 3 PLR 71
Laker Airways Ltd v *Department of Trade* [1977] QB 643; [1977] 2 WLR 234; [1977] 2 All ER 182
Office of Government Commerce v *Information Commissioner* [2008] EWHC
D 774 (Admin); [2010] QB 98; [2009] 3 WLR 627
Padfield v *Minister of Agriculture, Fisheries and Food* [1968] AC 997; [1968] 2 WLR 924; [1968] 1 All ER 694, HL
R (on the application of Cala Homes (South) Ltd) v *Secretary of State for Communities and Local Government* [2010] EWHC 2866 (Admin); [2010] 46 EG 116 (CS), [2011] EWHC 97 (Admin); [2011] JPL 887
R v *Braintree District Council, ex parte Halls* (2000) 80 P&CR 266; [2000] 3 EGLR 19; [2000] 36 EG 164; (2000) 32 HLR 770
Tesco Stores Ltd v *Secretary of State for the Environment* [1995] 1 WLR 759; [1995] 2 All ER 636; (1995) 93 LGR 403; [1995] 2 EGLR 147; [1995] 27 EG 154; [1995] 2 PLR 72, HL

E This was an appeal by the appellant, Cala Homes (South) Ltd, from a decision of Lindblom J, sitting in the Administrative Court, dismissing a claim against the respondent, the Secretary of State for Communities and Local Government, for judicial review of advice issued to planning authorities in May and November 2010 regarding the government's proposed abolition of regional spatial strategies as part of the development plan.

Peter Village QC, James Strachan and Sarah Hannett (instructed by Clyde & Co LLP) appeared for the appellant; Timothy Mould QC and James Maurici (instructed by the Treasury Solicitor) represented the respondent.

F Giving judgment, **Sullivan LJ** said:

Introduction

[1] The principal issue in this appeal against the order dated 7 February 2011 of Lindblom J dismissing the appellant's claim for judicial review is whether those responsible for making decisions under the Planning Acts (the respondent, local planning authorities (LPAs) and the Planning Inspectorate) are entitled to have regard to the government's proposal to abolish regional strategies, now embodied in clause 89 of the Localism Bill, as a material consideration for the purposes of section 70(2) of the Town and Country Planning Act 1990 (the 1990 Act) and section 38(6) of the Planning and Compulsory Purchase Act 2004 (the 2004 Act). The appellant contends that prior to royal assent being given to the necessary legislation, the proposal to abolish regional strategies is not capable of being a material consideration for the purpose of determining planning applications and appeals.

Statutory framework

Development plan

[2] Before turning to the facts, it is helpful to set out the statutory framework. The starting point is Part 5 of the Local Democracy, Economic Development and Construction Act 2009 (the 2009 Act), which came into effect on 1 April 2010, and which deals with "Regional Strategy". Section 70(1) provides that: "There is to be a regional strategy for each region other than London"; the regional strategy for each region must set out policies in respect of both "sustainable economic growth" and "the development and use of land" in the region: subsection 70(2).

[3] Section 38(3) of the 2004 Act, as amended by the 2009 Act, provides that for the purposes of any area in England other than Greater London the development plan is:

> (a) the regional strategy for the region in which the area is situated, and
> (b) the development plan documents (taken as a whole) which have been adopted or approved in relation to that area.

Development control

[4] Section 70(2) of the 1990 Act requires LPAs when determining planning applications to "have regard to the provisions of the development plan, so far as material to the application, and to any other material considerations." Section 70(2) also applies to the determination of appeals by the Planning Inspectorate and the respondent: see section 79(4).

[5] Section 70(2) of the 1990 Act must be read together with section 38(6) of the 2004 Act, which provides that:

> If regard is to be had to the development plan for the purpose of any determination to be made under the planning Acts the determination must be made in accordance with the plan unless material considerations indicate otherwise.

"Plan led" development control

[6] The combined effect of sections 70(2) of the 1990 Act and what is now 38(6) of the 2004 Act was explained by the House of Lords in *City of Edinburgh Council* v *Secretary of State for Scotland* [1997] 1 WLR 1447 (section 18A of the Town and Country Planning (Scotland) Act 1972 was the counterpart of section 54A of the 1990 Act, the predecessor of section 38(6) of the 2004 Act). The relevant extracts from the speeches of Lord Hope and Lord Clyde are set out in [28] of Lindblom J's judgment. In summary, section 38(6) creates a presumption in favour of the development plan. Two passages explain how that presumption is to be applied in practice:

> It requires to be emphasised, however, that the matter is nevertheless still one of judgment, and that this judgment is to be exercised by the decision-taker. The development plan does not, even with the benefit of section [38(6)] have absolute authority. The planning authority is not obliged, to adopt Lord Guest's words in *Simpson* v *Edinburgh Corporation*, 1960 SC 313, 318, "slavishly to adhere to" it. It is at liberty to depart from the development plan if material considerations indicate otherwise. No doubt the enhanced status of the development plan will ensure that in most cases decisions about the control of development will be taken in accordance with what it has laid down. But some of its provisions may become outdated as national policies change, or circumstances may have occurred which show that they are no longer relevant. In such a case the decision where the balance lies between its provisions on the one hand and other material considerations on the other which favour the development, or which may provide more up-to-date guidance as to the tests which must be satisfied, will continue, as before, to be a matter for the planning authority.

(Lord Hope, at p1450 B-D.)

> By virtue of section [38(6)] if the application accords with the development plan and there are no material considerations indicating that it should be refused, permission should be granted. If the application does not accord with

the development plan it will be refused unless there are material considerations indicating that it should be granted. One example of such a case may be where a particular policy in the plan can be seen to be outdated and superseded by more recent guidance. Thus the priority given to the development plan is not a mere mechanical preference for it. There remains a valuable element of flexibility. If there are material considerations indicating that it should not be followed then a decision contrary to its provisions can properly be given.

(Lord Clyde, at p1458 E-F.)

Local development documents preparation

[7] The role of the regional strategies in the preparation of the second tier of the development plan the development plan documents is somewhat more prescriptive. Section 19(2) of the 2004 Act lists in paras (a)(j) the matters to which the LPA must have regard when preparing a development plan document or any other local development document. Those matters include the regional strategy for the relevant region and the regional strategy for any adjoining region: paras (b) and (d). Section 24(1) provides that, outside Greater London "The local development documents must be in general conformity with… the regional strategy…".

Facts

[8] The facts are set out in some detail in the judgment of Lindblom J [2011] EWHC 97 (Admin)*, and a brief summary will suffice for present purposes. In a letter dated 27 May 2010 addressed to all LPAs in England, the respondent said:

ABOLITION OF REGIONAL STRATEGIES

I am writing to you today to highlight our commitment in the coalition agreements where we very clearly set out our intention to rapidly abolish Regional Strategies and return decision making powers on housing and planning to local councils. Consequently, decisions on housing supply (including the provision of travellers' sites) will rest with Local Planning Authorities without the framework of regional numbers and plans.

I will make a formal announcement on this matter soon. However, I expect Local Planning Authorities and the Planning Inspectorate to have regard to this letter as a material planning consideration in any decisions they are currently taking.

[9] On 6 July 2010, the respondent made a statement to parliament in which he said that he was revoking the regional strategies under section 79(6) of the 2009 Act. In judicial review proceedings before Sales J, the appellant challenged the lawfulness of that revocation on two grounds, namely that:

(i) the power of revocation under section 79(6) could not lawfully be used to abolish the entire regional strategy tier of the development plan; and

(ii) there had been no consideration as to whether this change in the development plan was likely to have significant environmental effects, in breach of Directive 2001/42/EC (the SEA Directive), as transposed into domestic law by the Environmental Assessment of Plans and Programmes Regulations 2004 (the SEA Regulations).

[10] In a judgment handed down on 10 November 2010, Sales J upheld the appellant's challenge on both grounds: [2010] EWHC 2866 (Admin)†. The respondent did not appeal against the judgment of Sales J, but he immediately issued a statement in parliament. The full text of that statement is set out in [17] of Lindblom J's judgment. The statement referred to a letter, also dated 10 November 2010, from the Department for Communities and Local Government's Chief Planner, which was addressed to the chief planning officers of all LPAs and to the Planning Inspectorate. That letter said:

ABOLITION OF REGIONAL STRATEGIES

I am writing to you today following the judgment in the case brought by Cala Homes in the High Court, which considered that the powers set out in section 79(6) of the Local Democracy, Economic Development and Construction Act 2009 could not be used to revoke all Regional Strategies in their entirety. G

The effect of this decision is to re-establish Regional Strategies as part of the development plan. However the Secretary of State wrote to Local Planning Authorities and to the Planning Inspectorate on 27 May 2010 informing them of the Government's intention to abolish Regional Strategies in the Localism Bill and that he expected them to have regard to this as a material consideration in planning decisions.

I am attaching the proposed clause of the Localism Bill that will enact that commitment. The Bill is expected to begin its passage through Parliament before Christmas, and will return decision-making powers in housing and planning to local authorities. Local Planning Authorities and the Planning Inspectorate should still have regard to the letter of the 27th May 2010 in any decisions they are currently taking. H

At that stage, there was no bill. The proposed clause in the draft bill attached to the Chief Planner's letter was as follows:

1 Abolition of regional strategies

(1) Part 5 of the Local Democracy, Economic Development and Construction Act 2009 (regional strategy) is repealed.

(2) The regional strategies under Part 5 of the Act are revoked.

[11] The claim form in these proceedings was issued on 19 November 2010. The appellant sought a declaration that it "is unlawful for the respondent (and for local planning authorities and the Planning J Inspectorate) to have regard to the government's stated intention to enact primary legislation in the future to abolish the regional strategies in England as a material consideration in making determinations under the Planning Acts". The appellant also sought a quashing order in respect of the respondent's statement and the Chief Planner's letter, both dated 10 November 2010, and the respondent's letter dated 27 May 2010, or a declaration that their contents were unlawful.

[12] At the hearing before Lindblom J on 17 and 18 January 2011, there were three grounds of challenge, namely that:

(i) the government's intention to abolish regional strategies was incapable in law of being a material consideration within the meaning K of section 70(2) of the 1990 Act;

(ii) the issuing of the statement and letter on 10 November 2010 was irrational; and

(iii) there had been a failure to comply with the requirements relating to strategic environmental assessment in the SEA Directive and the SEA Regulations.

Lindblom J rejected all three grounds. There is no appeal against his decision to reject grounds (ii) and (iii).

[13] The Localism Bill had its first reading in the House of Commons on 13 December 2010. On 10 March 2011, it concluded the committee stage, and we were told that the report stage was programmed to start L on 17 May 2011. On 5 April 2011, the parliamentary Under-Secretary of State at the Department for Communities and Local Government said in a statement to parliament that the government had decided to carry out a strategic environmental assessment of the regional strategies. He said that the assessment was being carried out on a voluntary basis, and explained:

We intend to compile an environmental report for each region and to consult on it in line with the process laid down in the Environmental Assessment of Plans and Programmes Regulations 2004. Local authorities and others should find this helpful in identifying issues relevant to their areas and policies or initiatives in the regional strategies which are no longer in effect, and it M should also help them decide how to proceed with preparing or reviewing their own plans.

This process of environmental assessment will be carried out during the passage of the Bill through Parliament. Subject to Royal Assent, the revocation of each individual regional strategy will be commenced after the assessment process has been completed.

[14] As amended in public bill committee, clause 97 of the Localism Bill now provides:

Abolition of regional strategies

(1) The following provisions are repealed

(a) sections 82(1) and 83 of the Local Democracy, Economic Development and Construction Act 2009 (effect of regional strategies), and

---

* Editor's note: Reported at [2011] JPL 887

† Editor's note: Reported at [2011] 46 EG 116 (CS)

A (b) the remaining provisions of Part 5 of that Act (regional strategy).

(2) Subsection (1)(b) does not apply to

(a) section 85(1) (consequential provision) of that Act,

(b) Schedule 5 to that Act (regional strategy: amendments) (but see Part 15 of Schedule 25 to this Act), or

(c) Part 4 of Schedule 7 to that Act (regional strategy: repeals).

(3) The regional strategies under Part 5 of that Act are revoked.

Schedule 8 contains a large number of consequential amendments. The provisions of section 97 are to come into force on such day as the Secretary of State may by order appoint: see clause 226(2).

B *Discussion*

[15] It is common ground that sections 70(2) of the 1990 Act and 38(6) of the 2004 Act confer a discretion, and that the planning decision maker (the respondent, the Planning Inspectorate or the LPA) must exercise that discretion so as to promote, and not so as to thwart or run counter to, the policy and objects of the legislation conferring the discretion: see *Padfield* v *Minister of Agriculture, Fisheries and Food* [1968] AC 997, *per* Lord Reid, at p1030B-D.

[16] At the heart of the appellant's case is the submission that the government's intention to abolish the regional strategies is not, as a matter of law, capable of being a material consideration for the purposes
C of sections 70(2) and 38(6) because taking into account an intention to abolish regional strategies would subvert or thwart the legislative purpose set out in section 70(1) of the 2009 Act that "*there is to be* a regional strategy for each region….". Mr Peter Village QC submitted that Lindblom J had not addressed this issue and had not answered the question: "How does taking into account the Respondent's aspiration to abolish regional strategies promote the object and purpose of the legislation that there should be regional strategies?"

[17] The *Padfield* principle was applied by the Court of Appeal in *R* v *Braintree District Council, ex parte Halls* (2000) 32 HLR 770*. Laws LJ said, at p779:

D The rule is not that the exercise of the power is only to be condemned if it is incapable of promoting the Act's policy, rather the question always is: what was the decision-maker's purpose in the instant case and was it calculated to promote the policy of the Act?

[18] Mr Village submitted that, in this case, the answer to the two-part question posed by Laws LJ was clear:

(a) the respondent's purpose was to abolish regional strategies; and

(b) that purpose was not calculated to promote (indeed, it was directly contrary to) the policy of the 2009 Act that regional strategies should exist as part of the development plan.

E [19] If the policy and objects of the legislation were defined solely by reference to section 70(1) of the 2009 Act, having regard to the proposed abolition of that which section 70(1) requires there to be in every region would be contrary to the policy and objects of the legislation. However, Mr Timothy Mould QC submitted that the policy and objects of the legislation are not defined solely by reference to section 70(1) of the 2009 Act, and that all of the enactments that comprise the legislative scheme section 70(1) of the 2009 Act, subsections 38(3) and (6) of the 2004 Act, and section 70(2) of the 1990 Act must be considered for the purpose of ascertaining the relevant statutory policy and objects. The proposition that there is to be a regional strategy for each region takes
F one only so far. In order to understand the role of that regional strategy in development control decisions one needs to appreciate that:

(a) the regional strategy is to be the upper tier of the development plan;

(b) the decision maker must have regard not only to the development plan but also to other material considerations when determining planning applications or appeals; and

(c) the decision maker may grant or refuse planning permission contrary to the development plan if material considerations indicate that it should not be followed.

[20] The issue between the parties is a very narrow one: what are the relevant legislative policy and objects? Mr Village did not dispute the general proposition that a prospective change to planning policy is capable of being a material consideration for the purposes of sections 70(2) of the 1990 Act and 38(6) of the 2004 Act. The weight to be given to any prospective change in planning policy will be a matter for the decision maker's planning judgment in each particular case. In principle, the means by which it is proposed to effect a change in policy, by new legislation, by amendment under existing legislation, or by administrative action such as the publication of a new planning policy statement (PPS), goes to the weight, not the materiality, of the prospective change. If the change is to be effected by legislation, will parliamentary approval be obtained, and if so in what form and within what timescale? Subject to the appellant's *Padfield* point, a change in policy that is proposed to be effected by legislation is not an immaterial consideration on the day before royal assent, and a material consideration on the day that royal assent is granted. The stage reached in the legislative process goes to the weight, not the materiality of the proposed change.

[21] Rather than considering the relevance for development control purposes of possible future changes to planning policy purely in the abstract, it is important to focus on the kind of policies that are in issue in this case. Regional strategies are typically concerned with the long term. The South East Plan 2009 sets regional housing requirements for a 20-year period: from 2006 to 2026. To meet those requirements, large-scale residential developments will typically be programmed for implementation over a lengthy period, perhaps 10-15 years. If the decision maker considers that there are valid site specific or "local" objections to the grant of planning permission, for example, landscape impact or loss of agricultural land, but these objections are said by the applicant for permission to be outweighed by the need to meet the long-term regional housing requirements in the development plan, the decision-maker would, subject to the appellant's *Padfield* point, surely be entitled to have regard to the prospect that the sole policy justification for permitting the development may well have ceased to exist long before the development has been completed, and perhaps before it has even been commenced, when deciding whether planning permission should be granted now for such a large-scale and long-term commitment to meet future needs.

[22] I accept Mr Mould's submission that in order to ascertain the policy and objects of the legislation conferring the discretion under section 70(2) of the 1990 Act, it is necessary to look beyond the requirement in section 70(1) of the 2009 Act that there "is to be a regional strategy for each region". The regional strategy is the first tier of the development plan; the development plan documents for each area within the region are the second tier. The 2004 Act requires each LPA to prepare and maintain "a scheme to be known as their local development scheme", and that scheme must specify "the local development documents which are to be development plan documents": see section 15(1) and (2)(aa). These local development documents must "set out the authority's policies (however expressed) relating to the development and use of land in their area": see section 16(3).

[23] In respect of any area in England outside Greater London, the legislative scheme requires there to be both a regional strategy for the region within which the area is situated and the development plan documents for the area. Together, they comprise the development plan for that area. If the legislative scheme provided that planning applications were to be determined in accordance with the development plan, having any regard to the prospect that the development plan might be abolished would be contrary to the policy and objects of the legislation, but *City of Edinburgh* (see [6]) above makes it clear that that is not the combined effect of sections 70(2) of the 1990 Act and 38(6) of the 2004 Act. The decision maker must consider not only the development plan, but also other material considerations. Those considerations may include the fact that the policies in the development plan have become outdated, or are no longer relevant because of a change of circumstances; and those considerations may indicate that the decision should not be in accord with the development plan.

* Editor's note: Also reported at [2000] 3 EGLR 19

[24] This "valuable element of flexibility" (see Lord Clyde's speech in *City of Edinburgh* cited in [6] above), given to the LPA when determining planning applications, is to be contrasted with the lack of flexibility when the LPA are preparing their development plan documents. They must have regard to the relevant regional strategy (among other specified matters), and whether they are precluded from having regard to other matters that are not listed in paras (a)-(j) of section 19(2) of the 2004 Act, the end-product, the local development documents, "must be in general conformity" with the regional strategy: see section 24(1) of the 2004 Act. Development plan documents must be submitted for independent examination by a person (in practice a planning inspector) appointed by the respondent, and one of the purposes of that examination is to determine whether the development plan document satisfies the requirement of general conformity in section 24(1). It would be unlawful for a LPA preparing, or a planning inspector examining, development plan documents to have regard to the proposal to abolish regional strategies. For so long as the regional strategies continue to exist, any development plan documents must be in general conformity with the relevant regional strategy.

[25] For these reasons, I do not accept the appellant's submission that the government's proposal to abolish the regional strategies is incapable, as a matter of law, of being a material consideration for the purposes of sections 70(2) of the 1990 Act and 38(6) of the 2004 Act. The prospect of a change in planning policy is capable of being a material consideration, and taking account of this particular prospective change would not be contrary to the *Padfield* principle because the policy and objects of the legislative scheme construed as a whole require those responsible for determining planning applications and appeals to look beyond the development plan, and to have regard to other material considerations. Contrary to Mr Village's submission, Lindblom J did deal with this issue what are the policy and objects of the legislation? in [47] of his judgment. The respondent's avowed purpose in promoting clause 89 of the Localism Bill is to abolish regional strategies, but the answer to the two-part question posed by Laws LJ in *Halls* is that:

(a) the purpose of the Chief Planning Officer's letter dated 10 November 2010 was to draw LPAs' attention to the proposed abolition as, potentially, a material consideration; and

(b) that purpose was calculated to promote the policy and objects of the legislation, that local planning authorities should have regard not merely to the development plan, but also to other material considerations.

[26] If the Chief Planning Officer's letter had advised LPAs to ignore the policies in the regional strategies, or to treat them as no longer forming part of the development plan, or to determine planning applications otherwise than in accordance with them because the government proposed to abolish them, or had it told decision makers what weight they should give to the government's proposal, such advice would have been unlawful. *Laker Airways Ltd* v *Department of Trade* [1977] QB 643, is an example of policy guidance that was unlawful because it was contrary to the statutory objectives laid down for the Civil Aviation Authority by section 3 of the Civil Aviation Act 1971: see *per* Lord Denning, at p704A-F. As Lindblom J explained in [58] to [62] of his judgment, it is necessary to look at what the Chief Planner's letter actually said. It expressly stated that the effect of the decision of Sales J was "to re-establish regional strategies as part of the development plan", and said that decision makers should have regard to the proposed abolition of regional strategies as a material consideration in planning decisions. No doubt the letter could have said more, but it must be remembered that the letter was addressed to the Chief Planner's fellow professionals: the Planning Inspectorate and the Chief Planning Officers of the LPAs in England. The Chief Planner did not need to remind those to whom he sent the letter that re-establishing regional strategies as part of the development plan meant that planning applications had to be determined in accordance with the regional strategies unless material considerations indicated otherwise.

[27] The advice that decision makers "*should* still have regard to the letter of 27th May 2010 in *any decisions* they are *currently* making" might well have been misleading if it had been addressed to a less G expert audience. Mr Mould accepted that in the great majority of cases, the letter dated 27 May 2010 will be of no relevance whatsoever because regional policy will not be in issue at all, or will at best be of marginal significance. The planning inspectors and chief planning officers who received the letter would have been well aware of this, and would have sensibly interpreted the letter as referring only to those decisions in which, in their judgment, regional policy was a significant issue. The letter says nothing about the weight to be given to the proposed abolition of regional strategies. Given the very early stage that the proposal has reached in the legislative process, and the fact that revocation of any individual regional strategy will be subject to the H SEA process, many planning inspectors and chief planning officers may well consider that they should give little, if any, weight to the proposed abolition of regional strategies in the decisions that they are *currently* taking. That position will change if the proposal progresses, or fails to progress, through the legislative and environmental assessment process, but those responsible for taking planning decisions are familiar with the general proposition that the weight to be given to emerging policy is contingent on its progress towards finality: see [52] of the judgment of Lindblom J. In summary, the Chief Planner's letter dated 10 November 2010 might have been better and more fully expressed, but what it did say was not unlawful. J

[28] However, the written submissions of Stevenage Borough Council that was given leave to intervene have identified one respect in which the reference in the letter to "any decisions" was potentially confusing. Stevenage are preparing their development plan documents. Those documents include a core strategy that has been submitted for independent examination. The planning inspectors' report is awaited. There has been correspondence between Stevenage and the Planning Inspectorate as to the effect of the purported "abolition" of regional strategies in July 2010, and their "reinstatement" in November 2010 following the decision of Sales J. The detail of that correspondence is not relevant for present purposes, but the fact that there was an K exchange of correspondence demonstrates the need to make it clear that the reference in the letter to "any decisions" is a reference to development control decisions; it is not a reference to decisions by LPAs and the Planning Inspectorate in the preparation and examination of development plan documents: see [24] above.

[29] I have referred to the terms of the Chief Planner's letter rather than the respondent's statement to parliament on 10 November 2010. It is unnecessary to consider the statement because it refers, and does not add anything of substance, to the Chief Planner's letter. More important, it would not be appropriate to examine the statement in any detail because to do so would run the risk of the court overstepping the line between the mere receipt of evidence of the proceedings of parliament and the L "questioning" of those proceedings: see the discussion of the ambit of parliamentary privilege in *Office of Government Commerce* v *Information Commissioner* [2008] EWHC 774 (Admin)*, *per* Stanley Burnton J (as he then was), in [30] to [50]. A quashing order in respect of the respondent's statement to parliament would have been out of the question.

[30] Mr Village submitted that if the proposed abolition of regional strategies was legally capable of being a material consideration, planning decision makers would be free to give a regional strategy no weight at all, and thus to subvert the statutory scheme by effectively "sidelining" the development plan. In my judgment, that concern is M overstated. Although the weight to be given to any particular material consideration is a matter for the decision maker, the decision maker must not "lapse into *Wednesbury* irrationality", see *Tesco Stores Ltd* v *Secretary of State for the Environment* [1995] 1 WLR 759†, *per* Lord Hoffmann, at p780F-G; see also Lord Keith, at p764H.

[31] In most cases, the constraint of *Wednesbury* rationality will be a very light rein because the courts normally give a very wide latitude to planners' judgments as to the weight to be given to planning

* Editor's note: Reported at [2010] QB 98

† Editor's note: Also reported at [1995] 2 EGLR 147

A considerations. However, it is proposed that this particular change of policy will be effected by legislation, so the proposal is subject to two legal obstacles:

(a) Parliamentary approval; and

(b) the SEA process.

[32] Although the point was not raised on behalf of the appellant, I asked the parties for their submissions as to whether it might be irrational for any decision maker to give any significant weight at this stage to the proposed abolition of regional strategies because to do so would require the decision maker to prejudge:

B (a) Parliament's acceptance of the proposal; and

(b) the outcome of the SEA process.

It would be inappropriate for any decision maker to consider in any detail what parliament's response to the government's proposal might be, because to do so might involve them in questioning the proceedings of parliament: see *Office of Government Commerce*: see [29] above. Moreover, even if clause 89 is enacted in its present form, it could not lawfully be assumed that revocation of any individual regional strategy is bound to occur regardless of the outcome of the process of environmental assessment, because to make such an assumption would be contrary to the requirement of the SEA Directive and the C SEA Regulations: that a decision to revoke may not be made until the process has been completed.

[33] Mr Village submitted that if the proposed abolition was a material consideration, it would be irrational to give it any weight at this stage. However, Mr Mould's submissions have persuaded me that where the issue is one of weight rather than materiality, "never say never" is the appropriate response to a submission that, as a matter of law, any decision maker in any case would be bound to give no significant weight to a potentially material factor. Mr Mould fairly acknowledged that even within the minority of cases in which the proposed abolition of regional strategies will be relevant, there may D well be very few cases in which it would be appropriate at this stage of the parliamentary and SEA process to give any significant weight to the proposal. However, the Chief Planner's letter is concerned with the whole of the period prior to the enactment of the Localism Bill (if it is enacted), and the position will change as it progresses, or fails to progress. Even now there might be finely balanced cases where the very slight prospect of a very substantial policy change might just tip the balance in favour of granting or refusing planning permission. Mr Mould gave the hypothetical example of a large-scale residential proposal (which he referred to as a "new town", but the point would equally apply to a proposed extension of an existing settlement), which E is proposed to be developed over the next 15-20 years, to which there are very strong site-specific objections, and where the sole justification for granting planning permission is the need to meet the requirement for residential development over the next 20 years in the regional strategy. In such a case, it would not be irrational for the decision maker to give some weight to the prospect, however uncertain, that the regional policy justification for granting permission for such a long-term proposal may cease to exist within the short term. In such a case, to give even very little weight to the prospect of a change in policy might be to give that factor "significant" weight, significant in the sense that it might tip the balance in favour of refusing permission. This hypothetical example may well F be an extreme case, but it does illustrate why it would not be safe for the court to assume that at this stage there are no circumstances in which any decision maker could rationally give some weight to the proposed abolition of regional strategies. In view of the uncertainty created by the legal obstacles referred to above (see [31]) any decision maker that does think it appropriate to give some weight to the government's proposal when determining an application or an appeal would be well advised to give very clear and cogent reasons for reaching that conclusion, but that does not mean that there could be no case whatsoever in which any decision maker might be able to give such reasons.

Conclusion

[34] I would dismiss this appeal.

Rimer LJ said:

[35] I agree

Rix LJ said:

[36] I also agree

Appeal dismissed.

R (on the application of Renaissance Habitat Ltd) *v* West Berkshire District Council

ADMINISTRATIVE COURT

[2011] EWHC 242 (Admin)

16 February 2011

Ouseley J

Estates Gazette 4 June 2011

[2011] 22 EG 104

Town and country planning — Planning agreement — Enforceability — Town and Country Planning Act 1990 — Developer entering into planning agreement under section 106 of 1990 Act — Terms of agreement requiring financial contributions to infrastructure improvements necessitated by development — Contributions calculated in accordance with supplementary planning guidance — Guidance changing in response to criticisms from planning inspectors on planning appeals — Whether local planning authority entitled to enforce payment of contributions under agreement

The claimant obtained planning permission from the defendant council for a residential development on land in Thatcham. By a planning agreement entered into by the parties under section 106 of the Town and Country Planning Act 1990, the claimant was to contribute towards infrastructure costs flowing from the development by making payments at specified times calculated according to the defendants' supplementary planning guidance (SPG); the matters to which the payments related included educational and healthcare facilities and improvements to open spaces, library services and transport.

Various sums fell due and were paid under the agreement. The defendants subsequently changed their SPG in response to criticisms of their methods of calculating contributions made by planning inspectors in their decisions on successful appeals by developers, including the claimant, against refusals of planning permission.

The claimant considered that it should not be required to pay further instalments under the section 106 agreement, on the grounds that no further contributions would be required on an application of the new SPG guidance and that a change in circumstances had rendered unnecessary any contribution towards educational provision. It brought judicial review proceedings to challenge the defendants' decision to bring proceedings to recover £47,375 in outstanding contributions as a debt due under the agreement.

The claimant contended that the defendants were seeking to enforce an unlawful obligation to the extent that they claimed more than could be justified under the new SPG and that they had failed to consider whether any planning purpose connected with the claimant's development would be served by enforcement.

***Held*: The claim was dismissed. It was not unlawful for the defendants to enforce the section 106 agreement. The agreement, which had been lawfully entered into, contained the specific sums that the parties had agreed rather than a method by which they were to be calculated. Accordingly, the basis for the amounts did not form part of the agreement and the enforcement of the agreement did not depend on the lawfulness of the method of calculation. The sums had been lawfully demanded because the claimant had**

agreed to pay them. Moreover, the claimant had not challenged the lawfulness of the SPG on which the contributions were based. The revision of the SPG in response to criticisms did not show that the earlier SPG was unlawful, and the inspectors' decisions had contained no finding to that effect. Further, where the section 106 agreement, on its true construction, required the contribution towards educational provision to be paid in circumstances where the effect that had originally justified its calculation in SPG terms no longer existed, the defendants were entitled to hold the claimant to its agreement. Enforcement of the agreement was not rendered unlawful by an absence of a useful planning purpose to which the "surplus" contributions could be put. Even assuming that the enforcement of the payments was lawful only to the extent that they would continue to serve a useful planning purpose, that purpose did not have to relate to the development in connection with which the section 106 agreement had been entered into. There was no reason why the useful planning purpose still being served should not be different from that which had led to the agreement in the first place; the statutory provisions contain no such limitation and none can be imported by necessary implication. The defendants' proposed use of the contributions could not be criticised.

The following cases are referred to in this report.

Pye (JA) (Oxford) Ltd v *South Gloucestershire District Council (No 1)*; *sub nom JA Pye (Oxford) Ltd* v *South Gloucester District Council* [2001] EWCA Civ 450; [2001] 2 PLR 66

R (on the application of Batchelor Enterprises Ltd) v *North Dorset District Council* [2003] EWHC 3006 (Admin); [2004] JPL 1222

R (on the application of The Garden & Leisure Group Ltd) v *North Somerset Council* [2003] EWHC 1605 (Admin); [2004] 1 P&CR 635

This was the hearing of a claim by the claimant, Renaissance Habitat Ltd, for judicial review of a decision by the defendants, West Berkshire District Council, to bring proceedings to enforce an agreement under section 106 of the Town and Country Planning Act 1990.

Richard Harwood (instructed by Horsey Lightly Fynn, of Bournemouth) appeared for the claimant; Thomas Jefferies and Charles Banner (instructed by the legal department of West Berkshire District Council) represented the defendants.

Giving judgment, **Ouseley J** said:

[1] Renaissance Habitat Ltd, the claimant, is a small developer carrying on business in West Berkshire. It obtained planning permission, on 14 December 2005, for the development of land for residential purposes at 69 Bath Road, Thatcham, West Berkshire. As was commonplace for the claimant and other developers in West Berkshire, it entered into an agreement with the defendants, West Berkshire District Council, under section 106 of the Town and Country Planning Act 1990 (the 1990 Act) on the same day as it was granted planning permission. Planning permission would not have been granted unless it had entered into that section 106 agreement.

[2] Under the agreement, the claimant agreed to pay money to the defendants by way of contributions towards the infrastructure costs that they would incur in dealing with the effect of the permitted development. The agreement specified the sums due and the time at which they were due. The basis on which those sums were calculated was set out in the supplementary planning guidance (SPG) produced by the council in April 2004. In the first week of September 2007, the claimant began development.

[3] Various sums became due under the section 106 agreement in 2008 but not all were paid to the defendants by the claimant. On 13 August 2009, the defendants issued an invoice for £57,792.88 as the balance of the sums due under the agreement. The claimant calculated that the sums it owed were all paid on 8 March 2010 by a payment of £20,236.90, which included interest and indexation. The defendants say that a further £47,375.81 plus interest is due. They issued proceedings in the Queen's Bench Division for the recovery of that amount as a debt due under the contract, that is, under the section 106 agreement. Those proceedings are due to be heard in a trial window commencing 21 March 2011.

[4] The claimant in these judicial review proceedings challenges the decision of West Berkshire District Council to bring those Queen's Bench proceedings, on the ground that, since the December 2005 agreement, the council have changed the SPG, on the basis of which they calculated what developers should pay as infrastructure charges. They have done so in response to decisions of inspectors on planning appeals against refusals of planning permission by the defendants on grounds, among others, that the developer had refused to pay proper infrastructure contributions. Inspectors have rejected some of the defendants' contentions as to what the proper infrastructure contributions should be. Two appeals by the claimant against refusals of planning permission by the defendants were allowed on those grounds on other sites. The disputes were not about the arithmetic but about the method of or inputs into the calculations. Circumstances have also changed since 2005 in respect of the need for educational provision, so far as this particular development is concerned, because there is greater capacity now than was realised or expected at the school in respect of which the educational contribution was calculated.

[5] The claimant contends that if the infrastructure contribution were calculated by reference to the changed SPG as at the defendants' final invoice date of 13 August 2009, no more would be due from the claimant than it had already paid.

[6] Hence, it challenges the defendants' decision to start proceedings on 1 April 2010 for the payment of the balance due under the section 106 agreement on the ground that it is seeking to enforce an unlawful obligation. The defendants were enforcing the agreement for more than they could now justify under their current SPG. It also challenges what it describes as the policy of the defendants not to agree to vary section 106 agreements in the light of changed SPG or changed circumstances, as set out in a letter dated 12 March 2010 sent to the local development industry forum.

[7] Mr Richard Harwood, for the claimant, sought at the last minute to raise a series of other points, not raised by counsel previously instructed. Consideration of many of these issues would have required the disclosure of documents and an adjournment of the hearing. This in turn would have jeopardised the trial window. I refused permission for the amendments, save to the extent that they involved no need for further research by the defendants; in reality, those surviving grounds were no more than illustrations or reformulations of the points already made. Those I refused did not seem likely to advance the essential point anyway and are likely to stand or fall for the most part with it.

Statutory framework

[8] Section 106 of the 1990 Act provides:

> (1) Any person interested in land in the area of a local planning authority may, by agreement or otherwise, enter into an obligation (referred to in this section and sections 106A and 106B as "a planning obligations"), enforceable to the extent mentioned in subsection (3)
>
> (a) restricting the development or use of the land in any specified way;
>
> (b) requiring specified operations or activities to be carried out in, on, under or over the land;
>
> (c) requiring the land to be used in any specified way; or
>
> (d) requiring a sum or sums to be paid to the authority… on a specified date or dates or periodically.
>
> (2) A planning obligation may
>
> (a) be unconditional or subject to conditions;
>
> (b) impose any restriction or requirement mentioned in subsection (1)(a) to (c) either indefinitely or for such period or periods as may be specified; and
>
> (c) if it requires a sum or sums to be paid, require the payment of a specified amount or an amount determined in accordance with the instrument by which the obligation is entered into and, if it requires the payment of periodical sums, require them to be paid indefinitely or for a specific period.

[9] There are two provisions for the discharge of these obligations. For the first five years after entry into the agreement, "the relevant period", it can be varied or discharged only by agreement. This may well reflect the total of five years during which a developer with the benefit of a permission must normally commence development. Thereafter, if the authority refuses the application to vary, the developer can appeal

to the Secretary of State, who can impose a variation or discharge. The current provisions are in section 106A and B:

> S106A (1) A planning obligation may not be modified or discharged except
>
> (a) by agreement between the appropriate authority (see section (11)) and the person or persons against whom the obligation is enforceable; or
>
> (b) in accordance with this section and section 106B
>
> …
>
> (3) A person against whom a planning obligation is enforceable may, at any time after the expiry of the relevant period, apply to the appropriate authority for the obligation
>
> (a) to have effect subject to such modifications as may be specified in the application; or
>
> (b) to be discharged.
>
> …
>
> (6) Where an application is made to an authority under subsection (3), the authority may determine
>
> (a) that the planning obligation shall continue to have effect without modification;
>
> (b) if the obligation no longer serves a useful purpose, that it shall be discharged; or
>
> (c) if the obligation continues to serve a useful purpose, but would serve that purpose equally well if it had effect subject to the modification specified in the application, that it shall have effect subject to those modifications.

Section 106B deals with appeals and provides:

> (1) Where an authority
>
> (a) fail to give notice as mentioned in section 106A(7); or
>
> (b) determine that a planning obligation shall continue to have effect without modification,
>
> the applicant may appeal to the Secretary of State
>
> …
>
> (4) Subsections (6) to (9) of section 106A apply in relation to appeals to the Secretary of State under this section as they apply in relation to applications to authorities under that section
>
> …
>
> (6) The determination of an appeal by the Secretary of State under this section shall be final.

[10] It was not disputed, at least for these purposes, that section 106A(6)(b) and (c) meant that the authority could discharge or vary the agreement if it no longer served a useful "planning" purpose, or could serve it equally well in a different form. That word, submitted Mr Harwood, was necessarily implied since the agreement could be made in the first place only for a planning purpose, which is correct, and could be enforced only by a public body acting for a public purpose under the Planning Acts. It was not exercising some private power or purely contractual power. Sullivan J had so held in *R (on the application of Batchelor Enterprises Ltd)* v *North Dorset District Council* [2003] EWHC 3006 (Admin)*.

[11] I am prepared for present purposes to accept that point, but I note that "planning", the word implied, very broad though it is, may lead to a debate concerning what constitutes a planning consideration for these purposes as opposed to some other useful public purpose that could be pigeonholed under some other head, or even a private purpose such as the protection of private views, which may show the implied restriction to be unjustified. Sullivan J also relied on ministerial guidance that in fact contradicts this interpolation since it said that an agreement should "normally", rather than "always", be discharged when there is no planning purpose to be served by its continuance. The learned editors of the *Planning Encyclopaedia*, without explanation, and both before and after *Batchelor*, have said that no planning purpose was necessary. Curiously, there is no express obligation to discharge if conditions section 106A(6)(b) or (c) are satisfied, though "may determine" could be so interpreted.

* Editor's note: Reported at [2004] JPL 1222

Section 106 agreement and facts

[12] The recitals record that the development can be permitted but that the defendants "desire" that certain planning obligations be entered into: contributions towards educational facilities as defined, towards healthcare "throughout the district", towards improvements to "existing open space in Thatcham" as defined, towards the costs of providing stock for the library service throughout the district and towards the costs of carrying out various transport improvements in the vicinity of the site and in West Berkshire more widely.

[13] The obligations bit if the development were commenced. The payment obligations in clause 1 were as follows:

> (a)… (i) to pay to the Local Authority the sum of… (£14,936.00) as a contribution to meet the impact of the Development on the secondary catchment school… such sum to be paid as to 50% thereof no later than fourteen days from the date of commencement of the Development and as to the remaining 50% thereof not later than six months from the date of commencement of the Development…
>
> (ii) to pay to the Local Authority not later than fourteen days from the date of commencement of the Development the sum of… (£1,320.00) as a contribution towards the cost of the Local Authority of carrying out the Healthcare Provision.
>
> (iii) to pay to the Local Authority not later than the date on which the Development or any part thereof is first used and occupied as such the sum of £33,064.00 as a contribution towards the costs of the Local Authority of carrying out improvements to existing public open space in Thatcham.
>
> (iv) to pay to the Local authority not later than fourteen days from the date of commencement of the Development the sum of £1,360.00 as a contribution towards the cost of the Local Authority of carrying out the Library Provision
>
> (b) pursuant to Section 106 of the Planning Act and Section 278 of the Highways Act and all other enabling powers:
>
> (i) to pay to the Local authority not later than the date upon which the Development or any part thereof is first used or occupied as such £14,400.00 as a contribution towards the costs of carrying out the Highway Works.

Each obligation involved further payments by reference to increases in RPI.

[14] When the claimant commenced development, the first half of the education contribution became due and was paid. The second indexed half (£8,168.00) was due on 1 March 2008 but was not paid. On 1 June 2008, the open space and highway contributions became due but were not paid. Index linking had raised these to £36,794.00 and £16,026.00 (rounded) respectively. The balance, including interest, as the claimant sees it was paid in March 2010.

[15] SPG changed in the following ways. In May 2006, the defendants removed a rounding-up process whereby forecast pupil numbers per household were rounded up to the next highest rather than the nearest whole number. On 3 September 2007, the defendants lost an appeal by the claimant against the refusal of planning permission. Their infrastructure calculations were criticised by the inspector for using an average occupancy rate for all dwellings, which produced a higher occupancy rate than was appropriate for the particular type of development, and for using an out-of-date and higher census figure for average occupancy anyway. This affected the library, health and open space contributions. The open space contribution was excessive and the calculation for equipped play space unreasonable in that case. SPG was changed in October 2007 to reflect those points. There was a very marked reduction in the amounts required for open space from developers.

[16] In August 2009, following another successful developer's appeal in March 2009 and after all the sums had become due in respect of 69 Bath Road, SPG was further revised to reflect the inspector's criticisms of the defendants' assumption that new housing would generate more children than old housing and would do so even where the development consisted of flats replacing family houses, contrary to the defendants' own data.

[17] If the 69 Bath Road contributions had been assessed at 12 August 2009 on the current SPG, the highway contribution would have increased from £15,000 to £18,000. The education contribution would have gone down from £14,936 to £405 because there was capacity at the primary school for the catchments and the situation

at the catchment secondary school had changed since the grant of permission; there was now capacity and so no contribution would be sought. The sum was not reduced to nil but to £405 because of an SPG requirement, new in 2008, that a sum be paid towards special needs education. The open space contribution would have gone down from £33,000 to £10,496.

[18] The claimant challenged the defendants in correspondence in December 2009 to explain what they would do with the "surplus" contribution that they would now receive. Although the highway contribution was lower than SPG would now require, the defendants explained the variety of transport improvements in Thatcham town centre to which the money received for highways would go: namely cycle, pedestrian, public transport and junction improvements. The sum received for education would contribute to the "re-provision and expansion of the secondary school's indoor sports provision or expansion of the sixth form provision". The open space contribution would be allocated to the programme of improvements at a local sports ground, subject to members' approval.

Nature of the case

[19] There was and is no challenge to the lawfulness of this agreement in the first place, nor to its materiality to the grant of permission. The claimant was well aware of the SPG when it entered into the agreement, and SPG had in fact been the subject matter of public consultation. The topic papers that explained the formulae that led to the specific sums in respect of education, open space, health and so on were publicly available, although they had not been the subject of public consultation. The claimant could have refused to enter into the agreement and appealed against the consequent refusal of planning permission on the ground that the infrastructure contributions it refused to pay were unreasonable and did not justify a refusal of permission. It did so, successfully, on two other occasions, but it did not go down that route on this development. The parties could have agreed provisions that enabled obligations to be adjusted with changes to SPG or in specified circumstances, but did not do so.

[20] This is not a challenge to any refusal of the defendants to discharge or vary the agreement under section 106A, although that would have tested directly the lawfulness of any such refusal. That route has not been followed. The issue is raised by the sidewind of a claim that enforcing the agreement, unvaried and undischarged, is unlawful. Its legal scope is therefore narrower than a challenge to a refusal to vary or discharge the agreement. It is simply that, since the sums due would not be calculated in August 2009, as they had been in 2005, because of what inspectors later thought were unreasonable inputs or because of changed circumstances, they could not lawfully now be demanded. Section 106B and the appeal provisions are not in play.

[21] The case does not turn either on whether the section 106 agreement on its true construction means that the sums claimed are due. This case assumes that some or all of those sums are due on the true construction of the agreement, but it is not for me to rule on that. I mention it because the way the different obligations are expressed may give rise to argument: does the educational contributions clause permit an argument that if the school in question had or already has capacity without more, it envisaged that no contribution would be payable?

[22] The date chosen by the claimant for the purposes of assessing how much is due does not necessarily reflect what the position was as at the various dates when sums were due under the agreement. The picture has been assessed at a single date and not at the dates when particular payments were due under the terms of the agreement, and there has been no examination of precisely how SPG stood at those particular dates. That does not affect the principle of the challenge, although that may be how any challenge should have been framed for which it is out of time; but it does illustrate a larger point. The sum that would have been due as a contribution to education, if it varied over time with SPG, is greater as at the August 2009 date selected, albeit not by much, than was due at an earlier stage. That works in the defendants' favour if they are not entitled to enforce the agreement as it stands. However, the highway contribution would have increased, and that has not been offset in the calculation undertaken by the claimant. So the date or dates at which **G** the enforcement of the obligation has to be judged for lawfulness can be important on the claimant's arguments and the lawfulness of any offset can illustrate the effect of confining the useful planning purpose to the purpose for which the contribution was payable.

Submissions

[23] Mr Harwood's case was that the SPG charges that were embodied in the agreement were based on "erroneous/unreasonable calculations, and therefore the charges themselves constitute unlawful demands for payment". He put it in other ways: the council had failed to consider whether any land-use planning purposes would be achieved by enforcement. **H** Enforcing the terms of the agreement had to achieve a useful planning purpose. The ability of the defendants to make some public use of the money was not the same as them having a planning use. As his argument developed, and in the light of how the defendants said that they would now use the money, he said that the planning purpose had to be related to the development to which the section 106 agreement was connected. There was no effect on schools from the development and so no effect to be absorbed by the claimant's contribution. He accepted, when pressed, that a useful planning purpose would be served by attributing a surplus contribution for one purpose to make good a deficient contribution for another.

J

[24] Mr Thomas Jefferies, leading for the defendants, submitted that there was no need for the section 106 agreement, when initially entered into, to relate to any development proposal so as to be a material consideration to the grant of permission. It needed to be entered into only for the purpose of restricting or regulating the use or development of land, and not for the purpose of any particular permission. It needed to have that relationship only were it in fact to be taken into account in the decision of whether to grant permission. He relied on the decision of the Court of Appeal in *JA Pye (Oxford) Ltd* v *South Gloucestershire District Council (No 1)* [2001] EWCA Civ 450*. That case was a stark example of the enforcement of an agreement that regulated the use of land, providing for the funding by the developer of the acquisition of **K** land for a road. It was assumed for the purposes of the challenge to the enforcement of the agreement that the provision of the road was immaterial to the development that had led to the agreement. The fact that it might have been unlawful for the defendants to have had regard to the agreement when granting permission did not mean that it was unlawful to enforce the agreement. On that basis, there was no scope for construing the agreement so as to require the contribution to be connected to any effect of the development. However, in any event, the sums were to be used for a planning purpose and for a planning purpose that would at least benefit the people who would live in the new dwellings. **L**

Conclusions

[25] There appear to be two related aspects to Mr Harwood's submissions: unlawfulness by reference to the cause of the "surpluses" and unlawfulness because of the purpose to which the "surplus" contributions will be put. I deal with those in turn.

[26] Is the decision to enforce the agreement unlawful because of the subsequent changes to SPG, the reasons for the changes or because of changes in circumstances? In my judgment, it is not unlawful. First, the defendants are simply seeking to enforce an agreement that was incontestably lawful when entered into. **M** The claimant could have contested the merits had it wanted before an inspector, so it was not forced to enter the agreement. It did not challenge its lawfulness before starting development, nor at any subsequent stage when payments were due nor when SPG changed or educational circumstances changed. It did not avail itself of the statutory means of seeking a variation or discharge and challenging as unlawful any refusal of variation or discharge. The parties could have stipulated for the changes that occurred. It would be very strange if enforcing the agreement were unlawful in those circumstances.

---

* Editor's note: Reported at [2002] 2 PLR 66

[27] Second, the lawfulness of enforcing the agreement does not, in my view, depend on the lawfulness of the calculations or method whereby a sum was put forward by the defendants and agreed by the claimant. The agreement simply contains the specific sums and not a particular method whereby sums are to be calculated. So the basis for the amounts, which is what is said to be unlawful, does not form part of the agreement at all. The sums demanded are lawfully demanded because the claimant agreed to pay them. The claimant in any event knew of the SPG and could see the topic papers. It could ask about the method if it did not know. It could appeal against a refusal of permission based on excessive demands.

[28] Third, no challenge has been brought to the lawfulness of the SPG that underlies the section 106 agreement, although a challenge could have been made in 2005 or 2007. The mere fact that SPG has changed in response to criticisms does not show that the earlier SPG was unlawful. The fact that inspectors found that the basis of the calculations of contributions led to sums that did not fairly or reasonably relate to the effects of the development, and were therefore contrary to government policy, does not and could not by itself amount to a finding that the SPG was unlawful. I am not prepared to accept that the inspectors' comments show them to be unlawful. Although the calculation methods as explained to me favoured the defendants unduly, I am not prepared to hold them unlawful on the limited material I have. SPG had to be changed only because inspectors would not support it. The contributions sought were characterised in argument as "erroneous or unreasonable"; that does not mean they were unlawful in public law terms.

[29] Fourth, if the section 106 agreement, on its true construction, does require the educational contribution to be paid in changed circumstances where the effect that originally justified its calculation in SPG terms no longer exists, the developer is simply being held to its agreement. Those changed circumstances do not make enforcement unlawful. This moreover invites the court to rewrite the agreement without going through an available statutory process, judicially reviewable, in which variations can be negotiated and refusals to change the terms justified in law.

[30] The fallacy underlying this aspect of Mr Harwood's submissions is that there is a necessary unlawfulness in enforcing an agreement that the defendants could not now force the developer to enter as the price for permission, since the developer would succeed in an appeal against the refusal of planning permission. It simply would not be unlawful for the defendants and claimant now to enter precisely this same section 106 agreement and for the defendants to enforce it. The SPG changes and changes in circumstance simply do not and would not make that agreement unlawful under section 106, nor its enforcement.

[31] The other aspect of Mr Harwood's argument as to unlawfulness concerns the absence of a lawful purpose within section 106 to which the "surplus" contributions will be put. On the assumption that enforcement of the payments due is lawful only to the extent that those payments would still serve "a useful planning purpose", Mr Harwood's arguments remain without substance.

[32] First, the useful planning purpose does not have to be related to the development in connection with which the section 106 agreement was entered into. The decision in *Pye* is quite clear, and I accept Mr Jefferies' submissions, set out above, as to its effect. Since there is no need for a connection between the agreement and the permission, in the first place, in order for the agreement to be lawful and enforceable, it is quite impossible to imply into section 106A(6)(b) and (c) on variation or discharge a requirement that the "useful planning purpose" must be one related to the permission itself. Since there is no need for the agreement to have any connection to a permission or a particular development, the variation or discharge power cannot be constrained in a way in which the power to enter the agreement in the first place is not. There is nothing in the 1990 Act that requires variation or discharge for want of useful planning purpose to be judged by reference to the development to which the agreement was related. There is nothing unlawful about enforcing an agreement in circumstances that would not warrant its variation or discharge.

[33] There is therefore, second, no reason why the useful planning purpose still being served should not be a different one from that which led to the agreement in the first place; the statutory provisions do not contain such a limitation and it is not there by necessary implication. Richards J took a similar view in *R (on the application of The Garden & Leisure Group Ltd)* v *North Somerset Council* [2003] EWHC 1605 (Admin)*, in [46]. There is nothing of course to stop the parties from agreeing to provisions that cover changes in circumstances, if they wish.

[34] Since there is no requirement that the useful planning purpose relates to the development itself, there can be no requirement either that it relates to an effect of the development at all, or to the same effect for which it was originally sought.

[35] There is no unlawfulness if the "surplus" contributions are applied for a useful planning purpose to the benefit of the area where the new residents will live, and from which they can benefit, even if they are not spent on the particular aspect of the development's effects for which they were originally sought. In so saying, I am not holding that to be a necessary test of lawfulness, merely that it cannot be unlawful to do so.

[36] Third, on the facts here Mr Harwood's own case is unsustainable. Each of the useful planning purposes on which the defendants are proposing to spend the "surplus" contributions is related to the development in at least this sense, that the people living in the development will benefit from the useful planning purposes to which the contributions will be put. That is sufficient, even on Mr Harwood's approach to the scope of section 106. However, the defendants' case is stronger than that.

[37] The defendants propose to spend the "surplus" open space contribution on improving open space provision locally in Thatcham. It is impossible to contend that that is not a useful planning purpose and one related to the development. It does not cease to be a useful planning purpose simply because the basis on which the agreed sum was calculated was found to be unfair or unreasonable by an inspector. (It seems, although it is not for me to decide whether this precludes a defence to the claim, that this proposed expenditure falls within the unvaried terms of the obligation anyway.). The educational provision on which the "surplus" is to be spent is undoubtedly a useful planning purpose related to the development; it is not for me to judge whether it falls outside the contractual provision so as to afford a defence to the claim as a matter of contract.

[38] The claim that the decision to enforce the section 106 agreement is unlawful is untenable.

Policy letter

[39] The policy challenged in the letter dated 12 March 2010 to the local development industry forum from the defendants' development control manager clarifies what he said at a meeting over a month earlier about how the defendants would approach applications to modify section 106 agreements. It said:

> For avoidance of doubt I would like to clarify what will be the Council's general position from now onwards. Where a developer applies to modify a section 106 planning obligation, on the basis that the Council's SPG relating to the relevant contribution(s) has changed since the date of that planning obligation, the Council's general position will be to reject such an application if the development has been implemented. By the implementation of the development, the developer will have obtained the benefit of the planning permission in question and it would not be appropriate to reopen the basis on which the associated s106 obligations were negotiated. Section 106A(1) makes clear that on an application to modify a planning obligation the test is not whether the obligation remains necessary to make the development acceptable but simply whether it still serves a useful purpose. The provision of infrastructure contributions plainly serves a useful purpose.
>
> Whilst this is intended to be a statement of the Council's general approach, it must be emphasised that all cases will be considered on their individual merits.

---

* Editor's note: Reported at [2004] 1 P&CR 635

[40] Mr Jefferies submitted that this letter was not reviewable because it did not even constitute a policy. I disagree.

[41] Mr Harwood criticised it as embodying the same refusal to revise agreements when the agreed infrastructure contributions have been based on formulae that have been rejected by inspectors and then changed; thus, it was a policy of refusing to vary agreements to the extent that they no longer served a useful planning purpose. I reject that argument; it merely traverses the same ground as I have already covered. A change in SPG or circumstances such that the original charge would not now be sought in a new agreement does not mean that no useful planning purpose is to be served by enforcing the agreement or anything unlawful in doing so.

[42] I do not think that the omission of the word "planning" as controlling the nature of the useful purpose can be criticised as showing an error of law; after all, the letter is quoting the wording of the 1990 Act. Nothing in the defendants' behaviour or submissions suggested that they were approaching modification without examining whether a useful planning purpose was still being served.

[43] Mr Harwood developed a further argument that the policy amounted to a general refusal to consider modifications after implementation of the permission. Yet the 1990 Act envisages C
that applications for modification or discharge will be made after implementation, that is, after the five-year period during which planning permission would normally have been implemented. There is nothing in this point. The letter is dealing with the defendants' approach to agreeing modifications in the light of changing SPG. It is dealing with a confined area. It will normally agree to a modification to reflect the change before implementation, but not after. The reason it gives is justifiable. During the five years after the agreement was entered into, it is entitled to adhere to the agreement, and for it normally to do so after implementation is obviously of itself not unlawful. After that five-year period, the defendants are unlikely to agree to a change to D
reflect changed SPG, but the developer has the remedy of appeal. If the developer's case is strong or of a type in which the Secretary of State has made his views known, the defendants will not succeed and may modify their views. If the complaint is that the developer has to exercise the statutory remedy, and take its chances, I see nothing remotely wrong with that.

[44] This application is dismissed.

Claim dismissed.

TRUST OF LAND

The Trustees Corporate Trustees Ltd and another *v* Capmark Bank Europe plc

COURT OF APPEAL

[2011] EWCA Civ 380

1 February 2011

LORD NEUBERGER MR, RICHARDS AND HUGHES LJJ

Estates Gazette 2 July 2011

[2011] 26 EG 86

Trust of land — Charge — Priority — Trustee's lien — Property unit trust formed as vehicle for acquisition of property — Purchase financed partly by bank loan secured by charge over trust property under deed of guarantee and debenture entered into by trustees — Trustees incurring substantial liabilities for rates on property — Bank appointing receivers to sell property — Insufficient funds realised to repay secured loan in full — Whether trustees' right to be indemnified out of trust assets for rates liability taking priority over bank's security as chargee — Construction of deed of guarantee and debenture

The appellants were the trustees of a property unit trust that had been formed as a vehicle for the sale and purchase of certain warehouse premises on behalf of a company. The purchase price of £28.1m was funded partly by the company and partly by a £21m loan from the respondent bank, secured by a deed of guarantee and debenture between the appellants and the respondent, which was registered as chargee of the property. The respondent was closely involved in the purchase transaction since it owned a 50% interest in the company.

At the time of purchase, the property was subject to a lease at a rent that was sufficient to meet all outgoings on the property and the administration of the trust. However, after the lease ended in 2008, the property remained largely empty and the trustees became liable for unoccupied property rates in excess of £1.3m. The respondent's loan was not repaid and it appointed receivers, who sold the property for £14.63m. The sale proceeds were insufficient to repay the loan in full.

The appellants applied for a declaration that their entitlement be indemnified out of the trust assets for all expenses reasonably and properly incurred, including their liability for the rates, and their lien over the trust property for that purpose took priority over the respondent's security under the deed of guarantee and debenture either by the express terms of the deed or by implication. They relied, in particular, on clause 17.6 of the deed, which stated, *inter alia*, that "Notwithstanding any other provisions of this Deed", the trustees had executed it solely in their capacity as trustees and with the intention of binding "the assets of the Unit Trust" and that, save for those trust assets, no liability should attach to or be enforceable against the assets of the trustees held in their capacity as trustees or otherwise. The claim was dismissed in the High Court: see [2010] EWHC 1605 (Ch). The appellants appealed.

***Held*: The appeal was dismissed. Clause 17.6 did not have the effect of giving priority to the appellants' rights of indemnity and lien under the trust instrument over the respondent's rights as chargee. Clause 17.6 had to be read in its legal and commercial context in the light of the purpose of the deed, which was to enable F trustee-owners to charge the property as security for a loan. It protected the appellants from the liability that a trustee, entering into a contractual obligation to another party in that capacity, would otherwise incur to enforcement of the obligation against itself and its assets as though it had entered into the contract on its own account. In the absence of any stipulation to the contrary, the respondent could have had recourse against each appellant under the relevant clauses of the debenture to the full extent of their assets, whether held by each of them as trustees under the trust, beneficially or possibly as a trustee of other trusts. The G purpose of clause 17.6 was to ensure that the respondent could not pursue the trustees beyond the trust assets, thereby providing protection to them in their personal capacity and in respect of assets held by them for their own benefit and for other trusts. It was concerned only with the appellants' liability under the debenture, not with protecting the appellants in respect of their liability to a third party. Particularly where the property was charged to the bank unconditionally, the clause could not properly be construed as confining the "assets of the Unit Trust" only to the net assets, subject to the trustees' right of indemnity and lien. Nor was it possible to imply a term to that effect. Such a term was not H necessary for business efficacy, was not obvious and conflicted with an express term of the debenture dealing with the order of priority in which any moneys received by a receiver should be applied. The fact that the appellants were bare trustees, and that the respondent was an indirect beneficiary of the trust as well as the secured lender to the trustees, did not justify departing from the clear meaning of the deed. Moreover, the absence of any provision for the appellants' lien to take priority was a significant factor against its implication, particularly where the terms of the trust instrument had been the subject of negotiation and agreement with the respondent, and the question of the trustees' expenses had therefore been considered.** J

The following case is referred to in this report.

Dominion Corporate Trustees Ltd v *Capmark Bank Europe plc (formerly GMAC Commercial Mortgage Bank Europe plc) (now Capmark Management plc)* [2010] EWHC 1605 (Ch)

This was an appeal by the appellants, The Trustees Corporate Trustees Ltd and The Trustees Trust Ltd, from a decision of David Richards J, sitting in the Chancery Division of the High Court, dismissing a claim against the respondent, Capmark Bank Europe plc, for declaratory relief as to the meaning and effect of a deed of guarantee K and debenture charging trust property as security for loan.

James Ayliffe QC (instructed by Teacher Stern LLP) appeared for the appellants; Raquel Agnello QC and Philip Hanks (instructed by Lawrence Graham LLP) represented the respondents.

Giving judgment, **Lord Neuberger MR** said:

[1] This is an appeal from a decision of David Richards J, brought with the judge's permission. It raises a point on the interpretation of a debenture granted to a bank over property held by trustees: the point is whether the judge was correct to reject the trustees' contention that their right to be indemnified in respect of expenditure under the trust deed

A creating the trust, had priority over the bank's rights as chargee under the debenture.

Introductory

[2] The appellants, Dominion Corporate Trustees Ltd and Dominion Trust Ltd (the trustees) are the trustees of a Jersey-based property unit trust called the Maylands Unit Trust (the trust). The trust was formed, for tax reasons, as the vehicle for the acquisition of warehouse premises at Hemel Hempstead (the property). The purchaser, Cantabria Investments Ltd (Cantabria), agreed to buy the property for £28.1m and the transaction was completed by the vendor transferring the property B to the trust in exchange for units, which it then sold to Cantabria for £28.1m.

[3] Cantabria raised £7.1m from its own resources, and the balance of £21m by way of a loan secured on the property and arranged by the respondent, Capmark Bank Europe plc (the bank). The security was created by a deed of guarantee and debenture (the debenture) between the trustees and the bank, as security agent. The lender was a different company in the same group as the bank, but, for present purposes, it can be treated as the bank.

[4] As trustees of the trust, the trustees were the registered proprietors of the freehold interest in the property. The bank was the C registered chargee of the property. It should be added that the bank was not simply a normal arm's-length secured lender, in the sense that it was closely involved in the entire transaction: it owned a substantial (around 50% or more) interest in Cantabria, and therefore it had a substantial, if indirect, interest in the units issued in the trust.

[5] At the time of the trust's acquisition of the property, it was effectively subject to a lease at a rent that was more than sufficient to meet all outgoings on the property and the administration of the trust. However, the lease ended in April 2008, whereafter the property remained empty, save between November 2008 and February 2009, when it was occupied by Royal Mail as a licensee.

D [6] A change in rating law that came into force on 1 April 2008 had the effect of imposing unoccupied property rates on premises such as the property after they had been unoccupied for six months. Liability for such rates falls on the persons entitled to possession. As a result, the trustees, as the registered freehold proprietors of the property, incurred a liability for rates (the rates) in excess of £1.3m, details of which are set out in the judgment below: see [2010] EWHC 1605 (Ch), in [4].

[7] Following the failure to repay the loan, the bank appointed receivers, who sold the property for only £14.63m in February 2010. There are therefore insufficient funds to repay the secured loan in full, even without taking account of the trustees' claim to be indemnified E against their liability for the rates.

[8] As trustees, the trustees are entitled to be indemnified out of the trust assets for all expenses reasonably and properly incurred by them (and they have a lien over the trust assets for that purpose), and it is not in dispute that the liabilities for the rates fall into that category. The issue is whether their right of indemnity ranks in priority to the bank's security under the debenture.

[9] The trustees' contention, that their right of indemnity ranks ahead of the bank's security, turns on the interpretation of the debenture. However, the trustees rightly accept that there is no "general principle that charges granted by trustees over trust property are subject to the F trustees' lien for their own expenses and they accept that their rights cannot prevail over the bank's rights as chargee unless the relevant contract, the debenture, so provides", to quote from the judgment below: see [7].

Trust instrument

[10] The trust instrument constituting the trust is dated 17 March 2006, and it provided in clause 2.1:

> The Trust Fund shall be constituted out of the proceeds of issues of Units in accordance with this Trust Instrument. All cash and other property for the time being held by the Trustees pursuant to this Trust Instrument… shall be held as a single common fund (the "Trust Fund") upon trust for the Holders [of the] Units… subject to the provisions of this Trust Fund. The Trust Fund shall be applied and otherwise dealt with by the Trustees in accordance with the provisions of this Trust Instrument. …

[11] Clause 13.7 empowered the trustees to create charges in terms that permitted the grant of the debenture. Clause 15 dealt with the payment of trust expenses. Clause 15.1 defined "Permitted Expenses" in terms wide enough to include the rates, and clause 15.2 provided that, subject to irrelevant exceptions, all permitted expenses should be payable out of the trust fund.

[12] By clauses 25.8 and 25.9, the trustees were given a right of indemnity out of the trust fund and a lien over the trust fund for "all liabilities and expenses properly incurred" as a result of, and "all actions, proceedings, costs, claims and demands" in respect of, their trusteeship, and clause 31 gave effect to this right of indemnity and lien in the event of termination of the trust. As mentioned, this indemnity extended to the trustees' liability for the rates.

[13] The trustees' fees were set out in clause 23.1: £3,500 was payable on the execution of the trust instrument, £6,500 for services provided during the first month after execution of the trust instrument and thereafter variable fees were payable in respect of each quarter at rates to be agreed.

Debenture

[14] The £21m loan was advanced under a facility agreement dated 21 March 2006, pursuant to which the loan was repayable on 21 April 2008 or, at the borrower's option, 20 March 2009. The debenture was executed next day.

[15] The definitions included the following:

(a) "Charged Assets":

> the assets and undertaking of [the Trustees] from time to time mortgaged, charged or assigned (or intended to be mortgaged, charged or assigned) by way of fixed and/or floating security as security for the payment or discharge of all or any part of the Secured Liabilities, and "Charged Asset" shall be construed accordingly.

(b) "Secured Liabilities":

> all present and future obligations and liabilities, whether actual or contingent and whether owed jointly or severally and whether as principal or as surety or in any other capacity whatsoever, of [the Trustees] to [the Bank] under or in connection with the Finance Documents (including this Deed)…

(c) "Security":

> a mortgage, charge, security assignment, pledge, lien or other encumbrance security interest securing any obligation of any person or any other agreement or arrangement having a similar effect…

(d) "Trust Fund" had the meaning given by clause 2.1 of the trust instrument.

(e) "assets" had a wide meaning, including "undertaking, assets, accounts, revenue, and rights of whatsoever nature… and every kind of interest…".

[16] Clause 2.1.1 contained a fixed charge over the property "for the payment, performance and discharge of the Secured Liabilities". Clause 2.1.2 created a "first fixed charge" over a variety of assets, including book debts, and "all other assets comprised in the Trust Fund not otherwise mortgaged or charged pursuant to the foregoing provisions…".

[17] Clause 11 was in these terms:

> 11.1 Any monies received by [the bank], any Receiver or any administrator after this Deed has become enforceable shall be applied in the following order of priority…:
>
> 11.1.1 in satisfaction of or provision for all costs and expenses incurred by [the bank], any Receiver or any administrator and of all remuneration due to any Receiver or administrator;
>
> 11.1.2 in or towards payment (in the order specified in… the Facility Agreement) of the Secured Liabilities or such part of them as is then due and payable; and
>
> 11.1.3 in payment of the surplus (if any) to [the trustees] or other person entitled to it.
>
> 11.2 Clause 11.1 overrides any appropriation made by the [trustees].

[18] By clause 12, the trustees agreed "to indemnify [the bank] against all… losses… incurred by or on behalf of [the bank]… or any Receiver in the execution or purported execution of any of the powers… pursuant hereto…".

[19] By clause 17.3, the trustees covenanted to "pay, perform or discharge all the Secured Liabilities on the due date therefore", and, by clause 17.4.1, they guaranteed the prompt payment, discharge and performance of the secured liabilities.

[20] Clause 17.6, on which the trustees particularly rely, provided:

> Notwithstanding any other provisions of this Deed:
>
> 17.6.1 [The trustees have] executed this deed solely in [their] capacity as trustee[s] of and with the intention of binding the assets of the Unit Trust… from time to time (the 'Trust Assets');
>
> 17.6.2 the aggregate of all liabilities of [the trustees] under this Deed shall at all times and for all purposes extend only to the Trust Assets;
>
> 17.6.3 in no circumstances shall any liability attach to or be enforced or enforceable against the assets of [the trustees] (held in their capacity as trustees of any other trust or in their personal capacity or in any other capacity whatsoever) other than the assets which comprise the Trust Assets; and
>
> 17.6.4 all representations, warranties, undertaking, obligations and covenants in this Deed are made, given, owed or agreed by or in relation to the Trust Assets and in [the trustees'] capacity [y] as trustees of the Unit Trust and, for the avoidance of doubt, shall not be construed to be made, given, owed or agreed by or in relation to [the Trustees] in their capacities as trustees of any other trust or in their personal capacity or in any other capacity whatsoever (other than in their capacities as trustees of the Unit Trust).

Summary of conclusions

[21] The trustees' case, that their right of indemnity and lien ranks ahead of the bank's charge, rests on the construction of the debenture. They base their case, primarily, on the express effect of clause 17.6 of the debenture (clause 17.6), but, alternatively, on a term to that effect to be implied into the debenture.

[22] In my view, the judge was right to reject both arguments raised by the trustees. I would summarise my reasons as follows.

[23] So far as the primary argument is concerned, it seems to me clear that clause 17.6 was concerned with the trustees' liability to the bank or, to put the same point another way, it was concerned with the trustees' liability under the debenture. The clause was not concerned with protecting the trustees in respect of their liability to a third party. Accordingly, the clause did not have the effect of giving the trustees' rights of indemnity under the trust instrument priority over the bank's rights as chargee of the property. As to the alternative argument, the alleged term is not necessary for business efficacy, it is not obvious and it appears to conflict with an express term of the debenture, namely clause 11.

Trustees' first argument: clause 17.6 of the debenture

[24] In the light of the opening seven words of the clause, I accept that clause 17.6 can be said to override other provisions in the debenture, and so the trustees' case would not be undermined simply because their interpretation of the clause appeared inconsistent with another provision of the debenture. However, the clause has to be read in its context. That context includes the legal and commercial circumstances in which the clause was agreed and, more particularly, the fact that it is in a document, which contains other provisions, and whose purpose is to enable trustee-owners of property to charge the property as security for a loan.

[25] In the absence of an express or implied provision (or, exceptionally, some other factor) to the contrary, where a trustee, acting as such, enters into a contractual obligation to another party, that party can enforce the obligation against the trustee, and the trustee's assets, in the same way as though the trustee had entered into the contract on his own account. In this case, the trustees, as trustees of the trust, accepted liabilities to the bank under the debenture for instance in clauses 12, 17.3 and 17.4. In the absence of any stipulation to the contrary, the bank could have had recourse against each of the trustees under those clauses to the full extent of the trustees' assets, whether held by each of the trustees *qua* trustees under the trust or beneficially or, conceivably, as trustee of other trusts.

[26] Hence, it is clear that clause 17.6 was intended to make it clear **G** that the bank could not pursue the trustees beyond the trust assets. As the judge put it, clause 17.6 "recognises that the trustees were contracting in a specific capacity, as trustees of the Trust, and is designed to provide protection to them in a personal capacity and to assets held by them for their own benefit or for the benefit of other trusts": see [31].

[27] However, the trustees contend that the clause does more than this and that, fairly read, it has the effect of giving the trustees' indemnity and lien priority over the bank's charge. By the end of his concise and clear submissions, it became clear that Mr James Ayliffe QC was resting the trustees' case on the definition of "Trust assets" in clause 17.6.1, the possible reach of clause 17.6.2 and the wording of clause 17.6.3, **H** together with the factual matrix in which the debenture was executed.

[28] So far as clause 17.6.1 is concerned, it appears, at any rate at first sight, simply to spell out that the trustees were entering into the debenture as trustees and were intending to bind only the assets of the trust. I find it impossible to accept Mr Ayliffe's argument that the expression "Trust Assets" means the "intention" was to "bind" only what he called the net assets of the trust that is, the assets of the trust subject to the trustees' lien.

[29] First, the natural meaning of "the assets of the Unit Trust", particularly in the light of the wide definition of "assets", is all the assets of the trust. As the judge said, "it is because they are trust assets **J** that the trustees are given a right of indemnity out of them and a lien to secure that right. They do not cease to be trust assets to the extent of the trustees' claim": see [38].

[30] Second, as pointed out by Richards LJ, the "Trust Assets" in clause 17.6.1 are not only the assets that are to be bound but also the assets of which the trustees are described as being trustees. They are clearly trustees of the gross assets of the trust, as Mr Ayliffe was constrained to accept.

[31] Third, by clause 2.1.1, the property was charged to the bank unconditionally, and it seems very hard to spell out of clause 17.6 that that should be read as being a reference to the property subject to the **K** trustees' right of indemnity and lien.

[32] Mr Ayliffe made much of the fact that clause 17.6 referred to, and indeed defined, "Trust Assets", whereas elsewhere in the debenture, the parties were concerned with the "Trust Fund". As with the judge, I consider that there is nothing in the point. The "Trust Assets" are defined in clause 17.6.1 as "the assets of the Unit Trust held by [the trustees] from time to time". It is true that there appears to be little if any difference there between "Trust Assets" and what is comprised in the "Trust Fund", but the fact that the parties have included an unnecessary new definition in clause 17.6 cannot justify giving the clause, or any subclauses of the clause, an unnatural meaning.

L

[33] As to clause 17.6.2, the notion that it is intended to go further than ensuring that the bank has no recourse against the trustees beyond the assets of the trust appears unarguable by the words "under this Deed". Those words seem to make it clear that clause 17.6.2 is concerned only with limiting the trustees' liability to the bank under the debenture, and is not concerned with giving the trustees' liability to third parties, or their right of indemnity or lien under the trust instrument, priority over the bank's rights under the debenture.

[34] Mr Ayliffe suggested that the judge's conclusion would not give effect to clause 17.6.2 if the trust assets were depleted by the trustees exercising their right by indemnifying themselves out of the trust assets before the bank brought a claim against the trustees based on an event **M** before the exercise of the trustees' right of indemnity. There is nothing in that argument, which appears anyway to based on rather unlikely facts. If the trustees' right of indemnity had been validly exercised, the "Trust Assets" would have been validly reduced and the bank could enforce its rights only against those assets, as reduced. That is clear from the words "from time to time" in clause 17.6.1 and the limitations in clause 17.6.3.

[35] Mr Ayliffe also suggested that if clause 17.6.2 had the effect that the judge accepted, it could result in arbitrary unfairness on the trustees. Thus, if, after clause 11.1 applied, the trustees discovered that they had had a liability that had arisen before clause 11.1 applied, they would

be unable to recover on their lien. However, that is the type of problem that arises in any case where there is a charge (such as a floating charge) that becomes crystallised on the happening of an event (often an event of default). In any event, the argument does not assist here, where the money against which the trustees are seeking to enforce their lien is the proceeds of sale of the property, which was the subject of a fixed charge from the inception of the debenture.

[36] That leaves clause 17.6.3. The trustees' case appears to involve reading the words "any liability" as extending to any liability of any nature irrespective of the identity of the person or the basis on which the liability arises, whether to the bank under the debenture, or to any other person on any other basis, subject to the requirement that it is a liability that arises from the trustees' trusteeship of the trust. In particular, of course, the trustees say that the words extend to the liability for the rates. There are a number of problems with that interpretation.

[37] First, it does not accord with the natural meaning of clause 17.6.3 in its context: to my mind, it plainly and naturally refers to liability under the terms of the debenture. Clause 17.6.1 identifies the basis on which the trustees entered into the debenture, namely as trustees of the trust assets. Clause 17.6.2 limits the extent of the trustees' liability under the debenture, namely only to the extent of the trust assets. Clause 17.6.3 then limits the extent to which any claims can be enforced against the trustees, namely only against the trust assets. There can be no doubt that clauses 17.6.1. and 17.6.3 are concerned only with the trustees' liability under the debenture: it seems to me to be unrealistic to suggest that clause 17.6.3 is not intended to be similarly limited in its extent.

[38] Quite apart from anything else, the reference to "any liability" in clause 17.6.3 must be treated as qualified in some way because it would otherwise appear to cover all liabilities of the trustees, whether as trustees of the trust or in any other capacity, which cannot possibly be right; the natural limitation would be to treat it as referring to any liability under the debenture.

[39] An ingenious argument, proffered by Hughes LJ and adopted by Mr Ayliffe, is that the right of lien enjoyed by the trustees is one of their "assets" in the light of the wide definition of that word, and that, accordingly, clause 17.6.3 precludes the bank from enforcing its charge to the extent that it deprives the trustees of their lien. In my view, the lien is not an asset "held in [the trustees'] capacity as trustees of any other trust or in their personal capacity or in any other capacity whatsoever" within the meaning of clause 17.6.3, since those words were not meant to extend to assets held in their capacity as trustees of the trust. Apart from the reasons already given, it seems to me that that conclusion is heavily reinforced by the closing words of clause 17.6.3, "other than the assets which comprise the Trust Assets".

[40] In any event, it seems very unlikely that clause 17.6 was entitled to have a meaning that would directly conflict with the order of priorities set out in clause 11.1.1, 2 and 3. As already mentioned, I accept that clause 17.6 plainly appears intended to take priority over any other provision in the debenture. However, one would have expected some express and clear signal if a fundamental provision such as clause 11 were to be overridden by another clause.

[41] Further, it would seem to follow from the trustees' argument that their lien would rank ahead of any charge or claim that any receiver administrator would have under clause 11.1.1 of the debenture. That seems unlikely, as a matter of commercial common sense, as well as in the light of clauses 11.1.1 and 12.

[42] Mr Ayliffe also relied on the circumstances surrounding the execution of the debenture. In particular, as both parties would have appreciated: (a) the trustees' role was limited to holding the property as bare trustees; (b) the bank was an investor with a significant indirect equity interest in Cantabria; (c) the trust structure was established for the benefit of the shareholders in Cantabria, including the bank; and (d) the trustees entered into the debenture at the request of Cantabria, ultimately for the benefit of those shareholders. In those circumstances, runs the argument, the parties cannot have been intended that the trustees would be exposed to any risk.

[43] The short answer to that point is that these circumstances cannot justify departing from the clear meaning of clause 17.6. There is nothing unusual about bare trustees charging trust property, and the Dominion group of companies, of which the trustees are members, is in a substantial way of business and has been providing offshore corporate services from Jersey since 1998. The fact that the bank was an indirect beneficiary of the trust, as well as the secured lender to the trustees, does not seem to me to assist the trustees' arguments. I cannot see how the various factors that I have just identified assist the trustees' arguments as to the meaning of clause 17.6 of the debenture. The only specific point that Mr Ayliffe made in this connection was that the bank would have been particularly well informed about the trustees' position, but, as the judge said in [28], all the relevant facts were "readily deducible from the relevant documents putting the arrangements in place".

[44] It is also relevant to mention that the trustees' liability for rates has arisen as a result of a combination of circumstances that were not anticipated at the time the trust acquired the property. The trust appeared to have substantial equity in the property, namely some £7m (the difference between the £28.1m purchase price and the £21m loan) and a very substantial drop in value, though obviously possible, would have been unexpected.

[45] Second, and connected to this, it was intended that the property would be held only for a short time by the trustees, so the risk of any exposure would not have been regarded as substantial. It is true that a new trustee would have been similarly exposed, but if such a trustee would have expected to hold the property for a long time, it may well have agreed to take the role only if insurance were provided.

[46] Third, the legislation imposing liability for sums in question came into force only some two years after the trust instrument and debenture were executed. I do not suggest that the liability that has transpired would have been regarded as wholly egregious in March 2006, but it would not have been expected. There is therefore not much force in the contention that the parties cannot have intended the trustees to be uncompensated for the rates liability, quite apart from the fact that there is at least as much force in the point that the parties cannot have intended the bank's security to be reduced by any such rates' liability.

Trustees' second argument: An implied term

[47] Having set out in some detail the arguments bearing on the contention that clause 17.6 gives priority over the bank's charge to the trustees' indemnity and lien under the trust instrument, I am bound to say that the implied term for which the trustees contend appears to me to be plainly unsustainable.

[48] Mr Ayliffe's argument, as developed orally on this aspect, amounted to this: that if it were apparent from clause 17.6 that the parties had intended the trustees' lien to have priority over the bank's charge, but that this had not been spelled out clearly enough to be an express term, one should give effect to the intention by way of an implied term. That is a hopeless argument. If the intention is clear from clause 17.6, one should give effect to it as an express term: the whole point of contractual interpretation is to give effect to the parties' intention as set out in the contract. If, on the other hand, the intention is insufficiently clearly set out on the clause, one cannot give effect to it unless, of course, it is made out on the basis of another clause or on the basis of an implied term established according to well-established principles.

[49] Subject to that point, I cannot do better than adopt the judge's reasons in [46] to [48] for rejecting this second argument based on an implied term. He said:

> 46. It is not difficult to accept the proposition that the circumstances in the present case are such that the parties might reasonably have agreed that the Trustees' lien should rank ahead of the mortgage. But in my judgment the circumstances are not such as to permit or require such a term to be implied.
>
> 47. There are a number of reasons for this. First, it would in my view contradict cl 11 of the debenture, which is directly in point as the sale of the property was effected by receivers. Clause 11 requires the receiver to apply the proceeds, after payment of his own expenses and remuneration and the [Bank's] costs and expenses, in or towards payment of the secured loan. Only the balance, if any, is paid to the trustees. Secondly, if there were to be some protection for

the trustees' position, it is far from obvious that it would be by giving priority to their lien. A lender might well say that such protection should take the form of a right of indemnity against the beneficial owners of the property… Thirdly, the very fact that no provision is made for the trustees' lien to take priority is a significant factor against its implication. As was said by Lord Hoffmann in *AG of Belize* v *Belize Telecom Ltd* [2009] 1 WLR 1988] at [17]:

"The question of implication arises when the instrument does not expressly provide for what is to happen when some event occurs. The most usual inference in such a case is that nothing is to happen. If the parties had intended something to happen, the instrument would have said so. Otherwise, the express provisions of the instrument are to continue to operate undisturbed. If the event has caused loss to one or other of the parties, the loss lies where it falls."

48. Among the circumstances of the arrangements on which the trustees relied in support of an implied term for priority of their lien, they particularly stressed that the terms of the Trust Instrument had been the subject of negotiation and agreement by the bank. The Trust Instrument conferred the trustees' rights of indemnity and lien, without the requirement contained in a number of the other clauses conferring rights or powers on the trustees for the C bank's prior consent. If anything, this seems to me to tell against an implied priority of the lien over the mortgage. It demonstrates that the question of the trustees' expenses was considered, but nonetheless no provision for priority was included in the debenture.

Conclusion

[50] For these reasons, I would dismiss the trustees' appeal and uphold the order made by David Richards J.

Richards LJ said:

[51] I agree.

D

Hughes LJ said:

[52] I also agree.

Appeal dismissed.

SUPPLEMENT

CASES NOT REPORTED IN "ESTATES GAZETTE"

CIVIL PRACTICE AND PROCEDURE

Middlegreen LP *v* Dominion Developments (2005) Ltd and another

COURT OF APPEAL

[2011] EWCA Civ 646

27 May 2011

RIX, RIMER AND STANLEY BURNTON LJJ

Civil Procedure Rules — Part 36 — Settlement of action — Offer and acceptance — Claimant offering to settle claim — Offer expressed to be made under CPR 36 and stated to be "open for 21 days" — Defendant purporting to accept offer after that time — Claimant obtaining declaration that offer no longer capable of acceptance — Whether time-limited offer capable of falling within CPR 36 Whether claimant's offer time-limited — Whether offer still open for acceptance

In December 2009, the respondent offered to settle proceedings that it had brought against the appellants for breach of contract in respect of a sale of land. The offer letter was headed "Offer to settle under CPR 36". It stated that the offer would be open for "the relevant period" of 21 days from the date of the letter and that if the appellants did not accept the offer and the respondent obtained a judgment that was equal to or more advantageous than that offer, it would seek a costs order in the terms of CPR 36.14.

The appellants did not accept the offer within the 21-day period; they eventually purported to accept it just before the trial date in November 2010. The respondent applied for a declaration that the offer could no longer be accepted and there was no binding settlement of the action, which should accordingly proceed.

Granting the declaration, the judge held that: (i) a time-limited offer, specifying a period for acceptance, could not be a Part 36 offer because such an offer had to be capable of acceptance unless and until withdrawn by the service of a notice under CPR 36.9(2); (ii) the respondent's offer letter, properly construed, created a time-limited offer that came to an end according to its own terms and therefore could not be a Part 36 offer; and (iii) that time-limited offer was none the less a valid offer under the common law and had expired, such that no offer capable of acceptance by the appellants had remained in existence as at November 2010: see [2010] EWHC 2940 (Ch); [2011] 1 EGLR xx; [2011] 03 EG 84. The appellants appealed.

Held**: The appeal was allowed. (1) There is a necessary inconsistency between an offer being time-limited and its being a Part 36 offer. It is an essential feature of the CPR 36 procedure, and the costs consequences thereof, that a Part 36 offer remains on the table until it is formally withdrawn by a notice of withdrawal. A time limit cannot form part of the subject matter of the offer over which an offeror within CPR 36 has autonomy but is instead the subject matter of CPR 36 itself. If an offeror wants to end a Part 36 offer, so that it cannot be accepted, it must do so by serving a formal notice of withdrawal. A time-limited offer, in the sense of an offer that lapses of its own accord at the end of a stipulated period, cannot be made under Part 36. The respondent's offer had to be construed against that legal context, of which both parties had to be taken to be aware. (2) An offer presented as a Part 36 offer, and otherwise complying with its form, will not readily be** F
interpreted in a way that will prevent it from being a Part 36 offer. An offer that is intended to be made and understood as a Part 36 offer should be construed as not being time-limited if it can reasonably be so construed. Where a purported Part 36 offer states that it is "open for 21 days", that can be construed to mean either that the offer is not open after 21 days or that it will not be withdrawn for 21 days. Since the former construction will prevent the offer from being a Part 36 offer, even though it is intended to be one, the latter construction should be preferred. Construing the respondent's offer letter in that way saved it as the Part 36 offer it G
was intended to be and also provided both parties with the clarity and certainty to which both CPR 36 and the offer letter aspired. (3) The respondent's Part 36 offer had not been withdrawn in e-mails sent subsequent to the offer letter. Those e-mails were concerned only with an extension of the 21-day period and did not amount to a withdrawal of the offer. The appellants' acceptance of the offer was effective since the respondent had made a Part 36 offer that it had not withdrawn.

The following cases are referred to in this report.

Charter Reinsurance Co Ltd (in liquidation) v *Fagan* [1997] AC 313; [1996] 2 H
WLR 726; [1996] 1 All ER 406; [1996] 1 Lloyd's Rep 261, CA

Gesellschaft Burgerlichen Rechts v *Stockholms Rederi AB Svea (The Brabant)*; *sub nom Gesellschaft Burgerlichen Rechts* v *Stockholms Rederiaktiebolag Svea, (SS Brabant)* [1967] 1 QB 588; [1966] 2 WLR 909; [1966] 1 All ER 961; [1965] 2 Lloyd's Rep 546, Comm

Gibbon v *Manchester City Council*; *LG Blower Specialist Bricklayer Ltd* v *Reeves* [2010] EWCA Civ 726; [2010] 1 WLR 2081; [2010] 3 EGLR 85; [2010] 36 EG 120

Investors Compensation Scheme Ltd v *West Bromwich Building Society (No 1)*; *Investors Compensation Scheme Ltd* v *Hopkin & Sons*; *Alford* v *West Bromwich Building Society*; *Armitage* v *West Bromwich Building Society* [1998] 1 WLR 896; [1998] 1 All ER 98, HL J

Mannai Investment Co Ltd v *Eagle Star Life Assurance Co Ltd* [1997] AC 749; [1997] 2 WLR 945; [1997] 3 All ER 352; [1997] 1 EGLR 57; [1997] 24 EG 122; [1997] 25 EG 138, HL

Middlegreen LP v *Dominion Developments (2005) Ltd* [2010] EWHC 2940 (Ch); [2011] 1 WLR 331; [2011] 2 All ER 404; [2011] 1 EGLR 53; [2011] 03 EG 84

Onay v *Brown* [2009] EWCA Civ 775; [2010] 1 Costs LR 29

Pagnan SpA v *Tradax Ocean Transportation SA*; *sub nom Tradax Ocean Transportation SA* v *Pagnan* [1987] 3 All ER 565; [1987] 2 Lloyd's Rep 342, CA, [1987] 1 All ER 81; [1986] 2 Lloyd's Rep 646, Comm

Rolf v *De Guerin* [2011] EWCA Civ 78; [2011] CP Rep 24

Sampla v *Rushmoor Borough Council* [2008] EWHC 2616 (TCC) K

This was an appeal by the appellants, Dominion Developments (2005) Ltd and A2 Dominion Enterprises Ltd, from a decision of Warren J, sitting in the Chancery Division of the High Court, granting a declaration to the respondent, Middlegreen LP, that an offer to settle proceedings had lapsed and could no longer be accepted such that the claim should proceed to trial.

Sue Carr QC and Jonathan Hough (instructed by Rawlinson Butler LLP) appeared for the appellants; Michael Barnes QC (instructed by SJ Berwin LLP) represented the respondent.

Giving the first judgment, **Rix LJ** said:

[1] The critical issue raised by this appeal is what it means in a purported Part 36 offer to say that the offer is "open for 21 days". On one view, it means that the offer is *not* open *after* 21 days. On another view, it means, in its context, that the offer will not be *withdrawn* for 21 days. That would reflect the terms of Part 36. Indeed, the judge, Warren J ([2010] EWHC 2940 (Ch)*), held that Part 36 does not allow for a time-limited offer, since a Part 36 offer can lapse only on service of a written notice of withdrawal. However, the judge nevertheless held that the phrase "open for 21 days" meant that the offer lapsed without express withdrawal, with the consequence that what was made as a Part 36 offer failed to be one, albeit it survived as an offer under the common law.

[2] In the present case, the offeree purported to accept the offer (well) outside the 21 days. The offeror submits that the offer had lapsed.

[3] On this appeal, the offeree submits that the judge was wrong in his construction of the offer. The offeror submits that the judge was wrong to say that Part 36 could not admit a time-limited offer.

[4] The judge also found that, had the offer been a Part 36 offer, it had never been withdrawn. The offeror submits that the judge was wrong about that as well. The offeror says that the offer was withdrawn by reason of the terms in which it was made, alternatively by subsequent e-mails.

Parties' dispute

[5] The offeror, the claimant in these proceedings, had sold development land to the offeree, the defendant, for a price of just under £12m. A dispute arose as to whether a condition precedent of their contract had been fulfilled. The defendant said that it had not, and declined to complete the transaction. The claimant originally sought specific performance of the sale, but has now given up the pursuit of that remedy and limited itself to a claim in damages. At the time of the offer, however, the claim for specific performance was still being maintained.

[6] The claimant is in this court the respondent, and the defendant is the appellant. I will refer to the parties as claimant or respondent or offeror and defendant or appellant or offeree, as appears most appropriate. The parties have been anonymised as *C* v *D* because of the possibility that a trial might have to take place.

Offer

[7] The offer was made in a letter dated 10 December 2009, written by the claimant's solicitor to the defendant's solicitor.

[8] The letter was headed "Offer to Settle under CPR Part 36". It began by commenting on the dispute. It said that the claimant was in discussions to sell the land to a third party, which would quantify a loss. Or, if the land was not resold, the defendant would have to buy it at the original contract price. In addition, the claimant was also incurring holding costs and interest. Those costs would fall on the defendant in any event. The letter then proposed a "Settlement Agreement" as follows:

> Our client's proposal is this:
>
> 1. [The defendant guarantor] purchase the property for £10,500,000; or
>
> 2. whilst there is no certainty as to the [third party] purchase of the Property, our client is willing to drop the action for specific performance and settle the claim for £2,000,000.
>
> Regardless of which offer your client chooses, the offer will be open for 21 days from the date of this letter (the "**Relevant Period**"). Your clients can thus walk away from the dispute by the year end having achieved an attractive settlement. [The judge referred to this paragraph as the "Offer Paragraph", which is a convenient device.]
>
> Both offers are intended to have the consequences set out in Part 36 of the Civil Procedure Rules and are to settle all matters raised in the proceedings.
>
> **Failure to Accept this Offer**
>
> If your clients do not accept this offer and our client obtains a judgment which is equal to or more advantageous than this offer, our clients will rely on CPR 36.14 to seek an order in the following terms:
>
> (i) that your clients will pay our client's costs up to the expiry of the Relevant Period;
>
> (ii) that your clients pay our client's costs on the indemnity basis from the date on which the Relevant Period expired, with interest on those costs of up to 10% above base rate and interest on the whole or part of any sum awarded at up to 10% above base rate for some or all of the period starting from the same date.
>
> We await your response.

[9] It is common ground that this offer was intended to be made pursuant to Part 36, as its language demonstrates. As will appear from the terms of Part 36 set out below, the concept of a "relevant period" and the consequences set out in CPR 36.14, are all part of the structure of Part 36.

[10] It is also common ground that the issue of construction as to the meaning of "open for 21 days" has to be considered in the context of the offer, and of Part 36, as a whole.

[11] It is therefore necessary to consider Part 36 in some detail.

CPR 36

[12] Part 36 provides in relevant part as follows:

> 1. Part 36 Offers to Settle
>
> *Scope of this Section*
>
> 36.1 (1)…
>
> (2) Nothing in this Section prevents a party making an offer to settle in whatever way he chooses, but if the offer is not made in accordance with rule 36.2, it will not have the consequences specified in rules 36.10, 36.11 and 36.14…
>
> *Form and content of a Part 36 offer*
>
> 36.2 (1) An offer to settle which is made in accordance with this rule is called a Part 36 offer.
>
> (2) A Part 36 offer must:
>
> (a) be in writing;
>
> (b) state on its face that it is intended to have the consequences of Section I of Part 36;
>
> (c) specify a period of not less than 21 days within which the defendant will be liable for the claimant's costs in accordance with rule 36.10 if the offer is accepted;
>
> (d) state whether it relates to the whole of the claim or to part of it or to an issue which arises in it and if so to which part or issue; and
>
> (e) state whether it takes into account any counterclaim…
>
> (3) Rule 36.2(2)(c) does not apply if the offer is made less than 21 days before the start of the trial…
>
> *Part 36 offers general provisions*
>
> 36.3(1) In this part
>
> (a) the party who makes the offer is the "offeror",
>
> (b) the party to whom an offer is made is the "offeree"; and
>
> (c) the "relevant period" means
>
> (i) in the case of an offer made not less than 21 days before trial, the period stated under rule 36.2(2)(c) or such longer period as the parties agree;…
>
> (3) A Part 36 offer which offers to pay or offers to accept a sum of money will be treated as inclusive of all interest until
>
> (a) the date on which the period stated under rule 36.2(2)(c) expires;
>
> (b) if rule 36.2(3) applies, a date 21 days after the date the offer was made…
>
> (5) Before expiry of the relevant period, a Part 36 offer may be withdrawn or its terms changed to be less advantageous, only if the court gives permission.
>
> (6) After expiry of the relevant period and provided that the offeree has not previously served notice of acceptance, the offeror may withdraw the offer or change its terms to be less advantageous to the offeree without the permission of the court.
>
> (7) The offeror does so by serving written notice of the withdrawal or change of terms on the offeree…
>
> *Acceptance of a Part 36 offer*
>
> 36.9(1) A Part 36 offer is accepted by serving written notice of the acceptance on the offeror.
>
> (2) Subject to rule 36.9(3), a Part 36 offer may be accepted at any time (whether or not the offeree has subsequently made a different offer) unless the offeror serves notice of withdrawal on the offeree…
>
> (3) [This rule lists four cases where the court's permission is needed to accept a Part 36 offer. None of them applies in this case. The fourth of them (d) is where "the trial has started". So permission is not needed to accept an offer after the "relevant period" has expired.]
>
> *Costs consequences of acceptance of a Part 36 offer*

* Editor's note: Reported at [2011] 1 EGLR 53; [2011] 03 EG 84

36.10(1) Subject to… paragraph 4(a), where a Part 36 offer is accepted within the relevant period the claimant will be entitled to the costs of the proceedings up to the date on which notice of acceptance was served on the offeror…

(4) Where

(a) a Part 36 offer that was made less than 21 days before the start of trial is accepted; or

(b) a Part 36 offer is accepted after expiry of the relevant period,

if the parties do not agree the liability for costs, the court will make an order as to costs.

(5) Where paragraph (4)(b) applies, unless the court orders otherwise

(a) the claimant will be entitled to the costs of the proceedings up to the date on which the relevant period expired; and

(b) the offeree will be liable for the offeror's costs for the period from the date of expiry of the relevant period to the date of acceptance…

The effect of acceptance of a Part 36 offer

36.11(1) If a Part 36 offer is accepted, the claim will be stayed…

Costs consequences following judgment

36.14(1) This rule applies where upon judgment being entered

(a) a claimant fails to obtain a judgment more advantageous than a defendant's Part 36 offer;

(b) judgment against the defendant is at least as advantageous to the claimant as the proposals contained in a claimant's Part 36 offer.

(2) Subject to paragraph (6), where rule 36.14(1)(a) applies, the court will, unless it considers it unjust to do so, order that the defendant is entitled to

(a) his costs from the date on which the relevant period expired; and

(b) interest on those costs.

(3) Subject to paragraph (6), where rule 36.14(1)(b) applies, the court will, unless it considers it unjust to do so, order that the claimant is entitled to

(a) interest on the whole or part of any sum of money (excluding interest) awarded at a rate not exceeding 10% above base rate for some or all of the period starting with the date on which the relevant period expired;

(b) his costs on the indemnity basis from the date on which the relevant period expired; and

(c) interest on those costs at a rate not exceeding 10% above base rates.

(4) In considering whether it would be unjust to make the orders referred to in paragraphs (2) and (3) above, the court will take into account all the circumstances of the case…

(6) Paragraphs (2) and (3) of this rule do not apply to a Part 36 offer

(a) that has been withdrawn;

(b) that has been changed so that its terms are less advantageous to the offeree, and the offeree has beaten the less advantageous offer;

(c) made less than 21 days before trial, unless the court has abridged the relevant period.

[13] It will have been observed that:

(1) the "relevant period" of at least 21 days before trial is critical to the operation of a Part 36 offer, and that the specification of such a 21-day period in which the acceptance of the offer will entail liability for the claimant's costs up to the date of acceptance of the offer is critical to the making of a Part 36 offer more than 21 days before trial: see CPR 36.2(2)(c), 36.3(1)(c), 36.10(1);

(2) Part 36 is prescriptive of the effect of a Part 36 offer as to costs and interest (see CPR 36.3(3), 36.10.1 and 36.14(2));

(3) a Part 36 offer cannot be withdrawn within the relevant period without the permission of the court (see CPR 36.3(5) and (6);

(4) a Part 36 offer can be withdrawn only by service of a written notice of withdrawal on the offeree (see CPR 36.3(7));

(5) an unwithdrawn Part 36 offer can therefore be accepted at any time, and that common law rules as to the lapsing of a Part 36 offer do not apply (see CPR 36.9(2));

(6) Part 36 prescribes the costs consequences of a judgment that betters or fails to better the Part 36 offer, subject only to the court's discretion to order otherwise if the prescribed consequences would be unjust, and subject also to a discretion as to the amount of the enhanced rate of interest to be awarded to a claimant on both any judgment sum and on costs (see CPR 36.14);

(7) it is an essential feature of the CPR 36.14 consequences (where the litigation has proceeded to judgment) that they apply only where the offer has not been withdrawn (see CPR 36.14(6)(a));

(8) an offer may be made outside Part 36, in which case, it will not have the prescribed consequences: see CPR 36.1(2).

Jurisprudence on Part 36 G

[14] Several of these matters have been commented or ruled on in decisions of the court.

[15] The most important decision in this respect is that of this court in *Gibbon* v *Manchester City Council* [2010] EWCA Civ 726; [2010] 1 WLR 2081*. The issue there was whether a Part 36 offer had been superseded by another revised offer, or by a rejection. At common law, an original offer would have been superseded in such circumstances, but that was not the way that Part 36 operated. Moore-Bick LJ began with some general observations:

> **4** It can be seen from Part 36 as a whole, as well as from the extracts cited above, that it contains a carefully structured and highly prescriptive set of rules dealing with formal offers to settle proceedings which have specific consequences in relation to costs in those cases where the offer is not accepted and the offeree fails to do better after a trial … [P]arties are not bound to make use of the mechanism provided by Part 36, but if they wish to take advantage of the particular consequences for costs and other matters that flow from making a Part 36 offer, in relation to which the court's discretion is much more confined, they must follow its requirements. H
>
> …
>
> **6** Basic concepts of offer and acceptance clearly underpin Part 36, but that is inevitable given that it contains a voluntary procedure under which either party may take the initiative to bring about a consensual resolution of the dispute. Such concepts are part of the landscape in which everyone conducts their daily life. It does not follow, however, that Part 36 should be understood as incorporating all the rules of law governing the formation of contracts, some of which are quite technical in nature. Indeed, it is not desirable that it should do so. Certainty is as much to be commended in procedural as in substantive law, especially, perhaps, in a procedural code which must be understood and followed by ordinary citizens who wish to conduct their own litigation. In my view, Part 36 was drafted with these considerations in mind and is to be read and understood according to its terms without importing other rules derived from the general law, save where that was clearly intended. J

[16] Moore-Bick LJ then continued by addressing the specific issue in the case. He said: K

> **16** In my view, attractive though these arguments are, they cannot be reconciled with the clear language of Part 36, or indeed with the scheme which it embodies. Rule 36.9(2) is quite clear: a Part 36 offer may be accepted *at any time* unless the offeror has withdrawn the offer by serving notice of withdrawal on the offeree. Moreover, it may be accepted whether or not the offeree has subsequently made a different offer, a provision which is contrary to the general position at common law. The rules state clearly how a Part 36 offer may be made, how it may be varied and how it may be withdrawn. They do not provide for it to lapse or become incapable of acceptance on being rejected by the offeree. That would be the case at common law, but it is inconsistent with the concepts underlying Part 36, which proceeds on the footing that the offer is on the table and available for acceptance until the offeror himself chooses to withdraw it. There are good reasons for that. An offer which appears unattractive when made, and which is therefore rejected, may become more attractive as the proceedings progress and the parties reassess the strength of their prospective cases. A defendant who chooses to leave his offer on the table may tempt the claimant into accepting it, with the benefit to himself of the consequences for costs of an offer made at an early stage. Part 36 allows a defendant (or for that matter a claimant) to decide whether to leave his offer open for acceptance or to withdraw it and make another offer later. To import into Part 36 the common law rule that an offer lapses on rejection by the offeree would undermine this important element of the scheme. It could give rise to disputes about whether the offer had been rejected in any given case, so as to render it incapable of acceptance. In *Sampla* v *Rushmoor Borough Council* [2008] EWHC 2616 (TCC) Coulson J held, largely for these reasons, that the rejection of a Part 36 offer does not render it incapable of late acceptance. In my view he was right to do so. L M
>
> **17** Nor do I think that the letter of 18 February 2009 [in which the claimant had rejected a counter-offer] can be read as containing an implied withdrawal of the Part 36 offer. Rule 36.3(7) provides that an offer is withdrawn by serving written notice on the offeree. In my view that leaves no room for the concept of implied withdrawal; it requires express notice in writing in terms which brings home to the offeree that the offer has been withdrawn. If justification for that

* Editor's note: Reported at [2010] 3 EGLR 85; [2010] 36 EG 120

A requirement is sought, it can be found once again in the need for clarity and certainty in the operation of the Part 36 procedure.

> **18** In some cases the demands of clarity and certainty in the operation of Part 36 may appear to produce injustice and I see the force of Dr Friston's submission that Mrs Gibbon's solicitors had made it clear that she was not willing to accept £2,500 to settle her claim. Her difficulty, however, is that a Part 36 offer had been made on her behalf and had not been withdrawn. It remained on the table and was therefore open for acceptance. We are not concerned with whether she has any ground for complaint against the council [the offeree] for accepting her offer despite the fact that she had made her intentions clear, since no suggestion has ever been made that she has a legitimate complaint about that.

B

[17] It may be said that this decision demonstrates that the need for clarity and certainty in the operation of Part 36 outweighs the pressure exerted on the court by thoughts that the offeree has somehow taken advantage of the error of the offeror in not formally withdrawing an offer that the offeree had reason to know was no longer acceptable to the offeror. An essential feature of the Part 36 procedure is that a Part 36 offer remains on the table until it is formally withdrawn by a notice of withdrawal. It is because the offer remains on the table (until formally withdrawn) that the Part 36 consequences of a Part 36 offer, whether under CPR 36.10 following acceptance prior to judgment, or under
C CPR 36.14 following judgment, can be prescribed in the form that Part 36 sets out. If a Part 36 offer has been withdrawn before judgment, the consequences of CPR 36.14(2) and (3) do not flow.

[18] *Sampla* v *Rushmoor Borough Council* [2008] EWHC 2616 (TCC), which Moore-Bick LJ in *Gibbon* had cited and approved, raised the issue of whether a Part 36 offer could be accepted: (a) after it had been rejected; and (b) after the trial had started. As to question (a), Coulson J ruled that it could be. He said:

> 29. First…CPR Part 36 is entirely silent as to offers that are rejected.
>
> 30. Secondly, CPR [36.9(2)] makes plain that an offer can be accepted "at any time". Those are wide words which are not qualified in any way. If an offer
> D could not be accepted because it had been initially rejected, then I would have expected Part 36 to say so in clear terms. …
>
> 31. Thirdly, in express confirmation of those first two points, are the words in brackets in CPR 36.9(2), to the effect that an offeree can accept an offer at any time, even if his original response had been to make a counter-offer. That is the direct opposite of the position in contract, where a counter-offer "kills" an offer just as completely as an express rejection.

[19] However, question (b) arose because CPR 36.9(2) was subject to CPR 36.9(3)(d), which required the court's permission for the acceptance of an offer after trial had begun. That rule gave the court the discretion to refuse permission where there had been a significant
E change of circumstances since the offer had been made; and Coulson J refused permission for that reason.

[20] It may be observed that a significant change of circumstances is not a general ground on which the court's permission is required to accept a Part 36 offer.

[21] In *Rolf* v *De Guerin* [2011] EWCA Civ 78*, the present issues did not arise. However, a Part 36 offer was there expressed to be "open" for a period of 21 days. After the 21 days had expired, the claimant offeror chased for a response. I commented, in [21]:

> It was thus clear that the Part 36 offer, and the offer of mediation etc, were
> F still on the table. There is in fact no practical limit on the time within which a Part 36 offer may be accepted: see CPR 36.9(2). There was no reply to that letter either.

Ultimately, the matter went for trial. Both the trial judge and this court treated the offer as a Part 36 offer. The judge used it against the offeror, because she had failed at trial to obtain as much as her offer. The judge proceeded as though she had failed at trial to obtain as much as a *defendant's* Part 36 offer (there was none) and because she had claimed so much more than her Part 36 offer. I said, in [34]:

> However, there is nothing about the procedure which states that an offeror is to be prejudiced as to costs because he has expressed his willingness to accept less than his open position. That would make the procedure a most dangerous one to use. The judge's ruling that the incidence of costs should change at the expiry of the relevant period of three weeks confirms the illogicality of his decision. The three weeks is given to protect the offeree, who has that period to make up his mind: the judge used it, however, (to its limited extent) to protect the offeror, Mrs Rolf.

Elias and Tomlinson LJJ agreed.

[22] Although the present issues were not before the court in that case, nevertheless it appears that no one regarded the offer as lapsing at the end of the 21 days, or as not capable of being a Part 36 offer at all.

[23] In *Onay* v *Brown* [2009] EWCA Civ 775; [2010] 1 Costs LR 29, the defendant made a Part 36 offer in, *inter alia*, the following terms:

> This offer is made pursuant to Part 36 of the Civil Procedure Rules and is intended to have the consequences of that rule.
>
> The relevant acceptance period is 21 days from the date of receipt of this letter, namely to 26 September 2008.

Within that period, the claimant accepted the offer. The offer and the acceptance said nothing about costs, and there was a dispute about how costs should be dealt with. It was submitted on the part of the claimant that the defendant's offer did not amount to a Part 36 offer because the wording of CPR 36.2(2)(c) was not expressly set out. This court rejected that submission. Goldring LJ said, in [27]:

> the reference to the *relevant* (my emphasis) acceptance period of 21 days plainly pointed to CPR 36.2.

[24] No one suggested that the stipulation of an "acceptance period" prevented the offer from being a Part 36 offer. Again, the precise point before us was not the point before the court. However, in a concurring judgment Carnwath LJ said this, which resonates in the context of our case:

> 32. I also agree. The moral of this story is that someone who writes a letter headed "Part 36 offer", and which is stated as "intended to have the consequence of that rule", should make sure that he knows what those consequences are. I agree with my Lord that those consequences in a case such as this are clearly set out in 36.2(2) and 36.10(1). If the party writing the letter does not want those consequences to apply, he should put his offer in some other way, as is expressly permitted by rule 36.2.
>
> 33. … it seems to me important, in the interests of certainty, that, when the Part 36 jurisdiction is expressly invoked, the court should generally take that at face value, and as far as possible give effect to the consequences as envisaged by the rules.

2007 amendment to Part 36

[25] This litigation has been conducted under the amended Part 36, which came into force on 6 April 2007. Prior to the amendments, the old Part 36 provided (*inter alia*) as follows:

> *Form and content of a Part 36 offer*
>
> 36.5…(6) A Part 36 offer made not less than 21 days before the start of the trial must
>
> (a) be expressed to remain open for acceptance for 21 days from the date it is made; and
>
> (b) provide that after 21 days the offeree may only accept it if
>
> (i) the parties agree the liability for costs; or
>
> (ii) the court gives permission…
>
> (8) If a Part 36 offer is withdrawn it will not have the consequences set out in this Part…
>
> *Time for acceptance of a claimant's Part 36 offer*
>
> 36.12(1) A defendant may accept a Part 36 offer made not less than 21 days before the start of the trial without needing the court's permission if he gives the claimant written notice of acceptance not later than 21 days after the offer was made.

If, however, the offer was accepted after 21 days and the parties agreed liability for costs, the offer could again be accepted without the permission of the court (rule 36.12(2)(b)(i)). That was consistent with CPR 36.5(6)(b)(i).

---

* Editor's note: Reported at [2011] CP Ref 24

[26] There was no defined concept of the "relevant period". There were no provisions relating to the withdrawal of an offer other than CPR 36.5(8), which recognised the possibility of withdrawal, but stipulated that the effect of a withdrawal would be that for Part 36 purposes it would be as though it had never been made.

[27] The provisions cited above demonstrate that under the unamended rules a Part 36 offer both had to be expressed as "open for acceptance for 21 days" and also survived for acceptance after 21 days. After 21 days, the offer could be accepted without the permission of the court if the parties agreed the incidence of costs, but otherwise needed the permission of the court and the court would then, if it gave permission, settle the incidence of costs (CPR 36.12(3)). The old rules therefore indicated that the expression "open for acceptance for 21 days" was not inconsistent with the offer remaining open for acceptance after 21 days, and that was emphasised by the required language of CPR 36.5(6)(b). Thus, the old rules proceeded on the basis that "open for acceptance for 21 days" did not in that context mean "not open for acceptance after 21 days". Consistently with that, it seems that an offer could be withdrawn after the 21 days for acceptance, but not, I would suggest, before, and that it could be accepted even after the 21 days, but any acceptance after 21 days lay either with the parties, where they could agree the incidence of costs, or with the court. These old rules were, in any event, inconsistent with the construction of the words "open for acceptance for 21 days" as meaning that the offer automatically lapsed at the end of the 21-day period.

[28] A consultation paper (CP 02/06) was published by the Department for Constitutional Affairs to assist in the amendment of the old Part 36. The "Policy Objectives" of the proposals contained this passage in para 6:

> to make it easier and more attractive to use Part 36 by removing unnecessary burdens/processes, in particular by:
>
> …
>
> allowing parties to accept offers after the initial time limit has expired without requiring the court's permission and, to balance that, allowing unaccepted offers and payments to be withdrawn after the time for acceptance has expired.

[29] Section 2 of the proposals (in paras 27ff) was concerned with "Acceptance and rejection of offers". The paper pondered the considerations as follows:

> 28. At present it is not entirely clear whether the court may exercise its general discretion to extend the prescribed time for acceptance of a Part 36 offer (so that the offeree will still enjoy automatic costs consequences).
>
> 29. Rule 3.1 probably allows it, but that rule refers to extending or shortening time limits for *compliance* with any rule, whereas Part 36 does not itself directly *require* acceptance (or explicit rejection) of an offer. The Department considers that it should be made clear, one way or the other, whether the court can amend the 21-day period for accepting offers.
>
> 30. Assuming the court's probable power to extend is to be confirmed…
>
> 31. … there may be circumstances in which the offeror may know that the offeree is likely to require longer than 21 days to assess the claim. To avoid unnecessary applications, and possible censure by the court, the offeror may prefer to express the offer as remaining open for more than 21 days. The Department proposes therefore that the offeror should be permitted to state a longer period of acceptance than 21 days.
>
> …
>
> 32. It is proposed to remove the requirement to obtain the court's permission to accept a Part 36 offer after the expiry of the time limit. This is because a permission requirement implies a possibility of refusal and it is unclear why the court should ever do this. Parties should be encouraged and assisted to settle cases, not hindered. (If the parties cannot agree costs, the court will have to agree those costs in any event). …
>
> …
>
> 34. It is proposed that:
>
> • the formal requirement for permission to accept an offer out of time should be removed;
>
> • an offer (if not withdrawn) should be acceptable at any time. If accepted out of time, it should not have the automatic costs consequences of Part 36 (see rules 36.13 & 14). If costs were not agreed, they would fall to be assessed by the court. The normal rule might be that costs incurred after the time for acceptance would not be allowed.

[30] Section 3 of the paper is headed "Withdrawal of offers". The **G** following is relevant to record:

> 36. As the rules stand, there is a lack of clarity and consistency in respect of the withdrawal of offers and payments.
>
> …
>
> 38. A Part 36 offer … may be withdrawn at any time but rule 36.5(8) provides that such offers that are withdrawn will not have the consequences set out in Part 36. However the Western Power judgment must throw doubt on this.
>
> …
>
> 41. It is proposed as a means to achieve such clarity to place litigants on an equal footing (removing the distinction in relation to offers and payments). The **H** approach favoured by the Department is that all offers (and payments in support of offers) may be withdrawn:
>
> • any time after the expiry of the period during which they are expressed to be open for acceptance, unless the offer has been accepted
>
> • before the expiry of the period during which they are expressed to be open for acceptance only with the permission of the court
>
> …
>
> 42. The Department considers that allowing a party who makes an offer to withdraw it (after the expiry of the period during which it is expressed to be open for acceptance) is a natural corollary of the offeree being able to accept it without permission after expiry of the same period (see paragraphs 32-34 above). **J**
>
> 46. However, we do not propose an absolute bar on withdrawal during that period, merely that there should be a requirement to apply to the court for permission.
>
> 47. The alternatives to what is proposed would be either an unfettered right to withdraw an offer at any time (by service of a notice), or a requirement for permission to withdraw at any time, and/or a complete ban on withdrawing during the period for acceptance (which might be expressed as more than 21 days or might have been extended by the court).

[31] Ms Sue Carr QC, counsel for the appellant, sought to draw some assistance from this paper in the light of the amendments to Part 36, but I am not sure that much assistance is to be obtained. The paper's discussion appears to be somewhat discursive, since it is not **K** focused on any particular wording. There is also some lack of precision about the terms deployed. Thus, a period (of 21 days) for acceptance is both spoken of as a "time limit" but also as a period that can be transcended for the purpose of acceptance (see para 41); and even after the end of that period, it may be that the offer can be withdrawn only with the permission of the court, even if the preferred proposal is that no permission should be required at that time. However, it is at any rate reasonably plain that it is not contemplated that language such as "open for acceptance" for 21 or any other number of days means that the offer automatically lapses (or is withdrawn) at the end of that period, and this may be important at several points of the arguments deployed **L** before us. There would of course be no need for withdrawal of an offer after the time for acceptance had expired if such language engendered an automatic lapse of the offer. That point would be more powerful, however, if the old Part 36's language (and the paper's language) of "open for acceptance" had survived into the amended Part 36, which it has not.

Issues

[32] Three issues were argued by Ms Carr on behalf of the appellant, and by Mr Michael Barnes QC on behalf of the respondent:

(1) Can a Part 36 offer be made in terms that limit the acceptance **M** of the offer to a stipulated period, such that the offer lapses at the end of that period? Such an offer was referred to as a time-limited offer. The judge said that such an offer could not be made as a Part 36 offer, although it could be made as an offer outside Part 36. Within Part 36, however, an offer must be withdrawn, otherwise it remains on the table. The appellant seeks to maintain that answer on appeal.

(2) What is the true construction of the respondent's offer being "open for 21 days" in the context of what was clearly intended to be a Part 36 offer? The judge said it was a time-limited offer, and that therefore it was not a Part 36 offer. The appellant submits that that was wrong. The respondent seeks to maintain that answer on appeal.

A (3) If withdrawal of a Part 36 offer is necessary, was the respondent's offer withdrawn: either by the time-limited terms of the offer itself, or by the e-mails that followed it? The judge did not formally decide that issue, but briefly expressed his view that, had a withdrawal been necessary, there was none. The appellant submitted that that opinion was correct, the respondent submitted that it was wrong.

[33] Ms Carr argued those three issues in that order. She did so because she submitted that it was important to come to the issue of construction with an understanding of the way in which Part 36 worked. That was the context in which offeror and offeree would reasonably understand an offer expressed to be a Part 36 offer. Mr Barnes, however, B was anxious to argue issue (2) first, for, although he recognised the context, he wanted to give to the disputed words "open for 21 days" the full benefit of what he submitted was their natural meaning.

[34] I will approach the issues in the order in which the judge did, as stated above.

Issue (1): Can a Part 36 offer be a time-limited offer?

[35] The judge concluded that it could not, for essentially three reasons. First, CPR 36.9(2) indicates that a Part 36 offer comes to an end by withdrawal rather than in accordance with its own terms, since otherwise the offer can be accepted "at any time". Second, the C CPR 36.14(3) sanction does not apply where the Part 36 offer is withdrawn (see CPR 36.14(6)(a)): that is, the policy of Part 36 is that, for the benefits and sanctions to work, the offer must be kept open. Third, however, there is no express loss of sanction where the offer lapses, as distinct from being withdrawn: but it would be unfair if a claimant could maintain the benefit of a Part 36 offer that lapsed, where it could not maintain its benefit where it was withdrawn.

[36] Ms Carr relieed on those reasons, but added two others, or at any rate built on the judge's reasons. First, she submitted that CPR 36.9(2) goes beyond recognising the importance of withdrawal. Combined with CPR 36.3(5), (6) and (7), it indicates that *only* the service of a written notice of withdrawal can bring a Part 36 offer, D together with its potential consequences, to an end. Moreover, that is the effect of what Moore-Bick LJ said in *Gibbon*, which is that Part 36 is a self-contained code to which the principles of the common law have to make way where inconsistent with it. Just as rejection of the offer, or the making of a counter-offer, do not bring a Part 36 offer to an end, so a Part 36 offer cannot lapse under its own terms. It can be brought to an end only by withdrawal: see especially *Gibbon* in [16], where Moore-Bick LJ said:

> The rules state clearly how a Part 36 offer may be made, how it may be varied and how it may be withdrawn. They do not provide for it to lapse or become incapable of acceptance on being rejected by the offeree. That would be the E case at common law, but it is inconsistent with the concepts underlying Part 36, which proceeds on the footing that the offer is on the table and available for acceptance until the offeror himself chooses to withdraw it.

[37] Second, Ms Carr submitted that the 2006 consultation paper, particularly in para 6, explains the purpose of the new proposals in a way that accords with withdrawal being crucial to the termination of the offer.

[38] In response, Mr Barnes accepted the force of these points as far as they go, but submitted first, that there is no express exclusion in Part 36 of a time-limited offer; second, that there is nothing in Part 36 F that necessarily, albeit implicitly, excludes, by reason of inconsistency, a time-limited offer; and third, that Part 36 as a matter of principle does not dictate to the offeror what may be contained in the offer as part of its subject-matter and that a time limit is to be regarded as part of the offer's subject-matter over which the offeror retains complete autonomy.

[39] As for the matters that the judge and Ms Carr relied on as indicating Part 36's inconsistency with a time-limited offer, the words "at any time" in CPR 36.9(2) were simply to be explained as pointing up the distinction with the old CPR 36.5 under which an acceptance after the time for acceptance required the permission of the court; any possible unfairness that might flow from imposing the CPR 36.14 consequences in the wake of the lapse of a time-limited offer could be avoided by the discretion of the court contained in the words "unless it considers it unjust to do so" within CPR 36.14(2) and (3); and, although an offer cannot be time-limited at less than 21 days, the requirement of a withdrawal in the absence of a time limit does not exclude the imposition of a time limit, for the functions of each are different.

[40] In my judgment, the reasons advanced by the judge and Ms Carr have more weight than those advanced on the other side. It is true that Part 36 does not contain an express exclusion of a time-limited offer. However, the essence of the matter is that a Part 36 offer, to have effect in terms of costs consequences after trial, has to be an offer that has not been withdrawn, but has remained on the table. The initial offer has to specify a period of at least 21 days during which the defendant remains liable for the claimant's costs until acceptance: CPR 36.2(2)(c) and 36.10(1). The offer cannot be withdrawn within that period without the permission of the court: CPR 36.3(5). After that period has expired, the offeror can withdraw the offer only by serving notice of withdrawal on the offeree: CPR 36.3(6) and (7). In the absence of withdrawal, the offeree can accept the offer at any time: CPR 36.9(2). The language of that rule's "unless the offeror serves notice of withdrawal on the offeree" states a pre-condition that has to be fulfilled in order to prevent the offeree having the right to accept "at any time". That critical rule is made subject to a few limited exceptions (CPR 36.9(3)), one of which is where the trial has commenced, but the expression of those exceptions serves only to emphasise that, in the ordinary way, unless the offer has been withdrawn before the expiry of the relevant period with the permission of the court, or the offer has been withdrawn after the expiry of that period by the service of a written notice of withdrawal, there is no room for an offer that is neither withdrawn before or after the expiry of the relevant period, but lapses as a matter of its own terms. It is, in any event, fanciful to think that there is a point of time that lies between before and after the expiry of the relevant period. Although the present issue was not in play in *Gibbon*, the logic of this court's decision there underwrites the conclusion that I, together with the judge below, would favour: and I agree that [16] of Moore-Bick LJ's judgment there comes very close to expressing the rule applicable in our case.

[41] I therefore do not accept Mr Barnes' submission that the availability of the costs advantages of CPR 36.14 to the offeror of a time-limited but not withdrawn offer is to be avoided, not by a rule, as occurs in the case where the offer has been withdrawn (CPR 36.14(6)(a)), but by the exercise of the court's discretion to avoid injustice. That would make no sense at all. If such a discretion was to operate in the case of a time-limited offer, it ought to operate in the case of an offer that was withdrawn. The discretion in CPR 36.14(2) and (3) is given for other reasons, to avoid injustice occurring by reason of a formula. Nor do I accept his submission that a time limit is a matter of the subject matter of the offer over which an offeror within Part 36 has autonomy: it is rather the subject matter of Part 36 itself, for the reasons stated above. Therefore, it is only outside Part 36 that a claimant has full autonomy over the means by which its offer may lapse in the absence of either the supervision of the court or the service of a notice of withdrawal.

[42] It would no doubt have been better if Part 36 had expressly stated that a time-limited offer could not be accommodated within its scheme. Against the backdrop of the old Part 36, the consultation paper, and the somewhat different terms in which the amended Part 36 was ultimately formulated from those that were being proposed and discussed in that paper, it is even possible that there was some uncertainty or lack of clarity on this point. After all, the old Part 36 had accommodated the appearance of a time limit in its formulation but had nevertheless imposed its own regime on that appearance. Be that as it may, in my judgment, the new Part 36 regime cannot accommodate a time-limited offer. The essence of a Part 36 offer is that it lies on the table until formally withdrawn. Only an offer that has not been withdrawn down to the commencement of trial is capable of having the scheme's costs consequences set out in CPR 36.14. A Part 36 offer can be accepted at any time unless withdrawn. Therefore, the scheme seeks to encourage offers that are not time-limited. The scheme nevertheless permits flexibility in permitting offers to be amended and withdrawn. The process, however, is strictly regulated in the interests of clarity and certainty.

[43] The difference between a time-limited offer and an offer that is formally withdrawn immediately after the end of the relevant period may not be a great one as a matter of substance or form. That, however, depends entirely on the service of such a timely withdrawal. There is a substantial difference between a time-limited offer and an offer that can be accepted at any time until it is formally withdrawn. The matter is illustrated by the facts of this case, as the e-mails cited below will demonstrate. At the end of a time-limited period, there is typically an exchange of correspondence (or even telephone calls) dealing with the possibility that the time limit is being extended. Strictly speaking, a time-limited offer will lapse at the end of the time limit. If there is a suggestion of an extension of the time limit on the next day, has the time limit been extended, or waived, or is this a new offer, and in any event when does the final limit occur? The scheme's rules are designed to lay such uncertainties, which would occur in nearly every case of a time-limited offer, to rest. If a Part 36 offer is to lapse, it must be formally withdrawn. That is what *Gibbon* said in dealing with another way in which at common law an offer might lapse, and in my judgment the same logic holds good for a time-limited offer.

[44] Therefore, the claimant's offer must be construed in the context of a Part 36 scheme that does not permit an offer within the scheme to be time-limited.

Issue 2: As a matter of construction, what does "open for 21 days" mean in context?

[45] It follows from my answer to the first issue that there is a necessary inconsistency between an offer being both time-limited and a Part 36 offer. An offer may be one or the other, it cannot be both. That is the objective context in which the offer in this case was made by the claimant's solicitor to the defendant's solicitor. Both the writer and the reader of that offer must be taken, objectively, to know the legal context. Of course, mistakes occur and must be allowed for. However, the question is how a reasonable solicitor would have understood the offer in that context, including the known context of the dispute as it stood at that time: see *Investors Compensation Scheme Ltd* v *West Bromwich Building Society* [1998] 1 WLR 896.

[46] It is common ground that the offer was intended to be made and understood as a Part 36 offer. It is disputed, however, what the meaning of "open for 21 days" means in that context. The claimant submits that it means that the offer lapses at the end of 21 days (in this case on 31 December 2009), that is, that the offer is not open for acceptance after 21 days. The defendant submits that it means that the offer is open for 21 days as an expression of the relevant period but that after those 21 days it may be withdrawn.

[47] It is accepted by Mr Barnes on behalf of the respondent that both meanings are feasible. That is a change from the submission that he made in his written skeleton argument. There he repeatedly submitted that no other meaning could be realistically given to the words other than that they amounted to a time limit that prevented the offer being accepted outside the stated period. He there submitted that the words "are not capable of having any other meaning"; and that the interpretation advanced by the appellant "is untenable". In my judgment, the concession made by Mr Barnes during the oral argument was correct. The issue therefore is which of two possible meanings is the objectively correct one.

[48] Mr Barnes also submitted that, if his meaning was not the only tenable meaning, it was at any rate a traditional and well understood way of expressing a time limit that prevented acceptance of an offer after the stated period had expired. If that is so, it is giving a negative meaning to a positive statement. It is not to my mind blindingly obvious that an offer that is stated to be open for a certain period cannot be accepted after that period has expired. It may be a way of stating that the offer will not be withdrawn for that period, although whether such a statement could be relied on by the offeree was a matter of conflicting submissions by Mr Barnes. However, Mr Barnes may be correct and for the sake of argument, I will assume that he is. I see force in what Rimer LJ has said in [72] below. However, that is not going further than saying that, outside the context of Part 36, that is the normal meaning of that expression. The question is whether that is the meaning that it G sustains within the context of a Part 36 offer, where *ex hypothesi* such an offer cannot be made.

[49] There is a general principle of construction that a document that falls to be construed should be read as a whole and its separate parts should be so construed, if that is possible, as to bring rational sense and consistency to that whole. Ms Carr relied on that principle and cited cases such as *Pagnan SpA* v *Tradax Ocean Transportation SA* [1986] 2 Lloyd's Rep 646 (Steyn J); [1987] 2 Lloyd's Rep 342 (CA) in support of it. The principle is not in doubt. One question there was as to the meaning of a term "Sellers to provide for export certificate…" Was it an absolute obligation or only one of reasonable endeavours? That H question had to be answered against the background of a prohibition of export clause that led to the cancellation of the contract where any executive or legislative act prevented fulfilment. In that situation, and if the certificate clause did impose an absolute obligation, which clause prevailed? The appeal arbitrators found in favour of the sellers, however Steyn J and this court found in favour of the buyers. Steyn J said that he found the arguments finely balanced. Was there a genuine inconsistency? If so, one clause might have to make way for the other. However, Steyn J cited *Gesellschaft Burgerlichen Rechts* v *Stockholms Rederi AB Svea (The Brabant)* [1967] 1 QB 588, and said, at p653:

> J This question must be approached on the basis that the Court's duty is to reconcile seemingly inconsistent provisions if that result can conscientiously and fairly be achieved.

[50] In this court, Bingham LJ cited Steyn J's statement of principle (at p349), cited the jurisprudence on inconsistent or repugnant clauses, and continued at pp349-350:

> It would in my judgment be quite wrong to approach this question of construction with any predisposition to find inconsistency between the special condition and cl 19. They are all part of the same contract … One should, therefore, approach the documents in a cool and objective spirit to see whether there is an inconsistency or not. K

He affirmed Steyn J's analysis, albeit saying that he did so with more confidence than Steyn J had expressed in his answer.

[51] The general principle is stated in *Lewison, The Interpretation of Contracts* (4th ed) 2004, in para 9.13, in this way:

> The court is reluctant to hold that parts of a contract are inconsistent with each other, and will give effect to any reasonable construction which harmonises such clauses.

[52] In the present case, it is at this stage of the argument a given that a time-limited offer and a Part 36 offer are inconsistent. Therefore, if the L claimant's offer was time-limited, it could not take effect as a Part 36 offer. In such circumstances, the court is obliged to consider whether there is a reasonable construction of the offer letter that does not involve it in being time-limited. It is not suggested that the offer can be regarded as other than intended to operate as a Part 36 offer. It states: "Offer to Settle under CPR Part 36" and "Both offers are intended to have the consequences set out in Part 36 of the Civil Procedure Rules".

[53] In my judgment, there is an entirely feasible and reasonable construction of the offer letter that avoids it being construed as a time-limited offer, and that is the meaning for which the appellant contends. M

[54] In the context of Part 36, it seems to me to be entirely feasible and reasonable to read the words "open for 21 days" as meaning that it will not be withdrawn within those 21 days. Part 36 permits withdrawal within the 21-day relevant period, but only with the permission of the court. It seems to me that "open for 21 days" is an obvious way of saying that there will be no attempt to withdraw within those 21 days. It is also a warning that after the expiry of those 21 days, a withdrawal of the offer is on the cards. Such a construction would save the Part 36 offer as a Part 36 offer and would also give to both parties the clarity and certainty that both Part 36 itself, and the offer letter with its reference to "open for 21 days", aspire to. It would leave the offeror entirely free to withdraw the offer immediately on expiry of the stated period, or to let

A it roll on for as long as it wished. At the same time, it would assure the offeree that it had 21 days to consider what it wanted to do, but was at risk if it had not accepted within that period. There might be an issue, had the offeror wished to withdraw within the relevant period, as to whether the court would permit it to do so where it had stated that it was open for 21 days: but that issue does not affect the current question.

[55] Another principle or maxim of construction that is applicable in the present circumstances is that words should be understood in such a way that the matter is effective rather than ineffective (*verba ita sunt intelligenda ut res magis valeat quam pereat*). If the words "open for 21 days" are given the meaning for which the respondent contends,
B then the offer, intended to take effect as a Part 36 offer, fails as such. If, however, the words are given the meaning for which the appellant contends, the intention of making a Part 36 offer is fulfilled. There are numerous instances of the application of this maxim. This is how *Chitty on Contracts* (30th ed) 2008, vol 1, in para 12-081, refers to this rule:

> If the words used in an agreement are susceptible of two meanings, one of which would validate the instrument or the particular clause in the instrument, and the other render it void, ineffective or meaningless, the former sense is to be adopted. This rule is often expressed in the phrase *ut res magis valeat cum* [*sc quam*] *pereat*. Thus, if by a particular construction the agreement would be rendered ineffectual and the apparent object of the contract would be frustrated,
> C but another construction, though by itself less appropriate looking to the words only, would produce a different effect, the latter interpretation is to be applied, if that is how the agreement would be understood by a reasonable man with a knowledge of the commercial purpose and background of the transaction. So, where the words of a guarantee were capable of expressing either a past or a concurrent consideration, the court adopted the latter construction, because the former would render the instrument void. If one construction makes the contract lawful and the other unlawful, the former is to be preferred.

[56] It seems to me that this construction is also supported by the offer's language that purports to define the 21 days for which the offer will be open as the "relevant period". This is not exactly the way in
D which Part 36 defines the relevant period, but the clear intention to assimilate the 21 days for which the offer will be open with Part 36's relevant period is another indication that a construction for "open for 21 days" must be found that is consistent with Part 36 and its concept of the relevant period. A point may perhaps have been taken that the offer did not comply with CPR 36.2(2)(c). However, no such point was taken, and the judge was satisfied that that rule had been complied with. No objection with his judgment in that regard has been taken for the purposes of this appeal. He said, in [12]:

> In the case of the Offer Letter, the specification of the relevant period can be found, if it is found at all, only in the Offer Paragraph. There is nowhere else in
> E the Offer Letter where the specification of a relevant period is to be found and the definition of "the Relevant period" makes the position clear. So one can see that at least one of the purposes of the Offer Paragraph is to ensure compliance with rule 36.2(2)(c).

If so, then that is a powerful pointer to the self-same words having to be interpreted in a way consistent with Part 36 as a whole and with the doctrine of the relevant period, something that on the respondent's construction becomes wholly irrelevant. Yet the same 21 days are involved.

[57] The solution to this issue is also, to my mind, demonstrated by
F the position that would occur if the offer said nothing about how long it was open for. Mr Barnes acknowledged that in the absence of any statement as to the time for which an offer is open, at common law the offer would lapse within a reasonable time. On that basis, every offer, if not expressly, then implicitly, is time-limited. That would make Part 36 impossible. That supports the view that when the Part 36 regime is invoked, the construction of the timing aspect of the Part 36 offer has to be construed with its regime well in mind. A Part 36 offer that says nothing as to the length of time for which it is open does not close within a reasonable time: it closes only with withdrawal.

[58] As it was, the judge favoured the claimant's construction. That construction was said by the claimant to be the only "admissible construction". The defendant's construction was said to be "simply not an admissible construction": in [31] in the judgment below. That, however, is to beg the question, and does not represent the respondent's present position. The judge appears also to have been influenced by the thought that the offer "shows clearly that it is drafted on the basis of the mistaken view" that a time-limited offer can be a Part 36 offer: see his discussion in [32] to [37]. However, the judge does not explain why he takes the view that the offer was drafted on that mistaken basis, for which purpose he limited himself quite properly to the terms of the offer itself. That, however, as it seems to me, is to beg the question, which is what the offer means as between the two possible meanings (even if the judge appears to have regarded only one as feasible). Or else it is to make an assumption about what the offeror subjectively intended to say, which is both to beg the question and to do so on an illegitimate basis, since what has to be sought is the objective and not the subjective intent.

[59] The court explored in the course of argument whether the terms of the offer itself showed that an answer was needed urgently, and whether this was additional context that supported the respondent's construction of a time-limited offer. After all, the respondent was saying in its letter that it was in negotiations for the sale of the property, in which case, if the property was sold to a third party, one of the two alternative offers, which involved the appellant buying the land at a discounted price, would have fallen away, with the attendant risk that if that offer was subsequently accepted, the claimant would have been unable to comply. That was not a point high-lighted by the respondent below or on this appeal. The answer to it, however, is that it is neutral between the two constructions. On any view, the offer letter is telling the offeree that it ought to make up its mind quickly, because the offer may be promptly withdrawn. The error, however, subject to issue (3), was in not withdrawing the offer.

[60] For similar reasons, the respondent's attempt to draw support for its construction from the subsequent e-mails fails. I shall set out those e-mails below, under issue (3), where they properly belong. A document cannot be construed by reference to the parties' subsequent conduct. In any event, as will appear below, the meaning of the e-mails themselves depends on the proper construction of the offer. If the offer letter is a time-limited offer, the e-mails can be construed in accordance with that premise. If, however, the offer letter is a warning of the imminence of a future withdrawal, then the e-mails can be read in accordance with that premise. What is not permissible is to assume the answer to the problem of construction, then read the e-mails accordingly, and then read back that reading to support the assumption made in the first place.

[61] For these reasons, I would respectfully disagree with the judge's construction of the offer. I prefer the construction of the appellant.

Issue (3): Do the terms of the offer itself, or the subsequent e-mails, amount to a withdrawal of the offer?

[62] Mr Barnes submitted that the terms of the offer itself amounted to a withdrawal. That submission, however, depended on his construction being adopted. And if it had been, the offer would not have been within Part 36, and withdrawal would not have been necessary. In the circumstances, it is unnecessary to say more than that I do not believe that an express time limit is the equivalent of a Part 36 withdrawal.

[63] Mr Barnes also submitted that the offer, if within Part 36 and to be construed as the appellant would construe it, was nevertheless withdrawn by subsequent e-mails.

[64] Those e-mails were as follows:

(i) Claimant's solicitor to defendant's solicitor, 18 December 2009:

> You will be aware that our client's Part 36 offer will remain open until 31 December. If your clients intend to revert after such time, we would expect a formal request detailing a specific date for extension of the deadline and the reason for this request. We would then revert to our client accordingly.

(ii) Defendant's solicitor to claimant's solicitor, 18 December 2009:

> Thank you for your email we are aware when the period for acceptance expires however to provide a formal reply our client has to go through its internal reporting procedures… this process may not have completed until 8th January.

(iii) Claimant's solicitor to defendant's solicitor, 23 December 2009:

It was not clear from your email that a response was required, apologies. We had understood your clients' internal reporting procedures necessitate a longer period for consideration of the settlement offer, until 8 January 2010. If this is still the case, our client is prepared to wait until that date for a full response to its offer.

(iv) Defendant's solicitor to claimant's solicitor, 23 December 2009:

Thank you for clarifying the position.

(v) Claimant's solicitor to defendant's solicitor, 8 January 2010:

Our client agreed to wait until today for a full response from your client to its settlement offer. We note that no response has been forthcoming and ask that you confirm by the end of the day when such response will be available.

(vi) Defendant's solicitor to claimant's solicitor, Friday 8 January 2010:

I have not been able to get into the office but will respond early next week.

(vii) Claimant's solicitor to defendant's solicitor, Monday 11 January 2010:

In relation to your clients' response on settlement, we request a response in the shortest possible order.

[65] In my judgment, none of these e-mails amounts to a withdrawal of the offer. On the contrary, they are all concerned with an extension of the 21-day period. The last e-mail is again extending the period, for some uncertain but short period. On the hypothesis of the respondent's construction, they may be read as dealing with a deadline after which acceptance would be ineffective. On the hypothesis of the appellant's construction, they may be read as dealing with the deadline after which the offer may be withdrawn. In any event, they do not amount to a withdrawal (and see, in this respect, Moore-Bick LJ in *Gibbon*, in [17]). The judge was of the same view, in [40].

[66] There was no suggestion on the part of the claimant that the terms of these e-mails gave it some additional right against the defendant to say that the offer had lapsed.

Conclusion

[67] For these reasons, I would allow the appeal. The defendant's acceptance of the alternative offer, to pay £2m, on 5 November 2010, only three and a half weeks before trial, was effective, for the claimant had made a Part 36 offer that it had never withdrawn.

[68] It is said that such an acceptance was opportunistic, for disclosure and the exchange of witness statements had led the claimant to think that its case had improved. However, whether that view is correct or not would have been revealed, if at all, at trial and has not been debated in these interlocutory proceedings. It is said that the offer had become less advantageous to the claimant, because it has continued to suffer additional holding costs of the property in the form of security and insurance costs. That may be, but such variables will always arise, as Moore-Bick LJ explained in *Gibbon*, in [16]. It is suggested that the defendant's case with regard to the offer lacks merits, but a similar suggestion failed to deflect the result in *Gibbon*. Ultimately, it is important for the security of the Part 36 scheme, in countless cases, that it should be clearly understood that if a claimant wishes to make a time-limited offer, in the sense that the offer is to lapse of its own accord at the end of a stipulated period, then such an offer cannot be made as a Part 36 offer; that an offer presented as a Part 36 offer and otherwise complying with its form will not readily be interpreted in a way that would prevent it from being a Part 36 offer; and that if an offeror wants to bring his Part 36 offer to an end, so that it cannot be accepted, then it must serve a formal notice of withdrawal. It seems to me that, although the precise point raised in this appeal is new, all the jurisprudence on Part 36 cited above contributes to these conclusions.

Giving the second judgment, **Rimer LJ** said: G

[69] I have had the advantage of reading in draft the judgment of Rix LJ and gratefully adopt his comprehensive explanation of the issues that have given rise to this appeal. I agree with Rix LJ that the appeal should be allowed. In deference to Warren J's carefully considered different view, I add my own reasons for so concluding, although I am conscious that they are based on ground already covered by Rix LJ.

[70] Rix LJ, in [32], has identified the three issues argued before us. As to the first, namely whether it is possible for a Part 36 offer to be a time-limited one, I agree with the judge and Rix LJ that it is not. Rix LJ has fully explained why the making of a time-limited offer is H inconsistent with the scheme of Part 36 and there is nothing on that topic that I can usefully add. It follows that if, on its true construction, the claimant's offer was a time-limited one, destined by its own terms to lapse at the expiry of 21 days from 10 December 2009, then: (i) it was not a Part 36 offer at all and the provisions of Part 36 had no application to it; and (ii) the defendant's purported acceptance of it well outside the 21-day period was ineffective.

[71] The crucial issue is the second one that Rix LJ has identified in [32]. In short, was the offer a time-limited one? The answer to that turns on its true interpretation in the context in which it was made, that context being a purported Part 36 offer. In advancing his submissions J on that question, and while rightly recognising that context is all, Mr Barnes asked that we should, at least in the first instance, simply consider the ordinary meaning of the key words of the offer paragraph divorced from their context. The judge, while of course also recognising that the context was crucial, engaged in such an exercise in [8] and [10] of his judgment. In his view it was obvious that, viewed in isolation, an offer declaring itself to be "open for 21 days" would be a time-limited one. As he put it, "What else, it might be asked, could those words mean?"

[72] There is, I consider, compelling force in that. If, following oral discussions between X and Y about a possible sale by X to Y of his K house, X writes to Y formally offering to sell it to him for £300,000 and stating that "this offer will be open for 21 days from the date of this letter", I consider that the objective interpretation of X's words, as Y ought reasonably to understand them, would be that acceptance within the 21-day period would secure the purchase, whereas a purported acceptance after that period would be too late. In the absence of any relevant context requiring a different meaning to be attached to that particular collection of English words (there being none in my example), that is the natural meaning of the offer and that is how Y would understand it.

[73] What, however, Y might not also understand unless he is a L lawyer is that he will not necessarily be able to bask in the certainty of actually having up to 21 days within which to accept the offer. That is because it would be open to X to withdraw it at any time before acceptance. X's offer therefore commits him to nothing unless and until it is validly accepted; and, until then, it likewise confers no legally enforceable rights on Y. It would be different if, for consideration moving to X from Y, X agrees to keep the offer open for 21 days. The arrangement would then be in the nature of an option and Y would be ensured the full 21 days in which to decide whether to exercise it. If he decides not to, the option will lapse. There is no question of any option having been granted in this case.

[74] Having devoted two paragraphs to a consideration of the M key words of the offer paragraph read in isolation, I suspect that Lord Hoffmann would regard them as an irrelevant distraction. He has perhaps done more than any other judge in recent years to focus judicial attention on the point that often the true sense of the words that people use can be assessed only by paying careful regard to the particular context in which they have used them. Thus, in *Charter Reinsurance Co Ltd (in liquidation)* v *Fagan* [1997] AC 313, at p392A, he observed that "[i]t is artificial to start with an acontextual preconception about the meaning of the words used and then see whether that meaning is somehow displaced". This is because, as he shortly afterwards explained in *Mannai Investment Co*

Ltd v *Eagle Star Life Assurance Co Ltd* [1997] AC 749* and *Investors Compensation Scheme Ltd* v *West Bromwich Building Society* [1998] 1 WLR 896 (and quoting from *Investors Compensation Scheme*, at p913B-C):

> The meaning which a document (or any other utterance) would convey to a reasonable man is not the same thing as the meaning of its words. The meaning of words is a matter of dictionaries and grammars; the meaning of the document is what the parties using those words against the relevant background would reasonably have been understood to mean. The background may not merely enable the reasonable man to choose between the possible meanings of words which are ambiguous but even (as occasionally happens in ordinary life) to conclude that the parties must, for whatever reason, have used the wrong words or syntax.

[75] In the present case, therefore, it is not of utility to consider the meaning of the offer paragraph in isolation from the context in which the offer was made. Whatever else may be in dispute, there is no dispute that the offer was intended to comply with Part 36. It was expressly stated to be an "Offer to Settle under CPR Part 36" that was "intended to have the consequences set out in Part 36… ." Of course, that does not mean that it did in fact comply with Part 36 and therefore must, come what may, somehow be shoehorned into the confines of its four corners: a stated bid to attain a particular goal does not also mean that the goal has been attained. The answer to the critical question still turns on how the reasonable man would read the offer. The relevance, however, of the claimant's expressed intention to make its offer a Part 36 offer is that, if there are any ambiguities in it raising a question as to whether the offer does comply with the requirements of Part 36, the reasonable man will interpret it in a way that is so compliant. That is because, objectively assessed, that is what the offeror can be taken to have intended. That is also in line with the principle of construction to which Rix LJ referred in [55].

[76] How, therefore, should the offer paragraph be construed? It must, I consider, be read and interpreted against both the requirements prescribed by Part 36 for the making of a valid Part 36 offer and also the effect and consequences of such an offer that Part 36 describes. So reading the offer paragraph, its description of the 21-day period as the "relevant period" is, I consider, important and is obviously a reference to the "relevant period" as defined in Part 36.3(1)(c). What does Part 36 tell us about the "relevant period"? One thing it tells us is in Part 36.3(5), which provides that "[b]efore expiry of the relevant period, a Part 36 offer may be withdrawn or its terms changed to be less advantageous, only if the court gives permission." Part 36.3(6) then provides that after the expiry of the "relevant period" an unaccepted offer can be withdrawn, or the offeror can change its terms, without the court's permission.

[77] Those are provisions of a Part 36 offer of which the reasonable offeree will be aware and that will be relevant to his interpretation of the critical words in the offer paragraph. He may perhaps read those words as, on one view, meaning that the offer is a time-limited one that will automatically expire after 21 days. However, that interpretation sits oddly with Part 36.3(6), which provides for an absolute right to withdraw the offer after 21 days. Why should the offer be read as such a time-limited one when the scheme of the provisions under which it is made expressly permits such a withdrawal and so provides the offeror with the almost exact commercial equivalent of an offer that is so time-limited? The reasonable man would, I consider, accordingly regard that interpretation of the critical words as unlikely, if not impossible; and that their more likely explanation is that they were simply intended to express a qualification of the provisions of Part 36.5(5) by way of a communication to the offeree that, during the 21 days during which the offer "will be open", the offeror will not seek the court's permission to withdraw it.

[78] That appears to me to be a feasible and reasonable interpretation of the critical words, and one that is congruent with the scheme of Part 36.3(5) and (6). Mr Barnes recognised it as a possible interpretation, although it was not one that he supported. In my view, it is an interpretation that, in the context of an offer expressed to be made under Part 36, fits comfortably with its terms. I regard it as the correct interpretation. It enables the offer to be read as the offeror expressly intended, namely as a Part 36 offer that would be governed by the provisions of Part 36.

[79] Once that conclusion has been arrived at, the force of the point that the offer was a time-limited one is, in my view, extinguished. If the sense of the offer paragraph can rationally be explained as serving the purpose just described, it is difficult also to ascribe to it the dual intention of conveying that, if unaccepted within the 21-day period, the offer will lapse. Nor is there any sensible or practical need to do so. The reasonable offeree will know that, once the 21-day period has expired, the offeror can withdraw the offer at any time and without the need for the court's permission. The offeror will, or ought to, know likewise.

[80] For these reasons, which essentially reflect part of Rix LJ's much fuller reasons, I agree with Rix LJ that, objectively interpreted, the offer was not time-limited but was an offer that complied, as it claimed to do, with Part 36.

[81] The third issue to which Rix LJ referred in [32] was as to whether, if so, the offer was withdrawn by the claimant before its purported acceptance. For the reasons given by Rix LJ, to which I cannot usefully add, I agree with him that it was not.

[82] I too would allow the appeal.

Giving the third judgment, **Stanley Burnton LJ** said:

[83] The first question to be considered by Warren J was whether CPR 36 permits a time-limited offer. It is regrettable that neither the CPR nor the practice directions 36A and 36B expressly address this issue. I agree with the conclusions of Warren J and Rix and Rimer LJJ on it, but I am left with the uncomfortable suspicion that when the offer letter was sent and received neither of the parties to these proceedings appreciated that a time-limited offer is not a Part 36 offer. However, their subjective assumptions on this issue are irrelevant to the construction of the offer.

[84] Any ambiguity in an offer purporting to be a Part 36 offer should be construed so far as reasonably possible as complying with Part 36. Once it is accepted that a time-limited offer does not comply with Part 36, one must approach the interpretation of the offer in this case on the basis that the party making the offer, and the party receiving it, appreciated that fact.

[85] I agree that the normal effect of the phrase "the offer will be open for 21 days" is that the offer is not open for acceptance after 21 days. However, Mr Barnes QC accepted that the use of that phrase is consistent with a warning that the offer will be withdrawn after 21 days. Given the clear express intention of the respondent to make an offer complying with Part 36, it should be so construed.

[86] There are other indications in the offer that it was intended to be available only for prompt acceptance. However, these indications are accommodated by the respondent's ability to withdraw the offer.

[87] It follows that, despite Warren J's carefully reasoned contrary conclusion, and my suspicion that he may well have correctly identified the subjective intention of the respondent, I agree that this appeal should be allowed.

Appeal allowed.

* Editor's note: Also reported at [1997] 1 EGLR 57; [1997] 24 EG 122; [1997] 25 EG 138

LANDLORD AND TENANT
GENERAL

Eastern Power Networks plc (formerly EDF Energy Networks (EPN) plc) *v* BOH Ltd and another

COURT OF APPEAL

[2011] EWCA Civ 19

26 January 2011

SEDLEY, RIMER AND BLACK LJJ

Landlord and tenant — Merger of interests — Split reversion — Leaseholder of electricity substation claiming rights over land of other parties for benefit of substation — Rights conferred under lease of substation site and other land — Leaseholder subsequently acquiring freehold of substation site only — Whether lease determined in respect of that site by merger of freehold and leasehold interests — Applicable presumptions in cases of split reversion

The respondent electricity supplier operated a substation on the Wembley Stadium trading estate. In order to access the substation, it was necessary to cross areas of land owned by the two appellants; underground electricity cables also crossed their land. The respondent claimed to have the necessary rights for those purposes under a lease of the substation site and other land for 42 years, granted to its predecessor in 1953 by a common predecessor in title of the parties. That lease had included a right of way over the lessor's land for all purposes necessary for the enjoyment of the demised premises, together with a right to lay and maintain electricity cables. Although the contractual term had expired in 1994 and the respondent had since acquired the freehold of the substation site, it contended that the lease, with its attendant rights, continued pursuant to Part II of the Landlord and Tenant Act 1954. The appellants disputed the respondent's rights and contended that the lease had come to an end by, *inter alia*, merger with the freehold acquired by the respondent.

In proceedings brought by the respondent to establish its rights, the trial judge held that no merger had taken place and that the lease continued, such that the respondent had rights of access over parts of the appellants' land and a right to maintain the underground cables. He held that although merger could occur in respect of part only of a leasehold interest, or where the reversion was severed, the appellants had not rebutted the equitable presumption against an intention of merger that applied in cases where it would benefit the leaseholder to keep the two interests separate: [2009] EWHC 3193 (Ch); [2010] 2 P&CR 3.

The appellants appealed. They contended that where a party had acquired the freehold of part of the land comprised in its lease, a presumption of merger should arise in respect of that part, so as to avoid the potential for tension between the respective interests of that party and the reversioners of the remaining part.

***Held*: The appeal was dismissed. Where a leaseholder acquires the freehold of part of the premises comprised in its lease, the two interests will merge only if the party in which the two estates are vested intends a merger. An equitable presumption arises against merger where the leaseholder does not expressly evince an intention to merge and a merger would be counter to its interests. That presumption applies regardless of whether merger would be of advantage to third parties, such as co-reversioners, with which the leaseholder is in no commercial relationship and to which it owes no duties; the focus of enquiry is exclusively on the leaseholder's interests.** F **It is not appropriate, in cases where a merger is in the interests of co-reversioners but against those of the leaseholder, to apply the contrary presumption that merger will occur in the absence of evidence that no merger was intended. That would involve an unwarranted departure from the established equitable principles, rather than a development of them. Such a presumption is not justified by any potential for tension between the respective interests of the leaseholder and the other reversioners. Accordingly, in the absence of any evidence of the respondent's intention on the issue of merger, and given that a merger would not be in its interests, it was to be presumed that no merger was intended.** G

The following cases are referred to in this report.

Capital & Counties Bank Ltd v *Rhodes* [1903] 1 Ch 631, CA
Dodson Bull Carpet Co Ltd v *City of London Corporation* [1975] 1 WLR 781; [1975] 2 All ER 497; (1974) 29 P&CR 311; [1975] 2 EGLR 62; 236 EG 484
Eastern Power Networks plc (formerly EDF Energy Networks (EPN)) v *BOH Ltd* [2009] EWHC 3193 (Ch); [2010] 2 P&CR 3; [2010] L&TR 14
Grice v *Shaw* (1852) 10 Hare 76
Ingle v *Vaughan Jenkins* [1900] 2 Ch 368, Ch
Jelley v *Buckman*; *sub nom Jelly* v *Buckman* [1974] QB 488; [1973] 3 WLR 585; [1973] 3 All ER 853; (1973) 26 P&CR 215; 228 EG 1512, CA H
Long (Nevill) & Co (Boards) Ltd v *Firmenich & Co* (1983) 47 P&CR 59; [1983] 2 EGLR 76; 268 EG 572, CA
Southport Old Links Ltd v *Naylor* [1985] 1 EGLR 66; (1984) 273 EG 767, CA
Thorne v *Cann* [1895] AC 11, HL

This was an appeal by the appellants, BOH Ltd and Layhawk Consultants Ltd, from a decision of Mr Edward Bartley Jones QC, sitting as a deputy High Court judge, allowing a claim by the respondent, Eastern Power Networks plc, for declaratory relief as to the continuation of a lease and certain rights conferred thereunder.

Mark Warwick QC (instructed by Colman Coyle LLP) appeared for the appellants; Christopher Stoner QC (instructed by Forsters LLP) J represented the respondent.

Giving judgment, **RIMER LJ** said:

Introduction

[1] This appeal, brought with the permission of Mummery LJ, is against an order dated 4 December 2009 made by Mr Edward Bartley Jones QC, sitting as a deputy High Court judge in the Chancery Division. The appellants (defendants to the claim) are BOH Ltd (BOH) and Layhawk Consultants Ltd (Layhawk). The respondent (the claimant) is Eastern Power Networks plc, the public utility supplier. It was formerly known, and sued, as EDF Energy Networks (EPN) plc K and, like counsel in argument, I will call it EDF. Mr Mark Warwick QC represented the appellants before us but not before the judge. Mr Christopher Stoner QC represented EDF both before us and below. The neutral citation number for the judge's judgment is [2009] EWHC 3193 (Ch) and it is reported at [2010] L&TR 14.

[2] The case concerns an approximately rectangular area of land bounded on the south by Fourth Way, a public highway, on the Wembley Stadium trading estate. Each party owns the freehold of part of the land. EDF owns plot 2, on which it operates an electricity substation. Plot 26 (owned by Layhawk) and plot 20 (owned by BOH)

A lie between plot 2 and Fourth Way. The issue in the claim was whether EDF has: (i) rights of access and egress between plot 2 and Fourth Way over plots 20 and 26; and (ii) a right to maintain in position in plot 20 various underground cables that pass through plot 26 to the plot 2 substation. BOH and Layhawk deny that EDF has any such rights and claim, by virtue of their ownership of plots 20 and 26 respectively, to be entitled to prevent EDF's claim to exercise such rights. The reality is that they wish to exploit their plots as ransom strips.

[3] If EDF does have such rights, it is only by virtue of its title as the tenant of plot 2 under a 1953 lease that created a 42-year tenancy of plots 2, 26 and 31. The critical question before the judge was
B whether EDF's tenancy of plot 2 was continuing in existence under the provisions of Part II of the Landlord and Tenant Act 1954 (the 1954 Act). That depended on whether, as BOH and Layhawk asserted, EDF's plot 2 tenancy had merged in the freehold of plot 2 that EDF acquired in 1993 and had so become extinguished. The judge accepted EDF's submission that it had not and that, as the tenant of plot 2, it continued to enjoy the appurtenant rights over plot 20 conferred by the 1953 lease, as well as its rights as the tenant of plot 26. He granted a declaration accordingly. By their first ground of appeal, BOH and Layhawk challenge that conclusion.

[4] By their second ground of appeal, BOH and Layhawk (which
C have changed their solicitor and counsel since the trial) have raised a new fall-back argument not argued below. It arises under the Convention for the Protection of Human Rights and Fundamental Freedoms (the Convention). It is said that if there has been no merger resulting in the extinguishment of EDF's plot 2 tenancy, the effect of the ordinary interpretation of section 44(1A) of the 1954 Act is potentially to disadvantage the appellants in a manner likely to impair their right to the peaceful enjoyment of their possessions under Article 1 of the First Protocol to the Convention. By this ground of appeal, the appellants assert that section 3 of the Human Rights Act 1998 (the 1998 Act) requires section 44(1A) so to be read as to avoid any
D such impairment.

Facts

[5] A fuller account of the facts is as follows. An understanding of them requires the reader to imagine an approximately rectangular area of land comprising four approximately horizontal plots each extending its full width. The northernmost plot is plot 31, a thin strip of just a few feet in depth. Plot 31 abuts plot 2, by far the largest plot and to its immediate south. Plot 2 abuts plot 26 (a strip of about 13ft in depth), which is to its south and which in turn abuts plot 20, a strip of rhomboid shape. Plot 20 abuts Fourth Way.

E [6] By a lease dated 20 February 1953 (the lease), the British Transport Commission (the BTC) leased plots 31, 2 and 26 (not plot 20) to the Eastern Electricity Board (the EEB, the predecessor of EDF) for a term of 42 years from 24 June 1952 and thus expiring on 23 June 1994. The rent was £25 pa, payable quarterly on the usual quarter days. Clause 2(7) restricted the use of the demised premises to that of an electricity substation in connection with the EEB's undertaking. Clause 2(6) prohibited the erection of anything other than an electricity substation on the demised premises and required the EEB to fence off the demised premises from the adjoining land. The EEB built an electricity substation on plot 2. Once built, the substation was
F enclosed by fencing around plot 2, with plots 31 and 26 remaining outside it (this is not what clause 2(6) required, but nothing turns on it). The demised premises were separated from the public highway by plot 20. The lease, however, conferred rights of way over plot 20 and also to lay cables, etc, under the highway and plot 20 (as they were), such cables passing under plot 26 to the plot 2 substation. The cables remain in position to this day. The BTC was the freehold owner of a substantial area of land, including plot 20 and what is now Fourth Way, and so was able to grant such rights.

[7] The substation is an important one. Without it there would be no power for a radius of around three miles. It serves as a back-up supplier to Wembley Stadium, but EDF's future plans are that the stadium should be fully supplied from it, a proposal that will involve the installation of more cables to the substation and a need for EDF to have a right of immediate access to it.

[8] The title to the lease is, and has been since 1953, a registered one (MX 269363). Subject to the exclusion from it in 2003 of a small, approximately square section in the north-eastern corner of plot 2 (referred to in the judgment below and in this one as the "X" land), EDF has been at all material times, and remains, the registered proprietor with a good leasehold title. The story relating to the X land is immaterial to the issues before us.

[9] In March 1984, Fastiron Ltd acquired from British Railways Board (the BTC's successor): (i) the freehold reversion expectant on the determination of the lease; and (ii) the freehold of plot 20. Fastiron thus became EDF's landlord, although it did not remain so for long.

[10] In November 1985, Fastiron transferred its freehold estate in plot 2 to Racal Properties Ltd and Racal became registered as proprietor with title absolute (NGL 541578). On 11 August 1987, Racal transferred that estate to Mr Leonard Dormer, who became the registered proprietor.

[11] Also in November 1985, Fastiron transferred to Racal its freehold estate in plot 26 and Racal became the registered proprietor with title absolute (NGL 541577). In March 1988, following two further transactions, Electrical Distributing Co Ltd (EDC) became the registered proprietor.

[12] The position by the spring of 1988, therefore, was that the title to the freehold reversion expectant on the determination of the lease was as follows: (i) Fastiron owned plot 31 (and plot 20, which was not demised by the lease but was subject to the appurtenant rights it created); (ii) Mr Dormer owned plot 2; and (iii) EDC owned plot 26. EDF remained the tenant.

[13] The property comprised in the tenancy created by the lease was or included premises occupied by EDF for the purposes of a business carried on by it. The tenancy was therefore one to which Part II of the 1954 Act applied: see section 23. Although the tenancy was, by its terms, due to expire by effluxion of time on 23 June 1994, the effect of section 24(1) was that (subject to section 24(2)) it would not then come to an end but would do so only if terminated in accordance with Part II.

[14] In 1992, EDF commenced negotiations with Mr Dormer for a consensual renewal of the tenancy (or perhaps only that part of it comprising plot 2: the explanatory material is not before us and the position was not explained). On 27 August 1993, he served a notice on EDF purportedly under section 25 of the Act. It related to "BEE Primary Substation and land" (the judge perhaps regarded as uncertain whether it purported to relate only to plot 2 or also to plots 31 and 26: see [50] of his judgment) and purported to terminate the tenancy on 24 June 1994. It notified EDF that any application by it to the court for the grant of a new tenancy would be opposed on the grounds specified in section 30(1)(*f*) and (*g*) of the Act. EDF gave Mr Dormer a counternotice on 3 September 1993 that, on the termination of the tenancy, it was not willing to give up possession of the premises referred to in the section 25 notice.

[15] The negotiations between EDF and Mr Dormer continued, resulting, on 23 December 1993, in the purchase by EDF for £237,500 and the transfer to it by Mr Dormer of his freehold estate in plot 2. EDF became and has since remained the registered proprietor of that estate (save, since 2003, for the X land). Thus, as from 23 December 1993, the freehold reversion immediately expectant on the determination of the tenancy created by the lease was split between: (i) Fastiron, which owned plot 31 (and also plot 20); (ii) EDF, which owned plot 2; and (iii) EDC, which owned plot 26. EDF was therefore in the relatively unusual position of having become the freehold owner of part of the land — plot 2 — of which it was also the tenant under the lease. Twenty-third June 1994 — the date specified in the lease for the termination of the tenancy — came and went. EDF had, in the meantime, made no application to the court for the grant of a new tenancy.

[16] Further dealings with the freehold title followed. In March 1999, Containerised Storage Co Ltd (Containerised Storage) acquired plots 20

and 31 from Fastiron and became registered as proprietor of both plots (NGL 770835). On 30 June 2002, it transferred that interest to BOH, which, on 24 December 2002, became the registered proprietor under a new title number, NGL 816821, having paid £750 for its acquisition.

[17] As for plot 26, in March 2002, EDC transferred it to Commercial Rental Ltd (Commercial Rental), which became registered as proprietor on 24 March 2002 (NGL 514577). Plot 26 was then transferred to Layhawk, which, on 19 January 2005, became its registered proprietor.

[18] In [29] of his judgment, the judge described how trouble started for EDF with the March 1999 sale, with a Mr Paul Todd engaging in increasingly aggressive correspondence. He was the sole witness at the trial for BOH and Layhawk. As I explain in [52] below, he is said to have been an agent for Containerised Storage and Commercial Rental; and they, BOH and Layhawk were or are apparently all connected in one way or another. In [30], the judge explained the plot 31 story, which resulted in Mr Avanzato also being joined as a defendant. He is not concerned with the issues before this court and was not represented before us. In [31], the judge explained the position with regard to the X land.

Issues before the judge

[19] The judge had before him more issues than we do. It was common ground at the trial that none of the rights of access or of cabling that EDF requires over and in respect of plots 26 and 20 is appurtenant to its *freehold* estate in plot 2. The only case that EDF sought to make was that despite its acquisition in 1993 of the freehold of plot 2, its *tenancy* of (at least) plots 2 and 26 created by the lease was still continuing under section 24 of Part II of the 1954 Act: that was because it had never been terminated in accordance with Part II. If that was correct, EDF would continue to enjoy the like appurtenant rights in respect of plot 2 as it had always done. As the judge explained, it did not matter to EDF whether its tenancy also continued in respect of plot 31; and leading counsel who appeared at the trial for BOH and Layhawk concentrated in particular on his submission that the tenancy in respect of plot 2 had terminated, to which end he argued that the section 25 notice that Mr Dormer had served in August 1993 had brought the tenancy of plot 2 to an end on 24 June 1994. That was the date specified in it for the termination of the tenancy and EDF had made no application to the court for the grant of a new tenancy. If the argument was right, it would hole EDF's case below the waterline.

[20] The judge, however, rejected it. He explained that the severance into three separate ownerships of the reversion did not create three separate tenancies: despite such severance, there remained but one tenancy: see *Jelly* v *Buckman* [1974] QB 488* and *Nevill Long & Co (Boards) Ltd* v *Firmenich & Co* (1983) 47 P&CR 59†. He explained further that, in order to be valid, a section 25 notice had to be given in respect of the whole of the property comprised in the tenancy: see *Dodson Bull Carpet Co Ltd* v *City of London Corporation* [1975] 1 WLR 781‡ and *Southport Old Links Ltd* v *Naylor* [1985] 1 EGLR 66§. Yet further, he pointed out that the conventional view had always been that, in the case of a split reversion, *all* the reversioners must join in the giving of a section 25 notice if it is to be valid, a principle now expressly affirmed with effect from 1 June 2004 by section 44(1A) of the 1954 Act. In this case, however, the notice had been given by Mr Dormer alone. The judge's conclusion, in [47], was that the section 25 notice he had served was invalid because: (i) it related only to plot 2 and not also to plots 31 and 26; and (ii) it was given by only one of the three separate reversioners. There is no appeal against those conclusions. (They were alternative conclusions, each fatal, and I have mentioned that, in [50], the judge perhaps evinced some uncertainty as to the extent of the land to which the notice purported to relate.)

[21] It was next argued that, because EDF had served a counternotice in response to the section 25 notice indicating that it was not willing to give up possession of the premises referred to in the notice, it became estopped, by representation or convention, from denying the validity of the section 25 notice. The judge dealt with that submission between [48] and [61] and rejected it. There is no appeal against his decision in that respect either. G

[22] The next issue argued before the judge was that on EDF's acquisition in 1993 of the freehold estate in plot 2, its tenancy in respect of plot 2 (although not in respect of plots 31 and 26) must be regarded as having merged in the freehold estate and so have come to an end. If so, even though its leasehold estate in plot 26 may have continued, EDF would not, as the freehold owner of plot 2, have continued to enjoy the rights over plot 20 that were formerly appurtenant to its tenancy interest in plot 2; nor could it use its rights as a tenant in respect of plot 26 to gain access to plot 2. H

[23] As to that argument, the judge rejected EDF's submission that a merger as to merely a part of a leasehold estate was not possible as a matter of law, and held that it was possible, including in a case where, as here, the reversion had become severed. As to the merits of the argument, he explained that, at common law, a merger occurred automatically, reflecting the principle that one person cannot be both landlord and tenant of the same premises, whereas in equity (which, by section 185 of the Law of Property Act 1925 (the 1925 Act), now prevails) the position was different. In equity, regard must be had to the intention of the parties. The judge referred to Cozens-Hardy LJ's observations in *Capital & Counties Bank Ltd* v *Rhodes* [1903] 1 Ch 631, at p652, that the courts of equity mitigated the rigours of the common law and: J

> had regard to the intention of the parties, and, in the absence of any direct evidence of intention, they presumed that merger was not intended, if it was to the interest of the party, or only consistent with the duty of the party, that merger should not take place.

K

[24] The judge's conclusion was that there had been no merger. Since that conclusion is challenged by the first ground of appeal, I will set out his reasoning. Having held that there was in principle no reason why there could not be a merger in respect of part of the premises comprised in a lease, he continued:

> 66. Out of this submission developed, however, a more sophisticated argument. Is it possible to merge as to part where the reversion is severed? The Lease might contain covenants which benefit ("touch and concern") other parts of the land comprised within the Lease. A lease may be granted of Plots 1, 2 and 3 (all of which are dwellinghouses) and contain a covenant, say, against keeping pigs on any part of the demised premises. The reversion then becomes severed so that there are individual reversions to Plots 1, 2 and 3. Following acquisition of the freehold reversion on Plot 1 by the tenant of Plot 1, the freehold reversioners to Plots 2 and 3 might well be concerned by merger on Plot 1 for, thereby, Plot 1 would be freed from the restraint against keeping pigs (to the detriment of the reversions on Plots 2 and 3). Equally the right to distrain for rent might be adversely affected. Thus, and taking the above example, if there be merger as to part in respect of Plot 1 the right of the severed reversioners of Plots 2 and 3 to distrain over Plot 1 would appear to be lost on merger in respect of Plot 1. And rent (absent formal apportionment on severance of the reversion) issues out of each and every part of the demised land. Ultimately, however, I could detect no rule of law or rule of equity that the existence of a severed reversion *of itself* prevents merger as to part. Rather it seems to me that if the other severed reversioners do have a genuine interest in the issue of merger then their interest has to be taken into account in ascertaining whether the equitable presumption against merger is rebutted. It may, perhaps, be that in an appropriate case the absence of the actual consent of the other severed reversioners to the merger as to part will be sufficient so as to conclusively prevent rebuttal of the presumption against merger. In other cases, perhaps, the absence of such consent may (because the interest of the other severed reversioners in opposing merger is, effectively, de minimis) not inform at all the issue as to whether the presumption against merger is rebutted. That there should be no absolute rule of law or equity that absence of the express or implied consent of the other severed reversioners prevents merger as to part seems to me to be in accord with the equitable principles which apply to merger (namely that the same is a matter of intent).

L M

* Editor's note: Also reported at (1973) 228 EG 1512

† Editor's note: Also reported at [1983] 2 EGLR 76; (1983) 268 EG 572

‡ Editor's note: Also reported at [1975] 2 EGLR 62; (1974) 236 EG 484

§ Editor's note: Also reported at (1984) 273 EG 767

67. I must, at this stage, comment on surrender as to part where there is [a] severed reversion. Can surrender as to part occur without the consent (express or implied) of all the severed reversioners? And does the answer to this question inform the issues which arise on merger as to part where there is a severed reversion? If, as EDF contend, the Lease has continued under Part II there has been a surrender of the "X" land without the consent of the severed reversioners. By such surrender the severed reversioners lost the right to distrain upon the "X" land and, perhaps more importantly, to enforce as to the "X" land the user restriction contained in clause 2(7) of the Lease. Although this issue was raised in submissions, no authority was cited to me which would assist on this question. As far as I am aware it is not considered in any textbook, indeed it may never have been considered at all. It may have no practical relevance on the facts of this case. Masterdent Ltd has the benefit of being registered under Title Number NGL 8273728 as Proprietor with Title Absolute of the freehold estate in the "X" land (entirely free of the Lease). It is difficult to see how the user covenant in clause 2(7) of the Lease confers any meaningful benefit on either Plot 31 or 26. The annual rent due under the Lease which should be apportionable to Plots 31 and 26 cannot be a matter of, at the most, a few pence or pounds and, in any event, the sub-station offers the ideal subject matter for distraint. But I do incline to the view that where there is a severed reversion a surrender as to part requires the consent (express or implied) of all the severed reversioners (save, perhaps, in a case where there has both been a formal apportionment of the rent *and* none of the covenants in the lease "touch and concern" the reversion of the other severed reversioners). To this extent, therefore, there may be a subtle difference between the principles applying on merger (as I have analysed them above) and surrender.

68. Against this background I turn to consider whether, in respect of Plot 2, the presumption against merger is rebutted. I do not intend to deal with this matter simply on the basis of the location of the burden of proof but it does seem to me that the burden of proof on this point is borne by [leading counsel for BOH and Layhawk]. Were it relevant I would have held that BOH and Layhawk had failed to discharge that burden (since, quite understandably, they were unable to advance any evidence whatsoever of any positive intent on the part of EDF to effect merger).

69. There is no direct evidence (either way) of the intent of EDF (or of the other two severed reversioners) in respect of merger following the acquisition of the freehold of Plot 2 by EDF on 23 December 1993. The reality, on the balance of probabilities, is that neither EDF (nor the other two severed reversioners who in all probability knew nothing whatsoever about the acquisition of the freehold of Plot 2) ever directed their minds to the question of merger. Whilst it is true that following the acquisition of the freehold of Plot 2 EDF stopped paying any of the rent due under the Lease I do not think that this evidences any intent to merge. On the contrary, it seems to me to be entirely neutral since EDF (absent any formal apportionment of the rent due under the Lease following initial severance of the reversion) could have paid the whole £25 rent as reserved by the Lease to itself (a pointless circular transaction). But if any evidence of intent to merge could be derived from the non-payment of rent this is more than balanced out by the fact that no application to merge was made by EDF when it became registered as proprietor of the freehold land comprised in Title Number NGL 541578. Thus Title Number MX 269363 continued in existence (without Plot 2 being taken out). This could not have occurred if there had been an express application for merger to HM Land Registry. On 6 November 2000 HM Land Registry wrote to EDF's solicitors indicating that, having inspected their files, HM Land Registry could not find any clear evidence as to merger one way or the other. This would clearly indicate that these files do not disclose any express application by EDF for merger. Over and above this, there is no further evidence on the issue of merger but merger was clearly to EDF's detriment granted the fact that the freehold carried no easements over Plot 20 whatsoever and that merger in respect of Plot 2 would give rise to the *Harris v. Flower* problem over Plot 26 (so far as use of Plot 26 for access to Plot 2 was concerned). That of itself, on the authorities, is sufficient to support the presumption against merger. Indeed it seems astounding, objectively analysed, that EDF should have paid £237,500 to Mr Dormer to acquire the freehold of Plot 2 if, thereby, it was going to lose its access and cabling rights over Plot 26 and Plot 20. To impute such an intention (absent direct evidence of intention) to EDF is absurd. When there is added into the equation the fact that the other severed reversioners in all probability knew nothing about what had occurred and, most certainly, did not give express or implied consent to merger the presumption against merger is only strengthened.

70. I can, therefore, find nothing which rebuts the presumption against merger and, so, merger did not occur.

[25] Given that conclusion, the judge held that the tenancy created by the lease was statutorily continued by Part II of the 1954 Act after the termination date of 23 June 1994 provided by the lease, which continuation applied equally to that part of the demised premises comprising plot 2. He rejected an argument that the lease had been forfeited so far as it related to plots 26 or 31. He also rejected an argument that the lease in respect of plot 26 had been surrendered, although he accepted that it had been surrendered in respect of plot 31. The result of that holding was that the lease was continuing solely in respect of plots 2 and 26. That conclusion was and is sufficient for EDF's purposes.

First ground of appeal

[26] The judge held that in a case in which the reversion becomes split between more than one reversioner, with the tenant becoming one such reversioner in respect of part of the premises, there can in principle be a merger (and consequential extinguishment) in its reversionary estate of its tenancy in that part. There is no cross-appeal by EDF against that conclusion. Mr Warwick's submission for the appellants was that where the judge fell into error was in finding on the facts that there had been no merger.

[27] For the purposes of this argument, Mr Warwick accepted, indeed asserted, that in the case of a split reversion of a business tenancy protected by Part II of the 1954 Act, it will be necessary for *all* the reversioners to join in the giving of a section 25 notice directed at terminating the business tenancy in respect of the entirety of the tenanted premises; and that the refusal of any of the reversioners to co-operate in giving such a notice will prevent its being given. The need for all reversioners to join in the giving of a section 25 notice is apparent from section 44(1A) of the 1954 Act, which has been in force since 1 June 2004 and which further explains the definition of "the landlord" in section 44(1). Section 44(1A) provides:

> The reference in subsection (1) above to a person who is the owner of an interest such as is mentioned in that subsection is to be construed, where different persons own such interests in different parts of the property, as a reference to all those persons collectively.

[28] In a case, however, in which there are, say, three reversioners of different parts of the demised premises — say R, S and T — with T also being the tenant of the whole, it is obvious that R's and S's commercial interests in giving a notice under section 25 may not be shared by T. The tenancy may, for example, be continuing under section 24 at a rent below the market rent, with the only prospect of achieving an increased rent being by the giving of a section 25 notice terminating the tenancy. If T then wants to preserve its security of tenure, it will have to apply to the court for the grant of a new tenancy at a market rent. T will or may, however, have no incentive to join in the giving of a section 25 notice that will bring its advantageous status to an end. It will, in practice, be able to block its giving and thereby protect itself from ordinary market forces. Although it is the purpose of Part II to confer security of tenure on business tenants, it is not its purpose to provide them with protection of that nature.

[29] The issue raised by this first ground of appeal turns, therefore, on whether the potential for tension between the respective interests of, in my example, R and S, on the one hand, and T, on the other, provides a basis for a conclusion that: (i) on the acquisition by T of its separate reversionary estate in respect of part of the demised premises; and (ii) in the absence of any evidence of an *intention* on T's part to effect a merger; there is (iii) a presumption of a merger in T's reversionary estate of its tenancy in the relevant part. If yes, it would follow that the tenancy of that part would determine and that R and S would become the sole remaining landlords in respect of T's tenancy of the other two parts and could alone combine in giving a section 25 notice in respect of such parts. They might of course then find that the market rent for such parts was no more or less than it had been for the whole. They might also want to avoid the grant of a new tenancy of those parts by relying on one or more of the grounds in section 30(1). They may of course anyway have divergent views as to the steps to be taken with regard to the tenancy: the possibility of conflict is not exclusively the preserve of situations in which one of the reversioners is also the tenant.

[30] In order to assess this issue, it is necessary to consider the principles as to when a merger of a lesser interest in a greater one will

take place. Where the lesser interest is a tenancy, the greater interest is the reversion immediately expectant on its determination and both estates become vested in the same person, the judge observed that, at common law, a merger would be regarded as occurring automatically. That is, I consider, correct, at any rate in cases in which the two estates become vested in the same person in the same right. The approach of the common law does not, however, provide a solution to the present case because a different principle applies in equity and the equitable rule now prevails. The rule in equity was that there would be a merger only if the party in which the two estates vested *intended* a merger. Section 185 of the 1925 Act, headed "Merger" (and substantially re-enacting section 25(4) of the Judicature Act 1873), provides that:

> There is no merger by operation of law only of any estate the beneficial interest in which would not be deemed to be merged or extinguished in equity.

It is, therefore, on the equitable principles relating to merger that attention must be focused. Mr Warwick's submissions raised an issue as to the equitable solution that the court should apply to the facts of the present case in circumstances in which there was (as the judge found) no evidence as to EDF's intention in respect of merger when the two estates became vested in it.

[31] We were referred to only two authorities on merger. The first was *Ingle* v *Vaughan Jenkins* [1900] 2 Ch 368, a decision of Farwell J. The point there was whether an equitable lease created by an uncompleted agreement for the grant of a lease became merged in the legal estate in the land that the equitable lessee later acquired or whether it remained unmerged so as to be enforceable by his executor. Farwell J held that there was no merger. He described the applicable principles thus, at p370:

> Whatever might have been the case at common law, as to which it is unnecessary that I should express an opinion, it is, in my opinion, clear that it was not merged or extinguished in equity. I think the proposition in Lewin on Trusts, 10th ed p889, is correct — namely, that "The principle by which the Court is guided is the *intention;* and in the absence of express intention, either in the instrument or by parol, the Court looks to *the benefit of the person in whom the two estates become vested*." The author goes on to point out that the chief importance of the doctrine of merger is with reference to charges, and the cases he cites are confined to charges.
>
> The defendants contended that a different principle applied in the case of a lease, but I am unable to follow that distinction. The principle being that the Court looks to the benefit of the person in whom the interests coalesce, I cannot see why there should be any distinction in this respect between a beneficial lease and a term to secure a charge. In either case the term is taken as an equivalent for money expended. Nor do I think it makes any difference whether the coalescence of the interests is brought about by operation of law or the acts of the parties. The principle of *Grice* v *Shaw* 10 Hare 76 is applicable to the present case. The head-note is as follows: "Where the tenant in fee or in tail of an estate becomes entitled to a charge upon the same estate, the general rule is, that the charge merges, unless it be kept alive by the party entitled to it; and where the merger of the charge would have let in other charges in priority, thereby rendering it the interest of the owner of the estate to keep alive his charge, the Court presumed that such was his intention, notwithstanding the absence of any other indication of such intention.

(Emphasis as in the original.)

[32] Farwell J cited from Turner VC's judgment in *Grice* v *Shaw* (1852) 10 Hare 76. That showed that, in that case, there was no evidence of any express intention to keep the charge alive and that the question was what intention should be *presumed* with regard to it. Since it was in the chargee's interest to keep it alive, the court presumed that that was his intention. Farwell J applied that principle to the case before him.

[33] The second authority was *Capital & Counties Bank*, a decision of this court. It is necessary to understand the material facts. In 1871, L granted T a 99-year lease of Whiteacre at an annual rent of £100. By May 1897, the first defendant, Rhodes, had acquired the lease. In that month, by way of mortgage to secure the payment of £16,700, he granted Flower & Sons (the third defendant) a sublease of Whiteacre for the residue of the lease less one day, such sublease not reserving any rent. The mortgage empowered Flower to dispose of the residue of the 1871 lease if it became necessary to enforce the security. In June 1899, Rhodes agreed to buy the freehold of Whiteacre for £3,650, the sale G particulars describing it as a freehold ground rent of £100 and making clear that the sale was subject to and with the benefit of the 1871 lease. In seeking to obtain, as he did, a mortgage loan of £3,000 from the plaintiff bank in order to assist his purchase, Rhodes presented the sale particulars to the bank. Rhodes's purchase and subsequent mortgage to the bank were completed by successive instruments executed on the same day in July 1899: first, a conveyance of the freehold to Rhodes subject to and with the benefit of the 1871 lease; second, an instrument (by way of mortgage) conveying the freehold to the bank and also demising to it the unexpired residue of the 1871 lease less the last day. Those instruments made no reference to the prior 1897 mortgage to H Flower, although the bank had actual knowledge of it. In April 1901, Rhodes executed a deed of arrangement transferring all his property to Mason (the second defendant) in trust for his creditors. The bank, to which the entire £3,000 was due, was not a party to the deed. By June 1901, Flower had taken possession of Whiteacre and contended that the 1871 lease had merged in the freehold on Rhodes's acquisition of it in 1899, so that it was entitled to remain in possession under its mortgage security without paying any rent. By its claim, the bank challenged that there had been a merger and claimed to be entitled to take possession of Whiteacre under the proviso for re-entry in the 1871 lease following the non-payment of rent. Kekewich J doubted, J without deciding, that there had been a merger, but held that if there had, the bank's mortgage interest would not take priority over Flower's and dismissed the claim against Flower. The bank appealed.

[34] Sir Richard Henn Collins MR summarised the issue succinctly, at p646:

> The principal question is whether on the purchase by Rhodes of the fee his term was merged in the reversion, the result of which would be that [Flower] would hold directly under the owner of the fee... without any obligation to pay rent, while, the term held by Rhodes being extinguished, the rent of £100 a year would be no longer payable, and the security held by the bank would be land subject to a term not expiring until 1968, during which no rent could K be received.

Collins MR observed, at p647, that it seemed to him "quite clear that, if the intention of the parties is that which determines this question, they certainly did not intend that a merger should take place". He regarded "the parties" as being Rhodes and the bank, his point being that it was the essence of Rhodes's wish to borrow from the bank on mortgage that he should be able to offer it the security of a freehold yielding a ground rent of £100 pa. He referred to that as supported by the July 1899 documents, and said, at p648, that:

> the transaction could not have been carried out at all unless the term was kept L alive, and it was clearly for the benefit of both parties to the transaction that it should be kept alive.

He rejected the argument that "the rule of equity that the question of merger must be decided by the intention of the parties was limited to the question of the merger of charges only, and did not apply to the merger of estates", and referred to Farwell J's decision in *Ingle* as deciding the contrary. He then held that there had been no merger, saying, at p651:

> It remains to consider whether the evidence of intention, as well as of benefit to Rhodes and the bank if the term were kept in existence, is not such as to prevent a merger as the law now stands. The law to be deduced from the M previous authorities was summed up by Lord Macnaghten in the well-known passage in *Thorne* v *Cann* [1895] AC 18, 19, and was applied to the case of leases by Fry J and Farwell J in the cases which I have already mentioned. I think those authorities clearly establish that in the present case, neither on the sale to Rhodes or on the mortgage by him to the bank, would "the beneficial interest in" the term have been "deemed to be merged or extinguished in equity," and that therefore the transaction falls within [what is now section 185 of the 1925 Act], and there is no merger at law.

[35] Romer LJ agreed both with Collins MR's judgment and that of Cozens-Hardy LJ, and did not add a substantive judgment of his own. Cozens-Hardy LJ explained that the first issue in the appeal was whether the 1871 lease had merged in the freehold. He said, at p652,

A that the rigid rule at common law was that if a term and its freehold reversion vested in the same party in its own right, the term merged in the freehold "whatever may have been the intention of the parties to the transaction which resulted in the union". He continued:

> The Courts of Equity, on the other hand, in many cases treated the interest which merged at law as being still subsisting in equity. They had regard to the intention of the parties, and, in the absence of any direct evidence of intention, they presumed that merger was not intended, if it was to the interest of the party, or only consistent with the duty of the party, that merger should not take place.

B He explained that the commonest application of the principle had been in respect of cases where a life tenant paid off a charge on the inheritance, but said that the principles applied equally to the merger of estates in land as to merger of charges on land. He said, at p653:

> It was well established that, according to the strict rules of the common law, there would be merger, notwithstanding that one of the two estates might be held in trust, and the other beneficially, by the same person, or one might be held on one set of trusts and the other on another set of trusts. But it was equally well established that equity would interfere, and would, if necessary, decree the execution of such deeds as would replace the parties in their proper position: see *Saunders* v *Bournford* (1679) Finchm 424, where Lord Nottingham LC decreed that, notwithstanding the merger of a term, the plaintiff should hold
> C possession of the premises during the remainder of the term, and that the defendant should make a further assurance of the remainder of the term. The merger was treated as an accident prejudicial to the real beneficial interests of the parties: see also *Attorney-General* v *Kerr* (1840) 2 Beav 420, a remarkable instance of the application of the equitable doctrine. I think the decision of Farwell J, or rather his dictum to this effect, in *Ingle* v *Vaughan Jenkins* [1900] 2 Ch 368 is consistent with principle and is supported by authority. A Court of Equity had regard to the intention of the parties, to the duty of the parties, and to the contract of the parties, in determining whether a term was to be treated as merged in the freehold.

[36] He explained, at p654, that it was obvious that it was the parties'
D intention that the bank should have a security on the freehold ground rent of Whiteacre, so that no merger that would destroy that ground rent ought to be allowed: "The intention of the parties was clear that there should not be a merger." He, too, plainly regarded "the parties" as being Rhodes and the bank.

[37] Mr Warwick, in his submissions in opening and reply, adopted a perhaps slightly inconsistent stance in respect of *Capital & Counties Bank*. I understood him in his opening to disavow the suggestion that it established a principle providing direct support for the appellants' argument that there was a merger in the present case. His point was, rather, that *Capital & Counties Bank* was not a case in which the court
E was concerned to take into account any third-party interests comparable to those of EDF's co-reversioners in the present case, and so it was not directly in point by way of guidance for its disposition. He said that EDF's claim, that there had been no merger, was one that was aimed at putting itself in a position to be able to abuse the legislative scheme of Part II of the 1954 Act, a position directly injurious to the interests of the co-reversioners. He said that it would be contrary to the latter's interests that EDF should maintain an unmerged interest in a tenancy interest of plot 2 because that would prevent the co-reversioners from serving a section 25 notice in respect of the demised premises without the co-operation of EDF, whereas, in all likelihood, EDF's interests would be
F against giving such a notice. The possible consequence of that would or might be that the tenancy could continue indefinitely. It followed that it could not therefore have been the intention of the co-reversioners that EDF's tenancy interest in plot 2 should survive unmerged. Had they been consulted as to whether there should or should not be a merger on EDF's acquisition of the freehold of plot 2, they would have responded with a resounding "yes", since that would enable them to bypass any reference to EDF when considering their interests as landlords.

[38] Mr Warwick's further submission was that a court of equity faced with this type of situation ought therefore to invoke its traditional flexibility and meet the potential for injustice by finding that the competing "pro-merger" interests of the co-reversioners must trump (his word) the different "no-merger" interest of EDF. The essence of his submission was that the present case involved circumstances that were quite different from any that had arisen in the reported authorities and that the court should therefore not feel constrained in developing the "merger" principles so as to meet such circumstances.

[39] In his reply, Mr Warwick focused on the last sentence of the passage from Cozens-Hardy LJ's judgment cited in [35] above, which he described as the *ratio* of the decision, and said that it supported the approach that he was urging. He sought to apply it to this case by submitting that, here, "the contract" is the 1953 lease and that, by 1993, when EDF acquired the freehold reversion in respect of plot 2, it had an obligation to consider the interests of its two co-reversioners as to the other two parts of such premises, namely plots 26 and 31. Mr Warwick submitted that, in [66] of his judgment, the judge had directed himself correctly as to the need, in considering the merger argument, to take account of the interests of the co-reversioners. By [69], however, the judge had, he said, forgotten such self-direction and had decided the question by applying the principle that absent any evidence pointing to a different intention, it should be presumed that EDF intended no merger if, as he found, it was contrary to its interest that there should be a merger.

[40] With respect to Mr Warwick's careful submissions, both in opening and reply, I regard them as mistaken at every point and would not accept them. The starting point is that, whereas the ordinary rule at law was that the coalescence of a lease and its reversion in the same person (A) in the same right would result in a merger and extinguishment of the lease, in equity it was open to A to form an intention, and declare accordingly, that there should be no such merger and extinguishment. Equity further developed the principle that in any case in which A did *not* expressly evince such an intention, or in which there was no other evidence of such an intention on its part, there was a presumption against any intention for a merger if such would be against its interest. In a case in which there was no express declaration or other evidence as to A's intentions, the focus of equity's inquiry was therefore exclusively on its interests, and if a merger would be against its interests, it was presumed to have intended against any merger. That is the principle that was applied in *Ingle*, and this court in *Capital & Counties Bank* made it clear that it regarded *Ingle* as having been correctly decided. The equitable principles now prevail over the principles applicable at law.

[41] This court in *Capital & Counties Bank* explained the principles, but I would not regard it as having developed them. *Capital & Counties Bank* differed in kind from *Ingle* in that it was not a case in which the court had to consider whether there was any *presumption* that Rhodes did not intend a merger. On the contrary, it was one in which the court considered that there was clear evidence as to what Rhodes (and the bank) had intended with regard to a merger or otherwise, and the evidence showed that neither of them could have intended a merger. Both Rhodes and the bank were closely bound up in Rhodes's acquisition of the freehold. Rhodes needed the bank's loan in order to finance the acquisition, and the documents showed that his deal with the bank was that its security was intended to be the freehold with the benefit of the rent yielded by the 1871 lease. It is those considerations that explain the references by Collins MR and Cozens-Hardy LJ to "the intention of the parties" and they pointed solidly against there having been any intention on the part of Rhodes to merge and extinguish the lease. It is perhaps worth citing the passage from the speech of Lord Macnaghten in *Thorne* v *Cann* [1895] AC 11, at pp18-19, to which Collins MR referred at p651: see [34] above) as summing up the applicable law. Lord Macnaghten said:

> Nothing, I think, is better settled than this, that when the owner of an estate pays charges on the estate which he is not personally liable to pay, the question whether those charges are to be considered as extinguished or as kept alive for his benefit is simply a question of intention. You may find the intention in the deed, or you may find it in the circumstances attending the transaction, or you may presume an intention from considering whether it is or not for his benefit that the charge should be kept on foot. Here, I think, the intention appears plainly on the face of the deed by which Miss Arnold purported to transfer her mortgage.

Ingle was a case in which the court presumed an intention that there should be no merger because that would have been against interest. *Capital & Counties Bank* was a case in which the court found positive evidence of an intention that there should be no merger.

[42] I disagree with Mr Warwick that the 1953 lease can or should be regarded as an example of the type of "contract" to which Cozens-Hardy LJ was referring in the last sentence of the passage quoted in [35] above. It is not. In making that reference, I presume that Cozens-Hardy LJ had in mind the mortgage contract between Rhodes and the bank, a transaction that made it self-evident that no merger was intended since otherwise the security that Rhodes purported to give would have been substantially worthless. The present case is miles away from *Capital & Counties Bank.* The 1953 lease did not constitute a contract between the three co-reversioners who were in existence by December 1993. It imposed no contractual obligations or other duties on any reversioner towards his co-reversioners. When EDF acquired the freehold reversion in plot 2 it was under no obligation to do other than consider its own interests.

[43] Mr Warwick's submission amounted ultimately to the proposition that because a merger is said to have been in the interests of the co-reversioners, EDF should be presumed to have intended a merger even though its own interest would be directly against it. That way of putting the case is not to invite a development of the principles that have so far been established by the courts of equity. It is to invite a departure from them and the adoption of quite contrary principles: namely that in the absence of evidence proving an intention one way or the other (and the judge found there was none, as to which there is no challenge), the court should presume an intention by EDF to merge, when to do so would be directly adverse to its interests but would or may be of advantage to third parties with which it is in no commercial relationship and to which it owes no duties. For the court so to presume in such circumstances appears to me to be contrary to the principles explained in the authorities, and wrong. Mr Warwick did not go the further distance of submitting that had EDF (as it might have done) made an express declaration of merger, the co-reversioners could have asked the court to set it aside, and any such submission would, I consider, have been hopeless. It is only, apparently, in the circumstance in which there is no evidence as to EDF's intention that the interests of the co-reversioners must be held to prevail. The argument appears to me to be an unprincipled one.

[44] In my judgment, the task for the judge in respect of this issue was to consider whether there was any evidence of any intention on the part of EDF as to whether there should be a merger. He found, in [69], that there was none. Having so found, he had next to consider whether it was or was not in EDF's interests that there should be a merger. He found that it was not, and that finding led him, correctly in my judgment, to the conclusion that he should presume that no merger was intended. That was the approach that *Ingle* (endorsed by *Capital & Counties Bank*) required him to adopt, and he was correct to do so. That it may perhaps have been, or might turn out to be, contrary to the interests of the co-reversioners that there should be no merger is irrelevant.

[45] I would dismiss the appellant's first ground of appeal.

Second ground of appeal

[46] This is a new ground of appeal that was not advanced to the judge. It is a fall-back argument that arises only if the "merger" ground of appeal were to fail, as in my view it should. The essence of the point, as originally advanced in the grounds of appeal, was that, on the footing, that: (i) plot 2 remains part of the premises demised by the 1953 lease and the tenancy thereby created is continuing in respect of (*inter alia*) plot 2; and (ii) EDF is a necessary party, together with BOH and Layhawk, to (for example) the giving of a section 25 notice in respect of the demised premises; then (iii) "the 1954 Act must be read so that the 'tenancy' held by EDF to which the 1954 Act applies does not include Plot 2".

[47] The claim that the 1954 Act should be so read was said to be required by section 3 of the 1998 Act, namely that:

> So far as it is possible to do so primary legislation and subordinate legislation must be read and given effect in a way which is compatible with the Convention rights. G

The need for a compatible reading of the 1954 Act was advanced on the basis that, otherwise, BOH's and Layhawk's enjoyment of their freehold interests in plots 31 and 26 respectively would be interfered with in a manner contrary to their rights under Article 1 of the First Protocol to the Convention, which provides:

> Every natural or legal person is entitled to the peaceful enjoyment of his possessions. No one shall be deprived of his possessions except in the public interest and subject to the conditions provided for by law and by the general principles of international law. H
>
> The preceding provisions shall not, however, in any way impair the right of a state to enforce such laws as it deems necessary to control the use of property in accordance with the general interest or to secure the payment of taxes or other contributions or penalties.

[48] In his written and oral argument, Mr Warwick materially qualified and refined this ground of appeal by confining the focus of his argument to the proposition that section 44(1A) of the 1954 Act should be read as excluding from such definition any party that is also the tenant of the relevant tenancy. Thus, applied to this case, a section 25 notice in respect of plots 2, 26 and 31 (and ignoring the judge's finding J that the tenancy of plot 31 had been surrendered) could validly be given by the freehold owners of plots 26 and 31 only. Such notice would, however, have to be given not just in respect of plots 26 and 31 but also in respect of plot 2.

[49] Mr Warwick fully developed this argument at the hearing, but the court did not, at that stage, call on Mr Stoner to answer it. That was because it was provisionally unconvinced that it ought to entertain this ground of appeal at all. In forming that view, the court had not overlooked that Mummery LJ had given permission for this ground of appeal to be argued as well as the merger ground. He expressed the view, when giving permission, that the "proposed appeal against [the K judge's order] has a real prospect of success on the basis of the grounds of appeal and the skeleton argument". Mummery LJ may, however, have overlooked that success on this ground of appeal could not justify any reversal or variation of the judge's order, as EDF pointed out in its skeleton argument and as Mr Warwick expressly conceded.

[50] The reason for that is simple. The statements of case before the judge put in issue the validity of the section 25 notice purportedly given by Mr Dormer in August 1993. That was apparently the only such notice in issue at the trial. At the time it was given, there were three separate co-reversioners, none of whom or which was also the tenant of L the demised premises. The judge held the notice to be invalid and the appellants do not challenge that decision. The point that they now want to make is not merely a point that was not advanced before the judge, it was one in respect of which no factual case was or, so far as I am aware, could have been made. It is a point that comes before this court entirely afresh and unsupported by any factual basis. Mr Warwick accepted that no section 25 notice had been served that purported to raise the issue. The court is therefore being asked to answer a question, and presumably to make some form of declaration, based on a hypothetical set of circumstances. In my judgment, this court should not do that. The point raised by this ground of appeal is potentially of some general M importance. If it is ever to be considered by this court, it should be only after a concrete set of facts raising it as a real issue has first been considered by a trial judge, who has then made findings on the facts and on the law applicable to them.

[51] The court reserved its judgment at the conclusion of the hearing. The following day, it received an e-mail purporting to have been composed by BOH and Layhawk. The identity of the human agent who was in fact its author was not disclosed, but the inference is that he was present in court at the hearing. He referred to the court's expressed unwillingness to deal with this ground of appeal in default of any section 25 notice having been served that could be said to have raised the issue. He continued:

A The Respondents [that is, EDF] failed to bring the attention of the Court the fact that they had indeed been served with Section 25 Notices. The Section 25 Notices were served by the appellants predecessors to the title, the appellants were given vacant possession and accepted the same in reliance on the said Section 25 notices and there was no requirement to provide the Respondents with a further opportunity as regards the Section 25 Notices.

We do feel that the three Lord Justices would have wanted to have their attention drawn to the three Section 25 Notices in order to provide clarity. The Judgment has yet to be delivered and we would therefore ask that you put the three attached notices before their lordships and Lady Black to enable them to address the issues.

The Respondent has been notified of our concerns regarding their decision
B to permit those representing the appellants, and the Court, to proceed without reference to these matters where the Court has decided there is a Human Rights element to the same. It remains open to the appellants to serve further Section 25 Notices but given these facts and matters the Court would wish to consider the same.

The three notices were acknowledged by the Respondent.

[52] The first attached notice was one on which the typed date of 2 April 2002 had been changed in manuscript to 29 April 2002. It was from Commercial Rentals Ltd, as landlord, to the EEB, as tenant: see [17] above. Mr Todd was described in it as the landlord's agent.
C The notice purported to terminate the tenancy on 11 November 2002. It was expressed as relating to:

Any part of the land included in freehold title NGL541577 in respect of which the Tenant asserts the registered title MX269363 should be included excluding any part of MX269363 where the Tenant has given up the protection of the Act and the landlord relied upon [sic] of which you the tenant.

The second notice had the same manuscript date change and was from Containerised Storage, as landlord, to the EEB: see [16] above. Mr Todd was again described as the landlord's agent. That notice was also expressed to terminate on 11 November 2002. It was expressed as relating to:

D

Any part of the land included in freehold title NGL770835 in respect of which the Tenant asserts the registered title MX269363 should be included excluding any part of MX269363 where the Tenant has given up the protection of the Act and the Landlord relied upon [sic] of which you are the tenant.

The third notice was dated 29 April 2002. It was given jointly by Commercial Rentals and Containerised Storage as landlords, with Mr Todd again being named as the agent. It was addressed to the EEB and purported to terminate its tenancy on 11 November 2002. It was expressed as relating to:

E Any parts of the land included in freehold title NGL541577 and NGL770835 in respect of which the Tenant asserts the registered leasehold title MX269363 should be included excluding that part of MX269363 in respect of which the Tenant is estopped by conduct and or merger and or the failure to commence proceedings after the service of a Section 25 notice dated 27th August 1993 and acceptance of the validity thereof by a Counternotice dated 3rd September 1993 which exchange and correspondence the Landlord has acted upon to its prejudice. [sic] of which you are the tenant.

[53] I interpret the notice separately given by Commercial Rentals as directed exclusively to (at most) plot 26 (the title to which was registered under title no NGL 541577) and the notice separately given by Containerised Storage as directed exclusively to (at most) plot 31 (title no NGL 770835, comprising plots 20 and 31). As neither notice appears to be directed to plot 2, I do not understand how they can severally or collectively satisfy the interpretation of section 44(1A) for which Mr Warwick argued. The notice served jointly by Commercial Rentals and Containerised Storage suffers from the same flaw.

[54] My view is, therefore, that these notices are irrelevant. They do not satisfy Mr Warwick's submission as to what section 44(1A) requires if read in the way he suggests would be compatible with the appellants' Convention rights. Even if my interpretation of them in this respect is incorrect, they were not pleaded in the statements of case before the judge, nor were they apparently the subject of any argument before him, let alone any findings by him. Had they been the subject of argument, it is inconceivable that he would not have dealt with them in the course of what was, if I may say so, a judgment of conspicuous meticulousness. It is not surprising that Mr Warwick did not refer us to them and the suggestion by the e-mail's author that it was Mr Stoner's duty to do so is astonishing. It is BOH and Layhawk that are the appellants and it was up to them to consider whether they wanted to draw the court's attention to any new evidence that might be thought to assist their appeal, and, if so, to seek (as would be necessary) permission to rely on it, explaining why permission ought to be given. Since the author was apparently present in court during Mr Warwick's submissions, it is surprising that he did not adopt the conventional course of instructing Mr Warwick about the notices at a convenient moment during the hearing. As it is, and having decided to assume the mantle of advocate on the appellants' behalf, he has provided no explanation as to why this court, whose function is to sit as a court of appeal, not as a court of first instance, should have regard to material that appears to have played no part at the trial, whereas had the appellants chosen to deploy it there they could have done so. I would disregard the representations made by the e-mail.

Disposition

[55] For reasons given, I would decline to express any view on the correctness or otherwise of the point argued under the second ground of appeal. Since I would dismiss the first ground of appeal, it follows that I would dismiss the appeal.

Black LJ said:

[56] I agree.

Sedley LJ said:

[57] I also agree.

Appeal dismissed.

LANDLORD AND TENANT
SERVICE CHARGES

Daejan Investments Ltd *v* Benson and others

COURT OF APPEAL

[2011] EWCA Civ 38

28 January 2011

SEDLEY, PITCHFORD AND GROSS LJJ

Landlord and tenant — Landlord and Tenant Act 1985 — Service charges — Qualifying works — Consultation requirements — Service Charges (Consultation Requirements)(England) Regulations 2003 — Landlord of block of flats carrying out qualifying works within sections 20 and 20ZA of 1985 Act — Landlord failing to comply with second stage of consultation requirements under 2003 Regulations in respect of those works — Whether appropriate to grant dispensation from consultation requirements — Whether financial consequences of refusal relevant to exercise of discretion to grant dispensation — Whether non-compliance causing significant prejudice to tenants

The five respondents each held a long lease of a flat in a building owned by the appellant landlord. In 2005, the appellant's managing agent gave notice to the lessees in the building of the appellant's intention to carry out qualifying works within sections 20 and 20ZA of the Landlord and Tenant Act 1985, in respect of which the consultation requirements of the Service Charges (Consultation Requirements)(England) Regulations 2003 applied.

The appellant made various errors in the second stage of the statutory consultation process. Although it provided the lessees with a report that set out the tenders received and weighed up the pros and cons of the two that were most favourable, it failed to provide copies of the actual estimates until August 2005. By then, the appellant had already informed the lessees that the contract for the works had been awarded.

The appellant sought to recover approximately £270,000 for the works from the respondents, who resisted payment on the ground of the appellant's non-compliance with the consultation requirements. The leasehold valuation tribunal (LVT) dismissed an application by the appellant, under section 20ZA of the 1985 Act, for dispensation from those requirements and accordingly limited its recovery from each lessee to £250. It held that the appellant's failure to provide the estimates, until after it had indicated that the contract had been awarded, had curtailed the consultation and led the respondents to conclude that further representations would be futile, thereby causing them significant prejudice. That decision was upheld by the Lands Chamber of the Upper Tribunal. The appellant appealed. It contended that the tribunals had erred in holding that: (i) the financial consequences to the appellant of allowing or refusing the application were irrelevant and should not be taken into account; (ii) the corporate nature of the landlord was relevant and justified a more rigorous approach to non-compliance with the consultation requirements; and (iii) the respondents had suffered prejudice from the appellant's non-compliance.

***Held*: The appeal was dismissed. (1) As a matter of statutory construction, with regard both to the context and the wording of the statute, the financial effect of the grant or refusal of dispensation is an irrelevant consideration when exercising the discretion under section 20ZA(1) of the 1985 Act. That section gives to the LVT the power to dispense with the consultation requirements themselves, not with the statutory consequences of non-compliance. Accordingly, in general circumstances, going to the integrity of the consultation process will be relevant while the consequences flowing from the grant or refusal of dispensation will be irrelevant. The fact that non-compliance might involve significant financial loss is by itself no more than an intrinsic part of the statutory scheme. (2) There can be no error in holding that the consultation requirements should be applied according to their terms in the case of a corporate or local authority landlord, even assuming that cases could be envisaged in which a less rigorous approach to the consultation requirements might be appropriate in respect of landlords that were owned or controlled by the lessees. There was no indication that either the LVT or the Lands Chamber had applied a more rigorous standard than that provided for in the statute. (3) Significant prejudice to the tenants is a consideration of the first importance in exercising the dispensatory discretion under section 20ZA(1). The appellant's non-compliance in curtailing consultation constituted a serious failing that had caused significant prejudice to the respondents. A proper consultation process is of the essence of the statutory scheme, which was devised to protect the interests of tenants, and its curtailment cannot be regarded as a technical, minor or excusable oversight. The loss of opportunity to make further representations had itself amounted to significant prejudice to the respondents. The LVT had been justified in refusing dispensation.**

The following cases are referred to in this report.

Camden London Borough Council v *Leaseholders of 37 Flats at 30-40 Grafton Way* LRX/185/2006 unreported 11 April 2008, LT

Daejan Investments Ltd v *Benson* [2009] UKUT 233 (LC); [2010] 2 P&CR 8

Edwards (Inspector of Taxes) v *Bairstow* [1956] AC 14; [1955] 3 WLR 410; [1955] 3 All ER 48, HL

Eltham Properties Ltd v *Kenny* LRX/161/2006 [2008] L&TR 14, LT

Majorstake Ltd v *Curtis* [2008] UKHL 10; [2008] 1 AC 787; [2008] 2 WLR 338; [2008] 2 All ER 303; [2008] 2 P&CR 2; [2008] 1 EGLR 44; [2008] 14 EG 102

Martin v *Maryland Estates Ltd*; *Seale* v *Maryland Estates Ltd* [1999] 2 EGLR 53; [1999] 26 EG 151; (2000) 32 HLR 116, CA

Paddington Basin Developments Ltd v *West End Quay Estate Management Ltd* [2010] EWHC 833 (Ch); [2010] 1 WLR 2735; [2010] 2 EGLR. 35; [2010] 27 EG 86

R (on the application of Cart) v *Upper Tribunal*; *R (on the application of MR (Pakistan)* v *Upper Tribunal (Immigration and Asylum Chamber)* [2010] EWCA Civ 859; [2011] QB 120; [2011] 2 WLR 36; [2010] 4 All ER 714

This was an appeal by the appellant landlord, Daejan Investments Ltd, from a decision of the Lands Chamber of the Upper Tribunal upholding a decision of the leasehold valuation tribunal, in favour of the respondent tenants, Jack Benson, David Lapes, Paul Wallder, Aldenspring Ltd and Alastair Gray, refusing dispensation from the consultation requirements of the Service Charges (Consultation Requirements)(England) Regulations 2003 relating to qualifying works under the Landlord and Tenant Act 1985.

Nicholas Dowding QC and Stephen Jourdan QC (instructed by GSC Solicitors) appeared for the appellant; Philip Rainey QC (instructed by K&L Gates LLP) appeared for the first to fourth respondents;

James Fieldsend (instructed by Jaffe Porter Crossick LLP) represented the fifth respondent.

Giving the first judgment, **Gross LJ** said:

Introduction

[1] This is an appeal, brought pursuant to section 13 of the Tribunals, Courts and Enforcement Act 2007 (the 2007 Act), by the appellant (Daejan), against the decision of the Upper Tribunal (Lands Chamber), known as the Lands Tribunal (LT), dated 27 November 2009 (the LT decision): see [2009] UKUT 233 (LC). In that decision, the LT dismissed Daejan's appeals against two decisions of the leasehold valuation tribunal (LVT) dated 11 March and 8 August 2008 (respectively the LVT's March and August decisions). Although the LT necessarily had regard to both the LVT's March and August decisions, the principal focus of the appeal before the LT, as indeed the appeal to this court, concerned the LVT's August decision.

[2] In a nutshell:

(i) By its March and August decisions, the LVT held that Daejan had failed to comply with the Service Charges (Consultation Requirements)(England) Regulations 2003, SI 2003/1987 (the consultation regulations), in respect of works to be done at Queens Mansions, Muswell Hill (the works and Queens Mansions).

(ii) By its August decision, the LVT declined to conclude that it was reasonable to dispense with the consultation requirements in the present case and so made no order for dispensation under section 20ZA(1) of the Landlord and Tenant Act 1985 as amended (the Act). The consequence is that Daejan failed in its claim to recover around £270,000 from the respondents, five long leaseholders of flats in Queens Mansions, in respect of the works. Instead, pursuant to the statutory scheme (of which more below), Daejan's recovery from the respondents was "capped" at £250 each. In reaching this conclusion, the LVT held that the financial effects of the grant or refusal of the application for dispensation on the landlord or tenant were irrelevant and were not to be taken into account. Further, the LVT held that Daejan's failure to comply with the consultation regulations had caused the respondents "substantial" or "significant" prejudice.

(iii) On appeal, the LT, whose constitution on this occasion included Carnwath LJ, senior president, upheld the decision of the LVT. The LVT had been entitled both to conclude that Daejan had committed a "serious breach" and to refuse dispensation with the consultation requirements under section 20ZA(1) of the Act.

[3] Before this court, Mr Nicholas Dowding QC, for Daejan, essentially contended as follows. First, the LT had erred in holding that the financial effects of granting or refusing dispensation were irrelevant. Second, the LT was wrong to hold that the nature of the landlord was relevant; that Daejan being a corporate landlord was irrelevant and did not justify the LVT adopting a more rigorous approach to non-compliance with the consultation regulations. Third, the LT had erred in law or, alternatively, had reached a conclusion no reasonable tribunal could have reached (*Edwards (Inspector of taxes)* v *Bairstow* [1956] AC 14) with regard to the question of whether the respondents had suffered prejudice flowing from Daejan's failure to comply with the consultation regulations; it was for the respondents to prove such prejudice; regardless of where the burden of proof lay, there was none; the only reasonable conclusion was that dispensation should have been granted.

[4] The respondents' case before us, presented by Mr Philip Rainey QC, for the first to fourth respondents, and Mr James Fieldsend, for the fifth respondent, was, in essence, that the decisions of the LT and LVT were correct for the reasons they had given. On a proper consideration of the statutory scheme, the financial effects of granting or refusing dispensation were irrelevant. Although the nature of the landlord could be relevant, in that a distinction could be drawn between lessee-owned landlords and others, there was no separate point here in this regard and there had been no error of law on the part of the LT. Regarding prejudice, the LT had not erred as to the burden of proof, but, in any event, there had been substantial prejudice to the respondents, flowing from Daejan's failure to comply with the requisite consultation process. If need be, the first to fourth respondents additionally relied on their respondent's notice (the respondent's notice), contending that the evidence of prejudice went beyond that found by the LT.

[5] Against this background, the principal issues on the appeal to this court can conveniently be grouped under the following headings:

(i) Are the financial consequences for the landlord or tenant relevant to the grant or refusal of dispensation under section 20ZA(1) of the Act? (Issue (I): Financial consequences.)

(ii) Is the nature of the landlord relevant to the grant or refusal of dispensation under section 20ZA(1) of the Act? (Issue (II): The nature of the landlord.)

(iii) What is the correct approach to prejudice allegedly suffered by a tenant in consequence of the landlord's failure to comply with the consultation regulations? (Issue (III): Prejudice.)

I return to these issues below and will deal with them in turn.

Statutory scheme

[6] It is next convenient to outline the statutory scheme with which this appeal is concerned.

[7] The background was helpfully explained by Mr Rainey in his skeleton argument:

> One of the singular aspects of most long leases of flats is that the landlord covenants to repair and maintain the block but the tenants covenant to meet the cost through a service charge. Often, service charges extend beyond repair to cover improvements. How, when and by whom and at what cost the work is done is decided by the landlord, despite the fact that where leases are long, the tenants effectively own the block. (A 99 year term is usually thought to be worth c99% of freehold value.) The landlord decides how to spend his tenant's money and at common law there is little control or basis for complaint by the tenants. Statutory consultation under… [the Act]… fills this gap. Its importance must not be understated.

[8] The particular position of long leaseholds merits additional emphasis. As observed by Baroness Hale of Richmond in *Majorstake Ltd* v *Curtis* [2008] UKHL 10; [2008] 1 AC 787*, in [22] to [23], by the 1980s long leaseholds had become an increasingly common form of tenure of flats. The relationship between leaseholders under such leases and the freehold owners was no longer akin to that of an ordinary landlord and tenant relationship. Long leaseholders not only faced what Baroness Hale termed "the wasting asset problem" but might also encounter poor management and high service charges.

[9] Accordingly, legislative policy has sought to strengthen the position of long leaseholders by regulating service charges. How the legislature has sought to do so was summarised in *Paddington Basin Developments Ltd* v *West End Quay Estate Management Ltd* [2010] EWHC 833 (Ch); [2010] 1 WLR 2735†, by Lewison J, as follows:

> 26. … there are two separate strands to the policy underlying the regulation of service charges. Parliament gave two types of protection to tenants. First, they are protected by section 19 [of the Act] from having to pay excessive and unreasonable service charges or charges for work and services that are not carried out to a reasonable standard. Second, even if service charges are reasonable in amount, reasonably incurred and are for work and services that are provided to a reasonable standard, they will not be recoverable above the statutory maximum if they relate to qualifying works or a qualifying long term agreement and the consultation process has not been complied with or dispensed with. It follows that the consultation provisions are imposed for an additional reason; namely, to ensure a degree of transparency and accountability when a landlord decides to undertake qualifying works or enter into a qualifying long term agreement . . .
>
> 27. One other general point needs to be made. The relevant provisions of [the Act] do not prohibit a landlord from entering into whatever contract he pleases for the carrying out of works or the supply of services. They merely prevent him from passing on the cost of the works or services to the lessees unless he has satisfied the statutory requirements about price, quality and consultation.

---

* Editor's note: Also reported at [2008] 1 EGLR 44; [2008] 14 EG 102

† Editor's note: Also reported at [2010] 2 EGLR 35; [2010] 27 EG 86

[10] Turning to the detail, section 18 of the Act defines "service charge" to mean:

(1)… an amount payable by a tenant of a [dwelling] as part of or in addition to the rent —

(a) which is payable, directly or indirectly, for services, repairs, maintenance, improvements or insurance or the landlord's costs of management, and

(b) the whole or part of which varies or may vary according to the relevant costs.

[11] Section 19 of the Act contains the first of the two strands of protection described by Lewison J (*supra*):

19. Limitation of service charges: reasonableness

(1) Relevant costs shall be taken into account in determining the amount of a service charge payable for a period —

(a) only to the extent that they are reasonably incurred, and

(b) where they are incurred on the provision of services or the carrying out of works, only if the services or works are of a reasonable standard;

and the amount payable shall be limited accordingly.

[12] Sections 20 and 20ZA of the Act deal with the second strand of protection spoken of by Lewison J and with which this litigation is concerned, namely consultation requirements.

20. Limitation of service charges: consultation requirements

(1) Where this section applies to any qualifying works or qualifying long term agreement, the relevant contribution of tenants are limited in accordance with subsection (6) or (7) (or both) unless the consultation requirements have been either —

(a) complied with in relation to the works or agreement, or

(b) dispensed with in relation to the works or agreement by (or on appeal from) a leasehold valuation tribunal.

20ZA. Consultation requirements: supplementary

(1) Where an application is made to a leasehold valuation tribunal for a determination to dispense with all or any of the consultation requirements in relation to any qualifying works or qualifying long term agreement, the tribunal may make the determination if satisfied that it is reasonable to dispense with the requirements.

[13] Section 27A of the Act makes provision for applications to an LVT (*inter alia*) to determine the amount of a service charge that is payable; such determinations may be sought either in advance of the costs being incurred or subsequently thereto. Paragraph 6 of the consultation regulations, read with section 20 of the Act, provides that the limit of the amount that can be passed on by a landlord to a tenant if the consultation requirements have neither been complied with nor dispensed with is £250.

[14] Pausing here, it may be noted that the terms of sections 20 and 20ZA of the Act differ from the predecessor statutory provisions. So, prior to amendment, section 20 of the Act had provided as follows:

20. Limitation of service charges: estimates and consultation

(1) Where relevant costs incurred on the carrying out of any qualifying works exceed the limit specified in subsection (3), the excess shall not be taken into account in determining the amount of a service charge unless the relevant requirements have been either —

(a) complied with, or

(b) dispensed with by the court in accordance with subsection (9);

and the amount payable shall be limited accordingly.

In proceedings related to a service charge the court may, if satisfied that the landlord acted reasonably, dispense with all or any of the relevant requirements.

It will be apparent that under these predecessor provisions, dispensation from the relevant requirements could be granted only if the landlord had acted reasonably: see, too, *Martin* v *Maryland Estates Ltd* [1999] 2 EGLR 53*, at p56, *per* Robert Walker LJ, as he then was. The current statutory provisions contain no such limitation; dispensation from the consultation requirements may be granted whether or not the landlord has acted reasonably.

*Editor's note: Also reported at [1999] 26 EG 151

[15] Coming next to the relevant consultation requirements, these are set out in Part 2 of Schedule 4 to the consultation regulations. With respect, I cannot improve on the summary contained in the LT decision in this case, itself drawing on a previous LT decision in *Camden London Borough Council* v *Leaseholders of 37 Flats at 30-40 Grafton Way* LRX/185/2006, unreported 11 April 2008 (*Grafton*). As the LT observed (in [4]), the "layout and drafting of the regulations leave something to be desired in terms of clarity". The LT, *ibid*, then followed the division of the requirements into "stages" adopted in *Grafton*: G

Stage 1

(1) **Notice of intention** Notice of intention to carry out qualifying works is given to each leaseholder and any recognised tenants' association ("RTA"). The notice must describe in general terms the proposed works, or specify a place and hours where the description may be inspected. The notice must state the reasons for the works, and invite written observations, specifying where they should be sent, over what period (30 days from the notice), and the end date. Further, the notice must contain an invitation for nominations of persons from whom the manager should obtain estimates. The landlord must have regard to written observations received during the consultation period. H

Stage 2

(2) **Estimates** The landlord must seek estimates. (There are detailed rules as to seeking estimates from nominees of the tenants or RTA).

(3) **The paragraph (b) statement** The landlord then issues a statement (free of charge) setting out the estimated cost from at least two of the estimates and a summary of the observations received during the stage 1 consultation period, and his responses to them. The statement is issued with a notice (see below). If any estimates have been received from leaseholders' nominees, they must be included in the statement. (The term "paragraph (b) statement" is used by the regulations themselves, by reference to sub-paragraph (5)(b) in which this requirement is found). J

(4) **Notice accompanying paragraph (b) statement** The statement must be sent out with a notice, detailing where and when all of the estimates may be inspected and inviting each leaseholder and any RTA to make written observations on any of the estimates, specifying an address where they should be sent, the consultation period (30 days from the notice) and the end date.

(5) **Regard to observations** The landlord must have regard to written observations received within this second 30-day consultation period. K

Stage 3

(6) **Notification of reasons** Unless the chosen contractor is a leaseholder's or RTA nominee or submitted the lowest estimate, the landlord must give notice within 21 days of entering into the contract to each leaseholder and any RTA, stating his reasons for the selection, or specifying a place and hours for inspection of such a statement . . .

[16] Matching these stages with the relevant paragraphs of Part 2 of Schedule 4 to the consultation regulations, stage 1 corresponds with paras 1 to 3; stage 2 with paras 4 to 5; and stage 3 with paras 6 *et seq.* As the LT observed, in [6]: L

The issues in the present case turn on the requirements of Stage 2: steps (3), (4) and (5) in the above sequence. They relate principally to the following paragraphs in Schedule 4 Part 2 of the Regulations: step (3) paras 4(5)(b) and 4(9); step (4) para 4(10); step (5) para 5.

Nature of hearing before this court

[17] As already noted, this appeal comes to this court pursuant to section 13(1) of the 2007 Act, which grants a right to appeal "on any point of law arising from a decision made by the Upper Tribunal". Section 14 of the 2007 Act goes on to provide as follows:

(1) Subsection (2) applies if the relevant appellate court, in deciding an appeal under section 13, finds that the making of the decision concerned involved the making of an error on a point of law. M

(2) The relevant appellate court —

(a) may (but need not) set aside the decision of the Upper Tribunal, and

(b) if it does, must either —

(i) remit the case to the Upper Tribunal or, where the decision of the Upper Tribunal was on an appeal… from another tribunal…, to the Upper Tribunal or that other tribunal…, or

(ii) re-make the decision.

[18] For reasons on which it is unnecessary to elaborate (and that may be of limited duration), it is common ground that this

A is not a "second appeal" to which CPR 52.13 and other related provisions apply.

[19] A question that is in dispute relates to the true nature of this appeal. Daejan contends that the decision that is reviewed pursuant to CPR 52.11 is that of the LVT. Weight may be given to the views of the LT if and to the extent that this court agrees with those views; special weight may be given to those views if and in so far as they reflect any special expertise of members of the LT, although it is also Daejan's contention that there was no relevant special expertise. Essentially, therefore, this court is concerned with the August decision of the LVT, rather than the decision of the LT.

B [20] By contrast, the respondents place much greater weight on the decision of the LT, submitting that, in the light of the failed first appeal, this appeal can only succeed "if the review by the… [LT]… can be vitiated on the usual grounds…", namely erroneous construction of the statutory provisions in question, failing to take account of a relevant factor and so on. The respondents further submit that despite how the matter is approached, the appeal can succeed only if this court is satisfied that the LT was wrong in law. It would seem to follow on the respondents' approach that even if this court was of the view that the LVT had erred in law, the appeal must be dismissed if the LT had reached a decision to which it was entitled to conclude.

C [21] In my judgment, the correct position lies somewhere between the submissions of the parties; both the August decision of the LVT and the LT's decision fall to be considered as follows:

(i) In the light of sections 13 and 14 of the 2007 Act, this appeal cannot succeed unless this court is satisfied that the decision of the LT involved the making of an error on a point of law.

(ii) That said, the starting point on this appeal is the August decision of the LVT. It is the LVT, not the LT, that is entrusted by statute with determining whether there has been compliance with the consultation regulations and, if not, whether dispensation should be granted; the LT's function is appellate only. I am thus unable to accept that this is D an appeal where this court's consideration is effectively confined to a "review" of the decision of the LT, so that provided only that the LT reached a decision it was entitled to reach, the appeal must be dismissed regardless of any flaws in the decision of the LVT. This court must instead explore whether the LVT had erred in law; but, in conducting that inquiry, it seems to me right that such weight as is appropriate should be given to the views of the LT as the designated and specialist appellate tribunal in this area. What weight is to be given to the LT's views must depend on the issue under consideration and all the circumstances of the individual case; this court cannot be prescriptive in that regard.

E [22] On reflection, there may be less in this particular dispute than first meets the eye or, put another way, the theory may suggest more complexity than will be encountered in practice. The examples that follow suggest why this is so:

(i) If this court is of the view that there was no error of law in the LVT's August decision, that would be an end of the matter, having regard to the dismissal of Daejan's appeal to the LT.

(ii) If, conversely, this court, having given such weight as it thought appropriate to the views of the LT, was none the less satisfied that the LVT had erred in law in its August decision, it would seem to me very difficult to resist the conclusion that the LT had itself erred in law in F deciding otherwise.

(iii) Should the situation arise where this court was of the view that a range of decisions was properly open to the LVT, this court would need to consider whether the decision to which the LVT had come was within the permissible ambit of that range. Once again, in considering that question, this court would no doubt pause before rejecting the views of the LT on a "discretionary" decision of that nature.

Facts and key conclusions of the LVT and the LT

[23] (1) Factual introduction: A very brief mention of the facts (largely taken from the "statement of agreed facts and issues" before the LT) is helpful to an understanding of the key conclusions of the LVT and the LT.

[24] Queens Mansions is a block of shops and flats in Muswell Hill. There are seven flats on the ground and upper floors. Five of the flats, nos 1, 3, 4, 5 and 6, are held under long leases that provide for the payment of service charges. Daejan is the freehold owner of the building and the landlord under those long leases. The respondents are the lessees of those flats. For completeness as to the other two flats, at all material times one was vacant and the other was occupied by a tenant under a statutory tenancy protected by the Rent Act 1977; no more need be said of them.

[25] The respondents are all members of the Queens Mansions Residents' Association (the QMRA), which is a recognised tenants' association under section 29 of the Act. Ms Marks was the chairman of the QMRA.

[26] The property was managed for Daejan by Freshwater Property Management (Freshwater), the trading name of Highdorn Co Ltd.

[27] By the start of 2005, it was clear that major works were needed to the property. The stage 1 notice of intention was sent on 6 July 2005. By December 2005, four priced tenders had been received by Freshwater. On the face of it, two contractors, Rosewood and Mitre, appeared to be the most competitive; as an indication of the scale of the works, all the tenders were priced in excess of £400,000. The respondents were, however, concerned that only Mitre's priced specification had been provided to them; in 2006, Ms Marks pressed Daejan for the opportunity to inspect all the priced tenders. By June 2006, this concern had not then been addressed but Freshwater, through a Mr Shevlin, was indicating a preference for instructing Mitre to proceed with the works. Mr Shevlin attempted to serve stage 2 notices from June 2006, but unhappiness as to the unavailability of the priced estimates to the respondents continued to simmer. A further stage 2 notice was purportedly served on 28 July. Daejan accepted before the LVT that the priced specifications were not made available to those of the respondents who wanted to see them until 11 August 2006.

[28] However, as the LVT recorded in its August decision:

> 56. … by 11th August 2006, the leaseholders had already been informed that the contract had in effect been awarded to Mitre, and the statutory consultation process was for all practical purposes ended.
>
> 57. It was not disputed that at the pre-trial review hearing in early August 2006, it was stated by the legal representative of Daejan, that Mitre Construction had been awarded the contract. This was confirmed in a letter from Mr Shevlin to Ms Marks dated 10th August 2006. In his oral evidence Mr Shevlin said that this was a mistake, but it had never been corrected.

[29] As already foreshadowed, the dispute centred on stage 2 and, in particular, on the events of June to August 2006.

[30] (2) The LVT: In its March decision, the LVT determined that Daejan had failed to comply with the provisions of Part 2 of Schedule 4 to the consultation regulations. Subsequently, the August decision both clarified the LVT's findings "in respect of the nature and extent of the landlord's non-compliance" and addressed the issue of dispensation.

[31] In [24] of the August decision, the LVT set out its conclusions on Daejan's non-compliance with the consultation regulations in the following terms: (note: paragraph numbers have been amended to conform to the version used by the LT):

> Having considered the evidence as a whole, the Tribunal considers that Daejan:
>
> 1] failed to comply with the requirements of paragraph 4(5)(b)(ii).
> A summary of the observations received and the landlord's responses thereto, were not properly included.
>
> 2] failed to comply with paragraph 4(5)(c), 4(9), 4(10) and 4(11)
> All the estimates were not available for inspection at a place, during the hours and for the period specified in the notice. The relevant period of thirty days was cut short as it was indicated to the leaseholders that the contractor for the major works had been decided by the landlord. As a result they concluded that further representations were futile. The consultation process was for all practical purposes curtailed, and the invitation to inspect the estimates and make observations was rendered ineffective at that point.
>
> 3] failed to comply with paragraph 5.
> The relevant period of thirty days was cut short before the leaseholders were provided with copies of all the estimates and had had an opportunity to make observations. The landlord did not have regard to the observations in respect

of the estimates which the leaseholders may have made, had they had the opportunity to do so within the relevant period.

[32] At [84] *et seq*, the LVT set out its conclusions overall. The LVT found that the failure by Daejan to comply with the statutory consultation requirements had caused substantial prejudice to the respondents. The LVT underlined that it had been a matter of great concern to the respondents that Daejan had not provided copies of all the estimates for their consideration. In [90], the LVT said:

> The Tribunal considers that the cutting short of the consultation period, by indicating to the leaseholders that the decision had been made to award the contract to Mitre, both at the pre-trial review and in Mr Shevlin's letter dated 10th August 2006, removed from the leaseholders the opportunity to make observations on the estimates to which [the] landlord was obliged to have regard. This opportunity to make informed comment on these matters was central to the consultation process. It had been stressed in correspondence how important this was to the leaseholders.

[33] The LVT acknowledged that there had been "extra-statutory" consultation carried out, notably through a Mr Harris, a surveyor originally engaged by the QMRA to advise the respondents and, at their suggestion, appointed by the appellant as contract administrator. The LVT did not, however, consider that this made good Dajean's failure to provide the estimates and an opportunity to make observations.

[34] Drawing on *Grafton* (to which reference has already been made), the LVT then addressed the submission advanced by counsel for Daejan, that Mr Lapes (one of the respondents) had been unable to identify the comments he would have made had he seen the Rosewood priced tender earlier. In the LVT's view (in [98]):

> it is a matter of speculation what comments may or may not have been made by Ms Marks, Mr Lapes and the other leaseholders and how this may have influenced the carrying out of the major works, had they had the opportunity to comment having seen all the estimates. The overall result of the landlord's failure to consult the leaseholders was that the Respondents lost the opportunity to make observations in respect of the estimates provided by contractors for the major works to Queens Mansions for which they would be contractually obliged to contribute substantial amounts of money. The Tribunal considers that the fact that they did not have this opportunity amounts to significant prejudice.

[35] The LVT next considered and rejected an offer on the part of Daejan to deduct £50,000 from the cost of the works; in the LVT's judgment, this offer did not alter "the existence of substantial prejudice suffered by the leaseholders".

[36] Counsel for Daejan submitted that it was not proportionate, in all the circumstances, for the LVT to refuse to make an order for dispensation; such refusal would result in Daejan having to bear all but £1,250 (£250 per long leaseholder) of the cost of the works. Once again guided by *Grafton*, the LVT held that:

> 104. … the financial effects of the grant or refusal of the application for dispensation on the landlord or tenant are not to be taken into account.

The LVT thus concluded that it did not consider it reasonable to dispense with the consultation requirements in this case and declined to make an order for dispensation under section 20ZA of the Act.

[37] (3) The decision in *Grafton*: Before the LT in this case, the decision in *Grafton* (*supra*) loomed large and must now be summarised, given its importance to the LT's reasoning.

[38] In *Grafton*, the landlord was the housing authority; there were 40 leaseholders, of whom 21 had formed a committee to represent them in discussions with the landlord. The dispute was concerned with stage 2 that, in *Grafton*, had been omitted altogether through an administrative error. In these circumstances, the LT in *Grafton* had upheld the LVT's decision not to dispense with compliance.

[39] The central reasoning of *Grafton* appears from the following passages:

> 32. Any process of consultation consists of giving information, inviting observations and taking those observations into account, and this is what paragraphs 1 to 6 make provision for. Information has to be given to tenants at three stages when there is an intention to carry out works, when estimates have been obtained and when a contract has been entered into. Observations from tenants are to be invited at the first two stages. Those observations must be taken into account, and the landlord's response to them must be given. This is the scheme of the provisions, which are designed to protect the interests of tenants; and whether it is reasonable to dispense with any particular requirements in an individual case must be considered in relation to the scheme of the provisions and their purpose. G
>
> 33. The principal consideration for the purpose of any decision on retrospective dispensation must… be whether any significant prejudice has been suffered by a tenant as a consequence of the landlord's failure to comply with the requirement or requirements in question. An omission may not prejudice a tenant if it is small, or if, through material made available in another context and the opportunity to comment upon it, it is rendered insignificant. Whether an omission does cause significant prejudice needs to be considered in all the circumstances. If significant prejudice has been caused we cannot see that it could ever be appropriate to grant dispensation. H
>
> 34. It was urged on us by Miss Holland that the consequences, for LBC and their tenants, was a material consideration, and indeed an important one. Also material, she suggested, was the unjustified benefit that the leaseholders here would receive in the event that dispensation was not granted. We can accept that the general nature of the provisions, with the £250 limit imposed as the consequence of section 20(1) and section 20ZA, forms part of the background to the consideration of reasonableness. We cannot accept, however, that the particular effects on the landlord or the tenant in the case in question are properly to be taken into account. It is in the very nature of the provisions that the landlord will suffer financially and the tenant will gain financially in the event that dispensation is not given. If it were material to take into account the degree to which the landlord might suffer or the tenant might gain, this would mean that a failure might achieve dispensation if the contract was a very large one but might not do so if the contract was small. We do not think that this could be the effect of the provisions. There would in any event be real practical difficulties for an LVT in dealing with a contention relating to the consequences for the landlord or other persons affected since the evidence relevant to these could be very far-reaching, time-consuming and costly to pursue and potentially inconclusive. J
>
> 35. The requirements relating to estimates are clearly fundamental in the scheme of requirements. The landlord must obtain estimates (in the plural), must include in the paragraph (b) statement the overall estimate of at least two of them and must make all of the estimates available for inspection. The purpose is to provide the tenants with the opportunity to see both the overall amount specified in two or more estimates and all the estimates themselves and to make on them observations, which the landlord is then required to take into account. In the present case stage 2 was completely omitted. It was a gross error, which manifestly prejudiced the leaseholders in a fundamental way. K

[40] (4) The LT: In the present case, and after a review of the facts, the LT turned its attention to *Grafton*. The LT underlined that it was not bound by *Grafton* but would "naturally treat it as persuasive". Having considered the matter, the LT saw no reason to depart from the approach taken in *Grafton*, supported, in the LT's view, by the statutory language. As the LT said: L

> 40. … The power given to the LVT is to dispense with the consultation requirements, not with the statutory consequences of non-compliance. The principal focus, therefore, must be on the scheme and purpose of the regulations themselves. If Parliament had intended to give a power to remove or mitigate the financial consequences, it could easily have done so… The potential effects — draconian on one side and a windfall on the other — are an intrinsic part of the legislative scheme. It is not open to the tribunal to rewrite it.

For the same reasons, the LT was not attracted to Daejan's alternative submission — or offer — of allowing a deduction from the charge rather than refusing dispensation in its entirety. M

[41] The LT went on (in [41]) to express agreement with *Grafton*, that the potential consequences for the parties "are relevant as part of the context in which the matter is to be considered". It was neither helpful nor accurate to describe the provisions as "penal":

> the tribunal should keep in mind that their purpose is to encourage practical co-operation between the parties on matters of substance, not to create an obstacle race. If the non-compliance has not detracted significantly from the purpose of the regulations and has caused no significant prejudice, there will normally be no reason to refuse dispensation.

Tribunals should use their own common sense to examine submissions as to the significance of non-compliance, given the parties' financial interests in playing such significance up or down, as the case may be.

[42] The LT added:

> 43. Finally, we emphasise the need to consider these issues having regard to the particular facts of each case, including the nature of the parties and their relationship. For example, the tribunal may reasonably take a more rigorous approach to non-compliance by a local authority or commercial landlord, than to a case where the landlord is simply a group of lessees in another form.

[43] Thereafter, the LT sought to apply these principles to the facts of the present case. It began with this observation:

> 44. … We have no doubt that the LVT was correct to find failures of compliance at Stage 2, but it is equally clear to us that for the most part these failures were minor and did not cause any significant prejudice to the lessees.

[44] The LT held that although Daejan's notice had not contained the requisite "summary of observations", it was "quite unable to see what prejudice was caused by the absence of such a summary". As to the stage 2 notices, these had not in terms specified the "place and hours" at which the estimates could be inspected and the estimates were not available for inspection at the time they were sent (whether on 14 June or 28 July); moreover, the notices had failed in terms to "invite" the making of written observations. The LT was, however, of the view that none of these failures to comply with the consultation regulations caused the respondents any prejudice:

> 50. … provided that, once they received the estimates, they had an appropriate opportunity to consider them and to raise any significant points that might emerge from them, and that they were taken into account before a final decision was made.

[45] The question of consultation being curtailed was different; this was the issue that had most concerned the LT. In its view, the LVT's conclusion that consultation had been curtailed was "difficult to criticise"; the LVT had been entitled to conclude that, by 10 August 2006, Daejan had "effectively closed its mind to any new observations on the choice of contractors, and that the lessees had reasonably formed the view that this was so".

[46] It remained for the LT to consider whether the respondents had been prejudiced by the curtailing of consultation:

> 57. … in particular whether the estimates did in fact disclose any new material, and whether the lessees would have had anything new to say.

The LT remarked on the LVT dismissing this inquiry as speculative, echoing the language used in *Grafton*. The LT, however, underlined that the circumstances in *Grafton* were different: there were many more lessees, not all represented by the main group, and it could not be said that all the same information had been made available in another form.

[47] After a somewhat critical review of the evidence and submissions as to prejudice, the LT expressed its own conclusions as follows:

> 61. We have not found this an easy case. Although the LVT was entitled to find a material breach of the regulations, the evidence of actual prejudice is weak. However, we remind ourselves that we are reviewing their decision, not substituting our own judgment. It is common ground that we can only interfere if the LVT has gone wrong in principle, or left material factors out of account, or its balancing of the material factors led it to a result which was clearly wrong. The LVT was in our view entitled to regard this as a serious breach, rather than a technical or excusable oversight. It involved a failure by a corporate landlord to ensure that those responsible in their office for the stage 2 consultation properly understood its requirements and its significance. The result was that the lessees' statutory right at stage 2, to make further representations following examination of the estimates, was nullified.
>
> 62. As to prejudice, the tribunal was entitled to start from the position that, given the seriousness of the breach, it was not for the lessees to prove specific lack of prejudice. It was enough that there was a realistic possibility that further representations might have influenced the decision. We bear in mind that, as is clear from the tender report, Mr Harris' decision to recommend Mitre in preference to Rosewood was a very close one. Although the issue was raised in Ms Marks' letter of 14 July, she made clear that it was an interim comment, pending access to full information. The lessees were entitled to proceed on the basis that they would have a further opportunity to present their case in its strongest terms and in the light of the full information; and that, given the marginal difference between the two tenders, they might be able to persuade the landlord to change its mind. In these circumstances, we are unable to say that [the] LVT has erred in principle, or that its decision was clearly wrong. The financial consequence may be thought disproportionately damaging to the landlord, and disproportionately advantageous to the lessees, but, as we have said, that is the effect of the legislation.

Issue (I): Financial consequences

[48] (1) The rival cases: For Daejan, Mr Dowding submitted that both the LT in *Grafton* and the LT (and LVT) in this case had fallen into error in treating the financial consequences of the grant or refusal of dispensation as irrelevant. The "strict" approach adopted in these cases resulted in "draconian consequences". The purpose of the dispensing power in section 20ZA(1) was to ensure that the landlord did not suffer the penalty imposed by section 20 except where it was reasonable that it should do so.

[49] Mr Dowding emphasised that the discretion to grant dispensation under the Act as amended was broader than under the predecessor provisions. Dispensation could now be granted even if the landlord had conducted itself unreasonably. The statute conferred a broad dispensing power and it was not for the court to restrict it. The consequence of doing so was that each tenant obtained the benefit of all the work done for its benefit in excess of £250 for nothing; yet service charges were the mirror images of the landlord's obligation to repair.

[50] Importantly, Mr Dowding said this: it would be impractical to have regard to the circumstances of the individual landlord and tenants; his submissions did not require it. Instead, he contended that the size of the sum itself (here hundreds of thousands of pounds) should be taken into account.

[51] The exercise of the discretion (namely whether it was reasonable to grant dispensation) required a balancing act. As a matter of fairness and principle, substantial financial consequences, albeit not decisive, had to be relevant and should be weighed in the scale. The weight to be given to those consequences depended on the circumstances of each case. As Mr Dowding put it:

> There is clearly a difference between a case where there is a modest degree of prejudice to the tenants and a huge financial loss to the landlord, and one where there is a modest degree of prejudice to the tenants and a modest financial loss.

[52] For completeness, Mr Dowding did not quibble with the outcome in *Grafton*. The complete failure to follow stage 2 in that case outweighed all other considerations.

[53] In the upshot, the LT and LVT had misdirected themselves in law by treating the financial consequences of the refusal of dispensation as irrelevant and this court should exercise the section 20ZA(1) discretion.

[54] For the respondents, Mr Rainey's submissions (supported by Mr Fieldsend) proceeded as follows. The LT and the LVT had rightly treated the financial consequences of the refusal of dispensation as irrelevant; there had been no error of law or misdirection. Daejan had approached the consultation requirements as "an annoying and artificial set of legal hoops to be jumped through prior to getting on with doing whatever works that a landlord had decided to do"; these submissions betrayed a misconception. Statutory consultation was a substantive right of a tenant: the reason was that X (the tenant) was being asked to pay for expenditure decided on by Y (the landlord); hence the legislative scheme. The landlord could do what it liked, but if it wanted to be reimbursed by the tenants, it was obliged to comply with the consultation requirements. That did not mean that the landlord was obliged to follow the views expressed by the tenants; but it did mean that the landlord had to pay due regard to the consultation. The consultation was an end in itself and the landlord's failure (if significant) to comply with the consultation requirements by itself gave rise to prejudice to the tenants. When, as here, consultation was curtailed, the tenants lost

the opportunity to consider making representations and to have those considered by the landlord.

[55] As to where the line was to be drawn, Mr Rainey submitted that *circumstances* rather than *consequences* were relevant to the decision as to whether to grant or refuse dispensation. Thus, the circumstances going to the integrity of the consultation process were relevant; by contrast, the consequences flowing from the decision to grant or refuse dispensation were not. It was common ground that the test for dispensation under the Act as amended was wider than the previous test and the respondents were not seeking to circumscribe it; but they did seek to identify what was relevant and irrelevant.

[56] Mr Rainey suggested the following as examples of when dispensation with the consultation requirements might be granted: (i) the need to react to an emergency; and (ii) a situation where (realistically) only one contractor could undertake repairs to (say) a particular type of boiler or lift. In such circumstances, compliance with the consultation requirements would not be practical.

[57] There was nothing in the point that if dispensation was refused, the excess of the cost of the works over £250 per tenant would fall to be borne by the landlord; the obverse was that if dispensation were granted, the tenants would be liable for that amount. Such consequences were simply what the legislative scheme provided.

[58] If wrong in these submissions, the respondents invited the court to remit the matter to the LVT, rather than seeking to exercise the section 20ZA(1) discretion afresh.

[59] (2) Discussion: Having introduced this issue at some length, my conclusions can be relatively briefly stated. With respect to Mr Dowding's forcefully presented submissions, I have a clear preference for those of the respondents. The issue is one of statutory construction. In my judgment, as a matter both of context and the wording of the statute, the financial effect of the grant or refusal of dispensation is an irrelevant consideration when exercising the discretion under section 20ZA(1). My reasons follow.

[60] First, it is instructive to identify the answer for which Daejan contends. It might have been expected that the attraction of the argument for weighing financial consequences in the balance lay in their effect on the individual parties, but that is not Daejan's case. Indeed, it is noteworthy that Mr Dowding drew back from contending that the financial effect of the grant or refusal of dispensation on the individual landlord and tenants should be considered. As a matter of practicality, Mr Dowding was right to hold back, even though, in logic, his submissions pointed to such an inquiry. The burden imposed on LVTs undertaking an inquiry into the financial circumstances of any particular landlord would be considerable. All the more so, in that any such inquiry would necessitate a similarly burdensome exploration of the financial circumstances of the tenants in question. However, Mr Dowding's pragmatic restraint means that he is left with the argument that the size of the sum is by itself relevant to the grant or refusal of dispensation. I am wholly unable to accept that submission, entailing as it does taking account of the sum involved, but not the effect on the parties. The submission must contemplate that the higher the service charge, the more readily dispensation should be granted. Whether Mr Rainey was right to characterise such an outcome as "perverse", it would, on any view, be a curious outcome and not one that I would be inclined to spell out of section 20ZA(1) unless driven to such a conclusion. With rather more plausibility, it could be argued that the higher the sum involved, the more necessary it is for consultation requirements to be observed.

[61] Second, regard must be had to the statutory language. Here, I respectfully agree with and adopt the LT's observations in [40] (set out above). Section 20ZA(1) gives the LVT power to "dispense with… the consultation requirements", not, as the LT put it, "with the statutory consequences of non-compliance". Here, too, the respondents' submission had force; the legislature did not employ the well-known drafting technique of granting relief from sanctions. Accordingly, and, as it seemed to me, plainly, the focus of the statute is on the consultation requirements, not the consequences of non-compliance. At least generally (it would be unwise to be unduly rigid), the distinction drawn by the respondents — between the relevance of circumstances G going to the integrity of the consultation process and the irrelevance of the consequences flowing from the grant or refusal of dispensation — seems well founded.

[62] Third, if perhaps a matter arising both under this issue and issue (III) (prejudice), the legislative focus on the consultation requirements serves to highlight that a proper consultation process comprises the substance of the scheme. Prima facie, therefore, curtailment of consultation (at least unless *de minimis*) involves substantial non-compliance with the consultation requirements. That is not to elevate *process* above *substance* — something I would be most reluctant to do and against which the LT properly cautioned in [41]. H This is not, as the LT remarked, about creating "an obstacle race". It is instead to identify, and from the outset, the true nature of the statutory purpose and substantive considerations in this area. Once again, this suggests that (other than as part of the context in which the matter is to be considered) the financial consequences of the grant or refusal of dispensation are unlikely to be relevant to the exercise of the discretion under section 20ZA(1); in any event, that non-compliance with the consultation requirements may involve significant financial loss is, by itself, no more than an intrinsic part of the statutory scheme.

[63] Fourth, some examples of when dispensation might be granted can readily be postulated. They are, it should be emphasised, no more J than examples; there is no closed list. So, all other things being equal, the following situations might commend the grant of dispensation:

(i) the need to undertake emergency works;

(ii) the availability, realistically, of only a single specialist contractor;

(iii) a minor breach of procedure, causing no prejudice to the tenants.

It may be noted that none of these examples undermines the integrity or importance of the consultation process.

[64] Fifth, I do not see this approach as in any way failing to give effect to the width of the dispensatory discretion conferred by the K Act as amended. For instance, under the current (amended) statutory provisions, the landlord may obtain dispensation in any of the examples given in the preceding paragraph, even if it has not "acted reasonably". However, granting that the reasonableness of the landlord's conduct is no longer a condition precedent for an order for dispensation, it does not follow that the financial effects of a refusal of dispensation are, or ought to be, a relevant consideration to the exercise of the statutory discretion.

[65] For all these reasons, I am amply satisfied that the LT in *Grafton* and the LVT and the LT in this case did not err in law or misdirect themselves in treating the financial consequences of the grant L or refusal of dispensation as irrelevant to the exercise of the discretion under section 20ZA(1). To the contrary, I would uphold their approach to this issue.

Issue (II): Nature of the landlord

[66] On reflection, this issue can be taken extremely briefly. In its decision, the LT referred to this question in [43] and touched on it in [61]. Rather seizing on these two mentions, Mr Dowding submitted that the nature of the landlord was irrelevant; there was no warrant for treating a corporate (or local authority) landlord more rigorously than a lessee-owned or controlled landlord. On the footing that the legislative M provisions were not penal and the focus was on the prejudice caused to tenants by the landlord's non-compliance with the consultation requirements, there was no basis for any such distinction turning on the nature of the landlord. The LT had accordingly erred in law and/or misdirected itself.

[67] With respect, I cannot agree.

(i) For my part, I do not think that the LT was saying more than (as Sedley LJ observed in argument) "context is everything"; that is certainly the impression created by the opening sentence of [43]. If so, the observations are unexceptionable and this issue simply falls away.

(ii) If it is necessary to go further, cases can be envisaged where a less rigorous approach may be justified in respect of lessee-owned/controlled landlords. So where the lessees are their own landlord, the consultation requirements have to be considered against the background that they are spending their own money; it may no longer be the case of X spending Y's money. Furthermore, in such a situation, there may be a greater likelihood of canvassing the relevant information by way of informal or extra-statutory consultation.

(iii) However, even if the LT was in error in contemplating a less rigorous approach in (some) cases where the tenants are their own landlord, the point does not go anywhere. It does not follow that the LT erred in holding that the consultation requirements were to be applied according to their terms in the case of a corporate (or local authority) landlord. There is certainly no suggestion in its decision that the LT applied a *more* rigorous standard than that provided for in the statute.

(iv) Finally, if there was an error in this regard, it was an error on the part of the LT — there is no hint of any such "error" in the August decision of the LVT. For this reason, too, as it seems to me, the point goes nowhere as a self-standing ground of appeal.

Issue (III): Prejudice

[68] (1) The rival cases: I focus here on the curtailing of consultation, the issue that most concerned the LT.

[69] For Daejan, Mr Dowding drew attention to the differences between this case and *Grafton.* In *Grafton*, stage 2 had been omitted altogether; here, it had not been. Moreover, in *Grafton* there had been a higher number of tenants, so (as the LT itself observed) there could be no question of all the information having been made available in another form. In those circumstances, Mr Dowding, as before, did not quarrel with the outcome in *Grafton.*

[70] In the present case, however, Mr Dowding contended that there was nothing further to be said or considered; in a letter dated 14 July 2006, Ms Marks had said all that could be said based on the information in Mr Harris's tender report. There was no basis for supposing that, if the consultation had not been curtailed, it would have made any difference. There were only five respondents. Accordingly, Mr Dowding submitted that the LVT (and LT) should have answered the "what if" question. If that question had been posed, it could not have been answered as the LT did in [62] (a conclusion in any event unsupported by any fact-finding of the LVT). The only conclusion that could reasonably have been reached was that the respondents had not proved any prejudice flowing from Daejan's non-compliance with the consultation requirements. It was, however, for the respondents not simply to assert but to prove prejudice: see, the decision of the LT in *Eltham Properties Ltd* v *Kenny* LRX/161/2006*, in [29] to [30], together with the observations of the LT in this case in [42]. In any event and whatever the incidence of the burden of proof, there had been no prejudice. If right thus far, there was no proper basis for the LVT or LT refusing to grant dispensation; their decisions to refuse dispensation were either perverse (within the meaning of *Edwards supra*) or disclosed an error of law.

[71] For the respondents, Mr Rainey submitted that if this issue turned on the burden of proof, then, since Daejan was seeking dispensation, it was for Daejan to prove that its non-compliance had not caused any prejudice so that it was reasonable for dispensation to be granted; it was not for the respondents to prove specific prejudice flowing from Daejan's non-compliance with the consultation requirements. In any event, the curtailment of the consultation process amounted to substantial prejudice; the respondents had been deprived of the opportunity to make representations and to have them considered. It was unnecessary to speculate the outcome of such further consultation had it taken place; a landlord could always say that nothing further said by the tenants would have made any difference. So far as concerned other communications between Daejan and the respondents (even assuming that they were capable of curing the position), these could not save the day for Daejan because the full facts had not been available until it was too late.

* Editor's note: Reported at [2008] L&TR 14

[72] (2) Discussion: For my part, I readily accept Mr Dowding's submission, as far as it goes, that significant prejudice to the tenants is a consideration of the first importance in exercising the dispensatory discretion under section 20ZA(1). I respectfully and entirely agree with the observations to this effect in *Eltham (supra)*, in [29] to [30], *Grafton*, in [33], and the LT in this case, in [41] to [42].

[73] I part company, however, with *Daejan* when it comes to determining whether the respondents did suffer significant prejudice in consequence of Daejan's non-compliance with the consultation requirements. In my judgment, Daejan's non-compliance in curtailing consultation constituted a serious failing and did cause the respondents serious prejudice.

(i) As already emphasised, a proper consultation process is of the essence of this statutory scheme, devised as it is to protect the interests of tenants such as the respondents.

(ii) In my judgment, the LT and the LVT were entirely right to treat the curtailment of the consultation process as a serious failing. It is striking that the observation of the Daejan legal representative at the LVT hearing in early August 2006, that Mitre had already been awarded the contract for the works, was never corrected; to the contrary, it was confirmed in Mr Shevlin's letter of 10 August 2006. Even assuming this failing to be the result of a lack of understanding or ineptitude, rather than a flouting of the consultation requirements, it is impossible to view it as a technical, minor or excusable oversight.

(iii) Against this background, I can detect no error of law or misdirection in the LVT's refusal to speculate (in [98] of its August decision) as to what might have been the outcome had the consultation been allowed to run its proper course. Indeed, given the seriousness of non-compliance in this case, I would endorse the LVT's approach of treating the respondents' loss of opportunity (to make further representations and have them considered) as itself amounting to significant prejudice. On any view, as it seems to me, that was a conclusion to which the LVT was entitled to come.

(iv) This view is reinforced by reflection on the rival contention advanced by Daejan. In many cases, a landlord could readily assert that further consultation would have made no difference. Disproving such assertions would inevitably give rise to an invidious exercise in speculation, quite apart from difficulties of proof (if and in so far as a burden rests on the tenants in this regard: see below). Although there will no doubt be some instances where a landlord may demonstrate that a failure to comply with the consultation requirements was, on the facts, such as to make no difference and to give rise to no prejudice to the tenants, arguments of this nature need careful scrutiny; there would otherwise be a risk of undermining the purpose of the statutory scheme or, as Pitchford LJ remarked in argument, a "premium on recalcitrance". Suffice to say that on the facts of this case, involving a serious failing on Daejan's part, I am not at all attracted to the argument.

(v) With respect to Mr Dowding's argument, I do not see a tension between the LVT's conclusion (that the curtailment of the consultation itself amounted to significant prejudice) and the observations of the LT already referred to, as to the importance of prejudice to the tenants, in *Eltham, Grafton* and in this case. The conclusion of the LVT involves a finding that there has been significant prejudice. Moreover, given that the LVT found as a fact that the extra-statutory consultation had not made good Daejan's failure to comply with the consultation regulations, nothing turns on the difference between the number of tenants in this case as compared with the much larger number in *Grafton.*

(vi) Accordingly, the LVT was amply justified in refusing dispensation in this case. Its conclusion betrays neither any error of law nor perversity.

[74] Having reached this view (which is sufficient to decide issue (III)), it is strictly unnecessary to decide whether the LT was correct to say (in [62]) that further representations might have influenced the *Daejan* decision as to the award of the contract for the works. For the avoidance of doubt, however, I am not at all persuaded that the LT fell into error. I acknowledge the difficulty of pointing in terms to a finding of fact by the LVT supporting the LT's observation.

That said, I am satisfied that the LT's conclusion involved no more than the drawing of a permissible inference from the LVT's findings.

[75] Accordingly, and whether by the route followed by the LVT or that taken by the LT in [62] (if and in so far as it differed), I entertain no doubt of the correctness of the LT's conclusion, namely that it was unable to say that the LVT had erred in principle on the question of prejudice or that its decision was clearly wrong.

[76] For completeness, it remains simply to mention a number of additional matters:

(i) First, in agreement with both the LT. in [40]. and the LVT, in [101], I do not think that Daejan's offer of a £50,000 "discount" off the price of the works provides a ground for the grant of dispensation. I incline to the view that, as the LT reasoned, the statutory scheme does not provide for such an alternative; however, even if it were open to Daejan to avert the refusal of dispensation by making a suitable offer of this nature, I agree with the LVT that the only offer on the table did not suffice.

(ii) Second, there was some debate as to the burden of proof with regard to prejudice suffered by the respondents. As will be apparent, it did not seem to me that the outcome in this appeal turned on the incidence of the burden of proof. In so far as it rested on the respondents, and as already discussed, they have satisfied the burden. I am, accordingly, reluctant to express a concluded view on a point, not without complexity, that does not require resolution in this case.

(iii) Third, in the light of the decision to which I have come on the curtailment of consultation, it is unnecessary to say anything of the additional points canvassed in the respondent's notice.

Overall conclusion

[77] For the reasons set out, Daejan's case fails on each of the principal issues. I would dismiss this appeal.

Pitchford LJ said:

[78] I agree with the judgments of both Gross LJ and Sedley LJ and would dismiss this appeal.

Giving the second judgment, **Sedley LJ** said:

[79] I agree with the reasoning and conclusions of Gross LJ and, like him, would dismiss this appeal.

[80] In [19] to [22], Gross LJ considers the arguments concerning the precise role of the appellate bodies below this one. On the face of them, the successive roles of the LT (that is to say, the LT, now reincarnated as the Upper Tribunal) and this court are identical: each is concerned with whether the LVT made an error of law, and there is no apparent reason why either appellate tier should defer to the view of the law reached by the tribunal below it.

[81] However, these are early days in the new tribunal system. For reasons touched on in *R (on the application of Cart)* v *Upper Tribunal* [2010] EWCA Civ 859* (in which the Upper Tribunal had resolved an issue of natural justice by asking simply whether substantial prejudice had resulted), it is distinctly possible that a body of administrative law will develop that is not entirely lawyer's law: see *Cart* in [42] to [44]. It will always, I hope, be bounded by principles of legality and fairness of which the High Court is the custodian, but within that perimeter individual chambers may develop their own understanding of the law and practice that are peculiar to their jurisdiction.

[82] This is why I would endorse the way Gross LJ resolves the argument in [21(ii)] in respect of the view reached by the tribunal. It may follow that the LT needed, in [44] of its decision, to explain why its view that most of the stage 2 shortcomings "were minor and did not cause any significant prejudice to the lessees" was not its own re-evaluation of the facts found by the LVT. On the other hand, its conclusion in [61] (see [47] above) seems to me a model of how the tribunal's appellate function is to be carried out.

[83] Turning to the substantive questions of law, I would emphasise (what must have been well to the forefront of the minds of the successive tribunals) that the tenants of a block of long-leasehold flats such as Queens Mansions have a real interest not only in the cost but in the quality of major maintenance works. As the sign in the Fleet Street shirtmaker's window used to say, there is nothing that someone else cannot do a little cheaper and a little worse. This is why it was no answer to say that Daejan, albeit prematurely, had accepted the lowest tender. As with Gross LJ (see [73] above), I would uphold the finding that cutting the tenants out of the consultation was a serious and a prejudicial failure.

[84] Lastly, it is relevant to examine the consequences of Daejan's casuistic submission that the magnitude of the sum involved should alone be the determinant of whether it is reasonable to dispense with compliance. Daejan, as it happens, is part of a large and prosperous group of property companies that would have difficulty in pleading poverty; but Mr Dowding would not be thanked by a small and struggling lessor if he were to succeed in this aspect of his argument.

[85] Beyond this, divorcing the cost from the means of the paying party would leave the LVT with the task, plainly outside the statutory intent, of setting a cash tariff. If it were to hold here, for example, that because the contract price was more than £250,000 consultation should be dispensed with, it would have in fairness to do the same in all other cases. By parity of reasoning, a cut-off point would come, somewhere down the monetary scale, at which the LVT declined to grant dispensation, again necessarily in all cases. What would then happen to cases falling in the gap between the two? The answer, happily, is that this adjudicator's nightmare has no foundation in legal reality. What may make dispensation reasonable, as Gross LJ has explained and as the tribunals below correctly understood, has to do with the circumstances in which the omission has occurred and its effect on the purposes of the consultation process.

[86] Lastly, I would add a word to what Gross LJ says in [76(ii)] about the burden of proof. It is common for advocates to resort to this when the factual case is finely balanced, but it is increasingly rare in modern litigation for the burden of proof to be critical. Much more commonly, the task of the tribunal of fact begins and ends with its evaluation of as much of the evidence, whatever its source, as helps to answer the material questions of law. In nine cases out of 10, this is sufficient to resolve the contest. It is only rarely that the tribunal will need to resort to the adversarial notion of the burden of proof in order to decide whether an argument has been made out, and tribunals ought, in my view, not to be astute to do so: the burden of proof is a last, not a first, resort.

Appeal dismissed.

---

* Editor's note: Reported at [2011] 2 WLR 36

Knights Construction (March) Ltd *v* Roberto Mac Ltd and another

LAND REGISTRY ADJUDICATION

9 February 2011

MR MICHAEL MARK, SITTING AS A DEPUTY ADJUDICATOR TO HM LAND REGISTRY

Land registration — Alteration of register — Rectification for mistake — Land Registration Act 2002 — Application for rectification — Part of applicant's land included by mistake in first registration of title to adjoining land — Para 5(a) of Schedule 4 to 2002 Act — Whether applicant entitled to rectification against purchaser of that title

The applicant held the paper title to an area of land that it had purchased in 1981 and on part of which it had constructed a block of flats and two shops. In 2007, the Land Registry allowed an application by the owner of an adjacent Salvation Army chapel for first registration of its title, proceeding on the evidence of a statutory declaration dated 1943. Owing to uncertainty in the statutory declaration concerning the relevant boundaries, the registered title to the chapel erroneously included land that fell within the applicant's title (the disputed land). The first respondent purchased the title to the chapel in 2009 for £150,000. It thereafter sought to enclose the disputed land and turn it into a private car park.

The applicant applied to the Land Registry for rectification of the register, under Schedule 4 to the Land Registration Act 2002, to remove the disputed land from the first respondent's registered title. That application was referred to a deputy Land Registry adjudicator. The central issue was the extent of the power, under para 5(a) of Schedule 4 to the 2002 Act, to alter the register for the purpose of "correcting a mistake" and whether it extended to restoring the applicant's title to the disputed land against the first respondent as an innocent purchaser from a vendor in whom the registered title had vested at the time of sale. The Chief Land Registrar was joined as a second respondent so that he would be bound by the adjudicator's decision as to the meaning and effect of the relevant statutory provisions.

An issue also arose as to whether the applicant was in occupation of the disputed land such that its interest overrode that of the first respondent. In that connection, there was evidence that: (i) the greater part of the disputed land was not fenced off from the adjoining highway, although the applicant had in the past tarmacked it to provide parking for tenants of the flats; (ii) the tarmacked area had become overgrown but continued to be used for parking by tenants of the shops and flats pursuant to rights conferred by their leases; and (iii) there was a fenced laundry-drying compound in one corner of the disputed land, built for use by occupants of the flats, although this had largely fallen out of use.

***Held*: The application was allowed. (1) To the extent that the disputed land was included in the registered title to the chapel, title to it vested in the first respondent: see section 58(1) of the 2002 Act. The applicant, although in possession of the disputed land, had not been in occupation of most of it at the time of the sale to the first respondent in a manner that would have been obvious on a reasonably careful inspection. Accordingly, its interest in that land did not override that of the first respondent pursuant to section 29(2)(a)(ii) of, and para 2 of Schedule 3 to, the 2002 Act. The exception was the drying compound, which, on a reasonably careful inspection, had clearly been occupied as part of the common parts of the flats at the time of the sale to the first respondent by reason of the fencing and other matters, such that the applicant's interest in it overrode that of the first respondent. The applicant was entitled to rectification of the register under para 5 of Schedule 4 to give effect to its absolute beneficial interest in the drying compound by removing it from the first respondent's title and adding it to the applicant's newly registered title. No exceptional circumstances justified a refusal to alter the register in that respect. (2) With regard to the remainder of the disputed land, rectification was available to correct a mistake pursuant to para 5(a) of Schedule 4. The relevant provisions should be construed so as to ensure that where parties are deprived by legislation of property to which they would otherwise have title, they should be compensated appropriately, at least if they are not at fault, and regardless of whether the party deprived is the original owner or the subsequent purchaser of registered land that finds that its title has been removed by rectification, and whether there is an error as to the amount of land included in the title. In light of the foregoing, a mistake had been made, capable of rectification, in: (i) the original registration of the owner of the chapel with title to the disputed land, which mistake should be corrected by removing the disputed land from the title; or (ii) flowing from, and as part and parcel of that original mistake, the registration of the first respondent as proprietor of the disputed land; or (iii) the registration of the transfer to the first respondent. Any other result would fall foul of Article 1 of the First Protocol to the European Convention on Human Rights unless the provisions of Schedule 8 were construed, in a manner that would stretch them almost to breaking point, so as to give the applicant, as the original owner, a remedy by way of an indemnity. Since the disputed land had continued to be in the applicant's possession since 1985, and had been used by its tenants in accordance with the terms of their leases, para 6 of Schedule 4, limiting the power to rectify under para 5 in the case of rectification affecting the title of a registered proprietor in possession, did not apply in the instant case. There were no exceptional circumstances to justify a refusal to order rectification.**

The following cases are referred to in this report.

Abbey National Building Society v *Cann* [1991] 1 AC 56; [1990] 2 WLR 832; [1990] 1 All ER 1085; (1990) 60 P&CR 278; 22 HLR 360, HL

Ajibade v *Bank of Scotland* REF/2006/0163/0174, unreported 8 April 2008

Argyle Building Society v *Hammond* (1984) 49 P&CR 148

Barclays Bank plc v *Guy* [2010] EWCA Civ 1396; [2011] 1 WLR 681, [2008] EWCA Civ 452; [2008] 2 EGLR 74; [2008] 25 EG 174, [2008] EWHC 893 (Ch)

Derbyshire County Council v *Fallon* [2007] EWHC 1326 (Ch); [2007] 3 EGLR 44; [2007] 45 EG 164

Epps v *Esso Petroleum Co Ltd* [1973] 1 WLR 1071; [1973] 2 All ER 465; (1973) 25 P&CR 402

Ghaidan v *Godin-Mendoza* [2004] UKHL 30; [2004] 2 AC 557; [2004] 3 WLR 113; [2004] 3 All ER 411; [2005] 1 P&CR 18; [2004] 2 EGLR 132; [2004] HLR 46

Guy v *Pannone LLP* [2009] EWCA Civ 30; [2009] 07 EG 90 (CS)

A *Kingsalton Ltd* v *Thames Water Developments Ltd* [2001] EWCA Civ 20; [2002] 1 P&CR 15
Lee v *Barrey* [1957] Ch 251; [1957] 2 WLR 245; [1957] 1 All ER 191, CA
Link Lending Ltd v *Bustard* [2010] EWCA Civ 424; [2010] 2 EGLR 55; [2010] 28 EG 86
Malory Enterprises Ltd v *Cheshire Homes (UK) Ltd* [2002] EWCA Civ 151; [2002] Ch 216; [2002] 3 WLR 1
Nevill Long & Co (Boards) Ltd v *Firmenich & Co* (1984) 47 P&CR 59; (1983) 268 EG 572, CA
Norwich & Peterborough Building Society v *Steed (No 2)* [1993] Ch 116; [1992] 3 WLR 669; [1993] 1 All ER 330, CA
Odogwu v *Vastguide Ltd* [2008] EWHC 3565 (Ch)
Pinto v *Lim* [2005] EWHC 630 (Ch)
B *Stewart* v *Lancashire Mortgage Corporation Ltd* REF/2009/0086/1556
Strachey v *Ramage* [2008] EWCA Civ 384; [2008] 2 P&CR 8
Thompson v *Foy*; *Mortgage Business* v *Foy* [2009] EWHC 1076 (Ch); [2010] 1 P&CR 16
Yaxley v *Gotts* [2000] Ch 162; [1999] 3 WLR 1217; [2000] 1 All ER 711; (2000) 32 HLR 547; 79 P&CR 91; [1999] 2 EGLR 181

This was an adjudication, before a deputy adjudicator to HM Land Registry, of an application by the applicant, Knights Construction (March) Ltd, against the first respondent, Roberto Mac Ltd, and the second respondent, the Chief Land Registrar, for rectification of the register to correct a mistake pursuant to para 5(a) of Schedule 4 to the C Land Registration Act 2002.

Andrew Gore (instructed by Bowser Ollard & Bentley, of March) appeared for the applicant; Stephen Jones (instructed by Fraser Dawbarns LLP, of Wisbech) appeared for the first respondent; Timothy Morshead QC (instructed by the Treasury Solicitor) represented the second respondent.

Giving his decision, **MR MICHAEL MARK, SITTING AS DEPUTY ADJUDICATOR TO HM LAND REGISTRY**, said:

Registered title

[1] The old Salvation Army Chapel in March lies between the High D Street and Chapel Street. Between the back of the chapel and Chapel Street is a roughly rectangular piece of land (the disputed land). It appears from a statutory declaration dated 6, 7 and 8 April 1943, by various officers of the Salvation Army and by solicitors acting for the Salvation Army, that the title deeds relating to the Salvation Army Chapel had been destroyed by enemy bombing during the second world war. The statutory declaration further indicated that the freehold title to the chapel had been acquired around 31 March 1888. The statutory declaration gave no indication as to the boundaries of the property so acquired.

[2] In 2007, the Salvation Army was contemplating selling the E chapel. It decided to apply for voluntary first registration of the property at the Land Registry. For this purpose, a further statutory declaration was sworn by Mr Andrew David Axcell. He had been working for the Salvation Army from 1979 to 1989 as chief surveyor within its property department. He had also rejoined the property department in September 2002. He claimed to be fully conversant with the records of the Salvation Army Trustee Co and of its property affairs since the date of his appointment in 1979. He went on to assert that the freehold premises known as The Salvation Army, High Street, March, Cambridgeshire, were shown in red on the plan marked ADA annexed thereto, were conveyed to Mr William Booth, the then General F of the Salvation Army, on 31 March 1888, and by virtue of section 6 of the Salvation Army Act 1931, became vested in The Salvation Army Trustee Co on 1 October 1931. He drew attention to the destruction of the title deeds by reference to the 1943 statutory declaration to which I have already referred and to a further statutory declaration dated 6 February 1984, which was not in evidence.

[3] The premises shown in red on the annexed plan included the disputed land.

[4] The statutory declaration of Mr Axcell went on to depose that from the completion of the "said Statutory Declaration on the dates aforesaid to the present time The Salvation Army Trustee Co has been in free undisturbed and undisputed possession of the said premises and in receipt of the rents and profits thereof". He stated that he deposed to the above from information obtained from the records of The Salvation Army Trustee Co and from his own knowledge of that company's property affairs since his appointment in 1979.

[5] Mr Axcell did not give evidence at the hearing, and no explanation has been provided by the Salvation Army as to how the disputed land came to be included in the land that he asserted that The Salvation Army Trustee Co had been in free undisturbed and undisputed possession of and in receipt of the rents and profits thereof. It is abundantly clear from the evidence to which I shall refer below that it never had title to the disputed land and had never, at least since the mid-1980s, been in possession of it or had anything to do with it, except pursuant to rights granted to it as underlessee of shop premises, 87 High Street, owned by the applicant, Knights Construction (March) Ltd.

[6] The Land Registry accepted Mr Axcell's statutory declaration, so far as the evidence before me appears, without question or further investigation. On 5 November 2007, it registered "The Freehold land shown edged with red on the plan of the above title files at the Registry and being The Salvation Army Hall, High Street, March" with title number CB328270. The land shown edged red on the title plan included the disputed land.

[7] The Salvation Army then sought to sell the property, which was surplus to its requirements, initially according to the evidence of Mr Roberto Divkovic, the managing director of the respondent, Roberto Mac Ltd, for £150,000. It appears to have been seriously overpriced, and was eventually acquired by Roberto Mac for £85,000. Roberto Mac was duly registered at the Land Registry as the proprietor of the land with title absolute on 13 March 2009, the same date that a registered charge dated 12 March 2009 was registered, the proprietor being Barclays Bank plc.

Paper title

[8] The unregistered paper title to the disputed land can be traced back to an indenture dated 24 February 1908. That indenture related to property to the south of the disputed land and of the chapel. That land was eventually acquired by Knights Construction in 1981, and it has built a block of flats and two shops on it. The disputed land, which is to the west of the chapel and to the north of the land on which the flats are built, is shown at that stage as including four buildings, two of which are shown as being in the occupation of Mrs E Spriggs. There follows an indenture dated 20 April 1920 between Ms Emma Spriggs, as vendor, and others, respectively as trustees and as purchaser, by which Ms Spriggs conveyed to the purchaser, *inter alia*: "All those Three cottages or tenements situate in and being Numbers 6, 7, and 8 Chapel Street in March aforesaid and the yard and premises adjoining Together with the outbuildings and strip of land on the opposite side of the said Street in the respective occupations of Mrs Williams, [illegible] and Robert Fitzjohn." The land and property conveyed by that conveyance included the whole of the disputed land, which was identified on the plan as including 6, 7 and 8 Chapel Street.

[9] The title deeds go on to show that, in 1937, the disputed land became vested in Mr William Sam Cartwright, and it was transferred by him to his wife in 1943. As described in the conveyance to Mrs Cartwright, the disputed land is described as being all those three cottages or tenements situate in and being 6, 7, and 8 Chapel Street in March aforesaid and the yard and premises adjoining, and they are identified further by reference to the plan on the 1920 indenture, to which I have referred.

[10] Mrs Cartwright then sold to Knights Construction, by a conveyance dated 13 August 1981, first the land conveyed by the indenture dated 24 February 1908, to which I have referred, and, second: "ALL THAT piece or parcel of land formerly forming the site of three cottages or tenements formerly being numbers and 6 7 and 8 Chapel Street March aforesaid together with the outbuildings and strip of land on the opposite side of the said street all which said property First and Secondly hereinbefore described is for the purpose of identification delineated and coloured pink and blue on the Location Plan annexed hereto." The location plan clearly includes almost all the disputed land, although for no apparent reason an irregular area in the north-east

corner of the land appears to have been omitted. That area includes part of the site of the former 7 Chapel Street. There is no evidence to suggest that that part of the site was ever transferred elsewhere or that anybody else has ever made any claim to it. I accept the evidence of Mr Rodney William Purser, a director of Knights Construction, that the land acquired was with the benefit of outline planning permission for the erection of eight one-bedroom flats, two bedsitter flats, seven garages and the provision of three parking spaces, and that the land he thought he was acquiring included the whole of the land in the 1929 conveyance.

Development and use of disputed land since 1981

[11] Subsequently, planning permission was granted for the erection of two shops and eight flats. The plan attached to the planning permission shares in the same area of the disputed land edged red as is shown in the 1981 location plan. It also shows the car-parking area with eight designated spaces for the flats and two further spaces for 87 and 89 High Street, the two shops. The flats and shops were duly constructed, and were completed in 1985. Their postal addresses are 1, 3, 5, 7, 9, 11, 15 and 17 Chapel Street. At the same time, almost all the disputed land, including the irregular area outside the plan on the 1881 conveyance but excluding the drying compound, to which I will refer, was tarmacked to provide a parking area for residents of the flats and tenants of the shops, together with a small area of additional land between the disputed land and the block of flats that had been part of the land included in the 1908 indenture. Knights Construction constructed a row of concrete bollards along the northern boundary separating the disputed land from a public footpath that led between Chapel Street and High Street. The development included a garage under flat 1 close to Chapel Street, access to and from which could only be obtained by crossing part of the disputed land.

[12] The disputed land was not fenced off from Chapel Street, thus making it easier for those with a right to park on the land. In the south-eastern corner of the disputed land, a drying compound was constructed for the exclusive use and benefit of the occupiers of the flats. This comprised a concrete base, surrounding which was a 1.8m close-boarded fence, with an ungated entrance leading to it on the southern side from a footpath forming part of the development that led from Chapel Street and High Street to the entrances to the flats. There was also space for four cars to park on land on the opposite side of Chapel Street owned by Knights Construction.

[13] Knights Construction still controlled, until recently, flats 1 and 15, which were let on assured shorthold tenancies, although flat 1 may have recently been sold on a 999-year lease. These tenancies contained no express rights over the disputed land, although Mr Purser stated, and I accept, that he had told the tenants of flats 1 and 15 when he handed over the keys that they could park on that land and use the drying area.

[14] The other flats were sold on 999-year leases following completion of the development, all of which leases contain similar terms. A copy of one of the leases, relating to flat 5, is included in the trial bundle. The demise refers to the 1981 conveyance and defines "the freehold" as the property shown on the location plan annexed thereto and thereon edged with red. The location plan includes the disputed land with the exception of the area not included in the plan on the 1981 conveyance. The lease includes the following rights:

(i) The right of free and unrestricted passage and running of electricity water and soil from and to the demised premises through the sewers conduits cables and soakaways on in or under the adjoining maisonettes shops and remainder of the freehold subject to the payment of a fair proportion of the expense of maintaining renewing altering rebuilding cleansing and repairing the same unless the same shall be the liability of the Local Authority Statutory Undertaker or any other Corporation.

(ii) The right for the Tenant to enter the remainder of the freehold for the purpose of repairing maintaining renewing altering rebuilding and cleansing the demised premises or any part thereof and the services thereto.

(iii) The right in common with Knights Construction and all other persons having like right with or without vehicles at all times and for all purposes connected with the demised premises to pass and repass to and from the demise premises over and along the drive paths and yard coloured green on the site plan G annexed — an area which included the disputed land — subject to payment of a fair proportion of the expenses of maintaining renewing altering rebuilding and repairing the same.

(iv) All the easements and similar rights now enjoyed by the demised premises in relation to the remainder of the freehold.

(v) The right in common with all others having like right to use the parking space as shown on the said site plan for the purpose of the parking of one private motor vehicle and/or motor or other cycles subject always to there being space available and also subject to the payment of a fair proportion of the expense of maintaining renewing altering rebuilding and repairing the same.

[15] The recent lease of flat 1, a copy of which is in the trial H bundle, contains similar provisions. The shops at 87 and 89 High Street were let to Mr Purser by a lease dated 21 June 1989, also for a term of 999 years. Once again, this lease contained similar provisions to those set out above in respect of flat 5. The shop at 89 High Street was underlet by Mr Purser to Ms Jacqueline Drewery by an underlease dated 5 December 1989 for a term of 25 years from that date. The underlease contains similar provisions to those set out above. The shop at 87 High Street was at one stage let to the Salvation Army Trustee Co by an underlease dated 23 February 1996, made between Mr Purser and that company, and which was terminated by that company as from 1 December 2004. The actual underlease was not in evidence, but the J letting particulars stated that there was car parking at the rear, and I am satisfied that the underlease would have contained similar provisions to those in the lease of 89 High Street. There was then a void at 87 High Street for a few months until it was relet.

[16] Evidence was given, which I accept, by Mr Neville Young, that he purchased flat 7 in 2004 and subsequently transferred it into the name of his company. He described the vegetation and the bins as being around the perimeter of the car-parking area, but not so as to prevent car parking. He let the flat on assured shorthold tenancies, and when he visited the flat, he would park on the disputed land. He spent two to three weeks decorating the flat when he first bought it. He would also K visit every two weeks when there were voids. He would see other cars parked there, which he knew belonged to the owners or visitors of the block of flats and the two shops. His right to park, dry clothes or keep a bin were never challenged. He would not have bought the flat without a right to park. He could not get to within around 4-5ft of the chapel wall because of the vegetation, but had no problem getting fully onto the land and to within a foot of that vegetation with the back of the car pointing towards the road, at which point his car would be several feet from the road in most places and would never be hanging over the road. He was not sure whether anybody used the drying compound as such and occasionally saw pieces of furniture there, although never, as was L suggested in cross-examination, faeces.

[17] Evidence was also given, which I accept, by Ms Stephanie Allen, the owner of the lease of 9 Chapel Street, which she purchased in October 2006. She used the bins and she parked her car on the disputed land overnight, from around 5.45pm to 8am, and also at weekends during the day. She was aware of the drying compound and had seen a washing line there. She did not use it, but thought that one family in the block used it around two to three times a month for part of the time she was there. Overnight and at weekends, the cars parked were mostly the same, and she believed that they belonged to the occupiers of the flats. Her visitors would also sometimes park there. She also confirmed M that the trees and shrubs were along the border and drying area. Before Roberto Mac came, six to seven cars could park facing inwards along the frontage. She identified the area in which she parked her car as that shown in a photograph at p226 of the trial bundle, in front of the chapel beyond the drying compound. The car would be 2-3m from the chapel wall facing it or with the back of the car towards it. There would still be some space behind and a reasonable distance to the vegetation in front of the chapel.

[18] Mr Jon Blyth, the owner of 17 Chapel Street since May 2004, gave evidence to similar effect and also confirmed that, on occasions, the shrubs and weeds were cleared by one of the owners of the flats on the occasion of their selling. During his period of occupation, the bins

A had been on the northern edge of the disputed land close to the footpath from Chapel Street to High Street. He had also not used the drying compound, but had seen clothes drying there, which he believed had been put there by a family occupying a ground-floor flat. He also had never seen faeces there.

[19] Ms Jacqueline Drewery, who had been atenant of 89 High Street since 1989 under a 25-year underlease from Mr Purser, gave evidence to the same effect with regard to car parking since that time. She would arrive to park between 8.30am and 8.45am Monday to Saturday and would leave between 3.30pm and 5.00pm. She would see the cars of some flat owners there during the day. The car park was busiest on B Saturday. She and the others used to park right across the area. She also saw no faeces in the drying area, which was occasionally used. Her estimate of the depth of the vegetation was 1-1.5m.

[20] None of the witnesses had ever seen anybody from the Salvation Army use the disputed land. All also gave evidence as to the problems caused by the work carried out by Roberto Mac to the land and as to the congestion and difficulties caused by the additional presence of his vehicles, including gangmaster minibuses, since April 2009.

[21] I accept the evidence of those witnesses and find that the disputed land has been used for car parking by several of the flats and by the shop at 89 High Street since 1985 or thereabouts. It has also C been used by Mr Purser to park his car when he has come into March, and to deliver and collect materials stored in the garage under flat 1. The garage was used to store these materials and has never been used as a garage. On rare occasions, when there was no space available on the disputed land, Mr Purser would park across the road. The drying area seems to have been infrequently used and does not appear to have been used at all in recent years, except that rubbish was dumped there. Council bins for refuse from the flats have also been kept on the disputed land in various places over the years, although principally, as Mr Purser confirmed, in front of the drying compound. The parts of the disputed land closest to the rear wall of the chapel have become D overgrown. No maintenance has been carried out on the disputed land by Knights Construction since 1985. It has occasionally been tidied up by tenants seeking to improve the appearance of the area when they were trying to sell the flats.

[22] Mr Divkovic confirmed in his evidence that he had inspected the land and that he was aware that cars parked there, including, at least in the area close to the flats, cars belonging to tenants of the flats. The Salvation Army had ceased to occupy the chapel around 2007 and he knew that the parked cars did not include cars belonging to the Salvation Army. He drew attention to an area close to the chapel wall that was raised by around 15-20cm and was too high and rough for cars to park E on. The raised area identified by him can be seen in the photographs at p59 of the trial bundle and barely extends beyond the limit of the vegetation as described by the witnesses for Knights Construction. It would not have inhibited parking as described by them. Mr Divkovic also knew that the bins were used by tenants of the flats.

[23] He was also able to see that the drying compound was fenced off and that the only practical access to it was from the path alongside the flats from which access had specifically been provided by the paving slabs, to which I have already referred. Although, theoretically, access might have been obtained from the remainder of the disputed land, this would not have been a practical way of obtaining access because of F the presence of the fencing and the narrow gap between the fencing and the wall of the chapel. It would also have been rendered wholly impracticable by reason of the vegetation close to the chapel wall and around the fence of the drying compound. Nevertheless, Mr Divkovic thought that the drying compound area belonged to the Salvation Army. He stated that it was derelict and contained faeces.

[24] The only issue of fact as to which there has been a serious dispute is as to the extent to which the disputed land had become overgrown by the time of the sale to Roberto Mac. Mr Divkovic contended that it had become so overgrown that most of it could not be used for car parking. Although cars could be seen on it, if they parked with their back to the road they were in fact parked partially on the road. The vegetation according to him extended across most of the site. The evidence of the tenants who gave evidence was that they had no problem in parking their cars on the site unless it was already fully occupied by other vehicles. The vegetation was only at the back of the site and immediately in front of the drying compound where there was a tree or large shrub, the base of which can still be seen in photographs in the trial bundle.

[25] I am satisfied that the vegetation was not nearly as extensive as Mr Divkovic suggests. The other witnesses were all clear that there was ample room for six to eight cars to park off the road in front of the vegetation. That is borne out by the photograph on p226 of the trial bundle and by that on p204. I appreciate that the photograph on p226 was taken in winter and that the perspective could be distorted. Nevertheless, it shows two cars parked with their rears pointing towards Chapel Street, and does not appear to show vegetation extending much beyond the extension to the chapel, apart from a tree or shrub of some age and size in front of the drying compound. The photograph at p204 also clearly shows left in front of the drying compound, and in that area cars could park well off the road. In addition, although other parking was available on the other side of Chapel Street, which had become a cul-de-sac for anything larger than a bicycle in around 2005, it was plainly advantageous to the tenants to be able to park off the road on their own side of it, and I doubt that they would have remained silent had the vegetation encroached to such an extent that they were no longer able to park.

[26] I accept that when Mr Divkovic saw the drying compound, there were faeces and rubbish there. Nevertheless, faeces were not normally to be found there, and although the area was not regularly used for drying clothes and could be used to dump rubbish, it was used as part of the common parts of the block of flats.

Sale of chapel to Roberto Mac

[27] The contract for the sale of the chapel is dated 15 January 2009. The solicitors for the Salvation Army confirmed by letter dated 15 January 2009 that they were not prepared to answer CPSE1 enquiries in this matter. They had, however, made it clear, according to the letter, that they were prepared to answer any specific enquiries that Roberto Mac's solicitor wanted to make and it appears that they had answered enquiries as the transaction proceeded. So far as appears from the evidence, however, there were no enquiries as to the boundaries of the disputed land or as to its occupation.

[28] Completion appears to have taken place on 12 March 2009. After completing its purchase in March 2009, Roberto Mac sought to enclose the disputed land and turn it into a private car park. In the process, it got rid of the vegetation and two of the bollards, took down the fencing of the drying compound and dug up the concrete foundation. Mr Purser was immediately advised of this by Ms Drewery. He met with an employee of Roberto Mac on the site and told him that the disputed land was owned by his company and that he had the title deeds. Mr Divkovic subsequently told him in turn that Roberto Mac had bought the land from the Salvation Army.

[29] There then appears to have been a stand-off. The new fencing was removed and the disputed land appears to have been used since for parking both by Roberto Mac and by the various tenants and undertenants of Knights Construction. On 20 April 2009, Knights Construction registered their title to their land other than the disputed land under title number CB344497. Their application, dated 1 April 2009, related to the whole of the land to which they had a paper title and also sought the removal of the disputed land from Roberto Mac's title.

Land Registration Act 2002

[30] Section 11 of the Land Registration Act 2002 (the 2002 Act) provides as follows in respect of the effect of first registration:

11 Freehold estates

(1) This section is concerned with the registration of a person under this Chapter as the proprietor of a freehold estate.

(2) Registration with absolute title has the effect described in subsections (3) to (5).

(3) The estate is vested in the proprietor together with all interests subsisting for the benefit of the estate.

(4) The estate is vested in the proprietor subject only to the following interests affecting the estate at the time of registration —

(a) interests which are the subject of an entry in the register in relation to the estate,

(b) unregistered interests which fall within any of the paragraphs of Schedule 1, and

(c) interests acquired under the Limitation Act 1980 (c 58) of which the proprietor has notice.

(5) If the proprietor is not entitled to the estate for his own benefit, or not entitled solely for his own benefit, then, as between himself and the persons beneficially entitled to the estate, the estate is vested in him subject to such of their interests as he has notice of.

[31] For the purposes of section 11(4)(b), the unregistered interests referred to in Schedule 1, so far as relevant to the present case, are as follows:

Leasehold estates in land

1 A leasehold estate in land granted for a term not exceeding seven years from the date of the grant, except for a lease the grant of which falls within section 4(1) (d), (e) or (f).

Interests of persons in actual occupation

2 An interest belonging to a person in actual occupation, so far as relating to land of which he is in actual occupation, except for an interest under a settlement under the Settled Land Act 1925 (c 18).

Easements and profits a prendre

3 A legal easement or profit a prendre.

[32] Section 26 deals with the transfer of a registered estate or charge as follows:

26 Protection of disponees

(1) Subject to subsection (2), a person's right to exercise owner's powers in relation to a registered estate or charge is to be taken to be free from any limitation affecting the validity of a disposition.

(2) Subsection (1) does not apply to a limitation —

(a) reflected by an entry in the register, or

(b) imposed by, or under, this Act.

(3) This section has effect only for the purpose of preventing the title of a disponee being questioned (and so does not affect the lawfulness of a disposition).

[33] Sections 28 and 29 deal with the effect of dispositions on priority as follows:

28 Basic rule

(1) Except as provided by sections 29 and 30, the priority of an interest affecting a registered estate or charge is not affected by a disposition of the estate or charge.

(2) It makes no difference for the purposes of this section whether the interest or disposition is registered.

29 Effect of registered dispositions: estates

(1) If a registrable disposition of a registered estate is made for valuable consideration, completion of the disposition by registration has the effect of postponing to the interest under the disposition any interest affecting the estate immediately before the disposition whose priority is not protected at the time of registration.

(2) For the purposes of subsection (1), the priority of an interest is protected —

(a) in any case, if the interest —

(i) is a registered charge or the subject of a notice in the register,

(ii) falls within any of the paragraphs of Schedule 3, or

(iii) appears from the register to be excepted from the effect of registration, and

(b) in the case of a disposition of a leasehold estate, if the burden of the interest is incident to the estate.

(3) Subsection (2)(a)(ii) does not apply to an interest which has been the subject of a notice in the register at any time since the coming into force of this section.

(4) Where the grant of a leasehold estate in land out of a registered estate does not involve a registrable disposition, this section has effect as if —

(a) the grant involved such a disposition, and

(b) the disposition were registered at the time of the grant.

G [34] For the purposes of section 29(2)(a)(ii), so far as relevant here, Schedule 3 provides that the following unregistered interests override registered dispositions:

Leasehold estates in land

1 A leasehold estate in land granted for a term not exceeding seven years from the date of the grant, except for —

(a) a lease the grant of which falls within section 4(1)(d), (e) or (f);

(b) a lease the grant of which constitutes a registrable disposition.

Interests of persons in actual occupation

2 An interest belonging at the time of the disposition to a person in actual occupation, so far as relating to land of which he is in actual occupation, H except for —

(a) an interest under a settlement under the Settled Land Act 1925 (c 18);

(b) an interest of a person of whom inquiry was made before the disposition and who failed to disclose the right when he could reasonably have been expected to do so;

(c) an interest —

(i) which belongs to a person whose occupation would not have been obvious on a reasonably careful inspection of the land at the time of the disposition, and

(ii) of which the person to whom the disposition is made does not have actual knowledge at that time;

J (d) a leasehold estate in land granted to take effect in possession after the end of the period of three months beginning with the date of the grant and which has not taken effect in possession at the time of the disposition.

Easements and profits a prendre

3(1) A legal easement or profit a prendre, except for an easement, or a profit a prendre which is not registered under the Commons Registration Act 1965 (c 64), which at the time of the disposition —

(a) is not within the actual knowledge of the person to whom the disposition is made, and

(b) would not have been obvious on a reasonably careful inspection of the land over which the easement or profit is exercisable.

K (2) The exception in sub-paragraph (1) does not apply if the person entitled to the easement or profit proves that it has been exercised in the period of one year ending with the day of the disposition.

[35] Schedule 3 is further widened by para 8 of Schedule 12 to include an interest that, immediately before the coming into effect of Schedule 3, was an overriding interest under section 70(1)(*g*) of the Land Registration Act 1925 by virtue of a person's receipt of rents and profits, except for an interest of a person of whom inquiry was made before the disposition and who failed to disclose the right when he could reasonably have been expected to do so. It is also widened by para 9 of Schedule 12 to include an easement or profit à prendre that L was an overriding interest in respect of a registered estate immediately before the coming into effect of Schedule 3, but that would not fall within para 3 of that Schedule if created after the coming into force of that Schedule.

[36] Section 58(1) of the 2002 Act provides that, subject to certain exceptions that do not apply here, if on the entry of a person in the register as the proprietor of the legal estate, the legal estate would not otherwise be vested in him, it shall be deemed to be vested in him as a result of the registration.

M [37] Schedule 4 to the 2002 Act contains provisions as to the alteration of the register. Paragraph 5 permits the registrar to alter the register for the purpose of: (a) correcting a mistake; (b) bringing the register up to date; (c) giving effect to any estate, right or interest excepted from the effect of registration; or (d) removing a superfluous entry. Paragraph 6 limits the power under para 5 so far as referring to rectification defined by para 1 of that Schedule, where used in that Schedule, as referring to alteration that: (a) involves the correction of a mistake; and (b) prejudicially affects the title of a registered proprietor. Such an alteration affecting the title of the proprietor of a registered estate in land may be made under para 5 without the proprietor's consent in relation to land in his possession unless: (a) he has by fraud or lack of proper care caused or substantially contributed to the mistake; or (b) it would for any other reason be unjust for the alteration not to be made.

A [38] Schedule 8 to the 2002 Act contains provision for a person to be indemnified by the Chief Land Registrar if he suffers loss by reason of, *inter alia*: (a) rectification of the register; or (b) a mistake whose correction would involve rectification of the register. Paragraph 11(2) of that Schedule provides that in that Schedule references to rectification of the register are to alteration of the register that: (a) involves the correction of a mistake; and (b) prejudicially affects the title of a registered proprietor.

Application of the law to the facts of this case

[39] It is clear that, whether or not The Salvation Army Trustee Co
B had title to the disputed land before first registration, to the extent that the disputed land was included in the land registered in title CB328270 title to it vested in that company by virtue of section 58(1) of the 2002 Act. At the hearing, I raised the question of whether it was in fact included, or whether all that was registered was the land conveyed to General Booth by the 1888 conveyance, the disputed land being shown on the general plan as included in error, as in *Lee* v *Barrey**, *Derbyshire County Council* v *Fallon* [2007] EWHC 1326 (Ch)† and *Strachey* v *Ramage* [2008] EWCA Civ 384‡. Neither counsel showed any enthusiasm for approaching the matter in that way, and in my judgment rightly so. The statutory declaration of Mr Axcell makes it
C clear that the application is to register title to the whole of the area shown on the plan annexed to the declaration, including the disputed land, and the registration was intended to give effect to that request. Counsel for Knights Construction further contended that the yellow line drawn by surveyors on the disputed land, on the basis of the plans provided to them, indicated the boundaries of the registered land. Near the boundary between the highway on Chapel Street, there was a significant area outside that yellow line that had been tarmacked in 1985 by Knights Construction in the same way as the land inside the line. It was submitted that this demonstrated that the registered land did not include a strip of land between the yellow land and the highway,
D thus providing Knights Construction with a ransom strip if its other claims were unsuccessful. I reject this contention. It is apparent from the statutory declaration of Mr Axcell that he was claiming the land up to the highway and if there was an error in the dimensions on the plan used by him, or an error in the measurements subsequently carried out by Roberto Mac's surveyors, that does not alter the fact that the land claimed, wrongfully, by Mr Axcell extended to the public highway. I am satisfied therefore that The Salvation Army Trustee Co was registered as proprietor of the disputed land.

[40] Knights Construction had no knowledge or means of knowledge that this registration had taken place. It may of course have registered
E a caution against first registration, but understandably did not do so, and probably had no reason to be aware of this procedure. It may also have voluntarily have registered its own title at an earlier date, but again understandably did not do so. There is no doubt that had it become aware at that stage of what had happened then it would have had a cast iron case for alteration of the register. There was a plain mistake, there were no exceptional circumstances, and the rectification provisions in para 6 of Schedule 4 to the 2002 Act would not have applied as it is clear that The Salvation Army Trustee Co was not in possession of that land.

[41] However, in March 2009, the land was disposed of to Roberto Mac, which I am satisfied took it in good faith and for
F valuable consideration. I am also satisfied that the parking facilities that the disputed land afforded were a significant factor in that company's decision to acquire the chapel. They needed the parking for their staff and occasionally needed the space for other purposes. There was no suggestion that Knights Construction had any knowledge of their land being included in this sale, and I am satisfied that they were not aware of it until contacted by Ms Drewery, at which point they acted very promptly.

* Editor's note: Reported at [1957] Ch 251

† Editor's note: Reported at [2007] 3 EGLR 44; [2007] 45 EG 164

‡ Editor's note: Reported at [2008] 2 P&CR 8

Is Roberto Mac in possession of the disputed land?

[42] I would add that I am satisfied that Roberto Mac is not in possession of the disputed land. It had been in the possession of Knights Construction since 1985, and although that company had allowed part of it to become overgrown, I do not consider that it was abandoned by it. It continued to be used by its tenants and subtenants in accordance with the terms of their leases and on occasions by Mr Purser as well. When Roberto Mac attempted to take possession of it, there was an immediate reaction from Mr Purser that led to that attempt being abandoned, and to the land being effectively shared since then. I do not consider therefore that para 6 of Schedule 4 has any application in this case even if it would have been relevant had Roberto Mac taken possession, an issue that would also depend on a determination that any alteration would involve the correction of a mistake, a matter to which I shall return. I would add that even if Roberto Mac were to be treated as in possession of the disputed land between the yellow line drawn by their surveyor and the chapel, they could not be treated as in possession of any part of it between the yellow line and the highway, as Mr Divkovic understood the yellow line to represent his boundary and acted accordingly.

Position of Barclays Bank as chargee

[43] I do not deal in this decision with the position of Barclays Bank as chargee, as it is not before me.

Actual occupation by Knights Construction

[44] The next factual issue is whether at the time of the disposition: (a) Knights Construction was in actual occupation of any part of the disputed land; and (b) its occupation would not have been obvious on a reasonably careful inspection of the land at the time of the disposition. Apart from the drying area, it was obvious to Mr Divkovic that the land was being used for parking on by cars and for tenants' bins. It was also clear to Mr Divkovic that, at least close to the garage, it was tenants' cars that were being parked and he stated in evidence that he did not think that he was buying the land right up to the garage but only up to 2–3m from it. He also drew on the photograph at p58 of my copy of the trial bundle approximately where he thought that the boundary came to allow enough space to enable a car to get in and out of the garage without trespassing on the land that he thought he was buying. The line ran from the south-western corner of the drying compound to the highway going away from the garage at an angle of around 45 degrees, covering half, diagonally, of around one additional parking space beyond the car shown in the photograph parked outside the garage. He never enquired as to the exact boundary either of his solicitor or of the vendors. There is no difference between the surface on one side of the line drawn by him and the other side. Nor did he make any enquiry as to the reason for the fencing around the drying compound or the access provided to it from the path outside the flats.

[45] It has been submitted by Mr Andrew Gore, on behalf of Knights Construction, that that company was in actual occupation of the car park, including the drying compound. He has pointed out that the rights granted to the various tenants did not render the disputed land part of the physical demise in the sense of giving any of the tenants exclusive possession of the disputed land. Knights Construction had laid the tarmac and erected the bollards on the land, and erected the drying area, concrete pad, the fencing around it and the washing line. He also placed reliance on the presence of the bins on the land for the use of the tenants, although it appeared from the evidence that these bins were the property of the local authority and there was no evidence that they were in fact placed there by Knights Construction. Mr Gore referred me to the test as to actual occupation set out by Lord Oliver in *Abbey National Building Society* v *Cann* [1991] 1 AC 56, at pp93D-94B, and to the recent decision of the Court of Appeal in *Link Lending Ltd* v *Bustard* [2010] EWCA Civ 424*, and in particular to the judgment of Mummery LJ, in [17] to [31].

* Editor's note: Reported at [2010] 2 EGLR 55; [2010] 28 EG 86

[46] Mr Gore also submitted that I should give a purposive construction of the relevant provisions of the 2002 Act and treat the occupation of the lessees as occupation by Knights Construction. Alternatively, it is submitted by him that policy reasons dictate that the sufficiency of a landlord's actual occupation should be judged against the fact that a number of its tenants have been granted rights over part of the landlord's land.

[47] Mr Stephen Jones contended that the mere parking of vehicles is not actual occupation, and he refers to the decision of Templeman J in *Epps* v *Esso Petroleum Co Ltd* [1973] 1 WLR 1071. He also drew attention to a need for a degree of permanence and continuity that he contended has not been present in this case.

[48] I do not find it necessary to set out at any length the passages referred to in *Abbey National* and *Link Lending* cited by Mr Gore. Those cases were concerned with actual possession of dwelling houses. As Mummery LJ pointed out, in [27], in *Link Lending*:

> Whether Ms Bustard was in "actual occupation" of the Property at the relevant date was an issue on which the trial judge had to make an evaluation based on his findings of primary fact. As for the law he considered the relevant authorities on the concept of a "person in actual occupation" of land in the earlier Land Registration legislation and now found in the 2002 Act. The construction of the earlier equivalent provisions by the House of Lords is binding on this court. The trend of the cases shows that the courts are reluctant to lay down, or even suggest, a single legal test for determining whether a person is in actual occupation. The decisions on statutory construction identify the factors that have to be weighed by the judge on this issue. The degree of permanence and continuity of presence of the person concerned, the intentions and wishes of that person, the length of absence from the property and the reason for it and the nature of the property and personal circumstances of the person are among the relevant factors.

[49] Taking into account all factors, including those identified by Lewison J with reference to earlier decisions in *Thompson* v *Foy* [2009] EWHC 1076 (Ch)*, in a passage cited and approved by Mummery LJ, I have still to apply them to the wholly different circumstances and property in the present case, and I must take into account that it is not every person in actual occupation whose interest will be protected under the 2002 Act. That must to some extent depend on the nature of the occupation, as Schedule 3 expressly excludes an interest that belongs to a person whose occupation would not have been obvious on a reasonably careful inspection of the land at the time of the disposition and of which the disponee does not have actual knowledge.

[50] With the exception of the drying compound, whether Knights Construction was in actual occupation of the disputed land, I am unable to see how such occupation could have been obvious on a reasonably careful inspection. Knights Construction had done nothing to the land for more than 20 years, it was unfenced and open to the road, and was apparently being used for parking and to keep waste bins. There was no sign, the bollards separating it from the public footpath could as well have been put there by the Salvation Army as by anybody else, and the area was in a dilapidated condition. I can see no basis on which I can find that such occupation as there may have been by Knights Construction was obvious on a reasonably careful inspection.

[51] The position is different with regard to the drying compound. This was fenced off from the rest of the disputed land, and a reasonably careful inspection would have shown that for all practical purposes it was accessible only from the adjoining land that formed part of the flats development, and that it was not tarmacked in the same way as the other land. That inspection would also have shown that there were paving stones specifically providing access to that area from that development. Although the washing line had gone and the area within the drying compound was derelict, it appears to me to have been in the actual occupation of Knights Construction in the same way as the other common parts of the development. The fact that there is no caretaker or other person present on-site does not mean that the landlord is not in occupation of the common parts, or that that occupation cannot be G
obvious if there is no caretaker present. Further, it should be obvious to anybody inspecting the common parts that there may be areas within the common parts that are not being used and are derelict.

[52] In my judgment, it should have been obvious on a careful inspection that the drying compound was being occupied as part of the common parts of the flats and not by the Salvation Army. In those circumstances, I also find it surprising that no enquiry about this area was ever addressed to the solicitor for the Salvation Army, and that Mr Divkovic should have assumed that it was included in the title he was acquiring even though part of the land between it and the public highway was, as he thought, not included. H

[53] I therefore conclude that Roberto Mac acquired title to the disputed land subject to the overriding interest of Knights Construction in the drying compound. Accordingly, the Chief Land Registrar may alter the register under para 5 of Schedule 4 to the 2002 Act to give effect to the beneficial interest of Knights Construction excepted from the effect of registration. Since that is an absolute beneficial interest, the alteration that is appropriate is one that removes the drying compound from Roberto Mac's title and adds it to the newly registered title of Knights Construction. Despite the contentions of Mr Jones, I can see no exceptional circumstances that could justify not altering the register in this respect. J

[54] I note that the conclusion to which I have come in respect of the drying compound is in accordance with *Malory Enterprises Ltd* v *Cheshire Homes (UK) Ltd* [2002] EWCA Civ 151; [2002] Ch 216, which was not cited to me, where the question was as to actual occupation of land on which there was a half constructed and derelict block of flats. At first instance, the trial judge, Judge Maddocks, made the following findings, quoted in [11]:

> In this case the land remained in a state which did not admit to any serious use or occupation. The rear flats were incomplete, derelict and uninhabitable. The land and the buildings were not appropriated to any alternative use, such as, for example, the parking of cars or the drying of clothes. As it is the only use K
> relied upon is the temporary deposit of refuse items from the Hometel Flats — old mattresses and beds awaiting collection and removal in a skip, and the deposit of fencing panels to be broken up and used as hardcore on site. Such casual use and intermittent activities could not as themselves be viewed as "actual occupation". I attach greater significance to the secure fencing on all three sides, coupled as it was with the unfenced boundary with the Hometel site and with the access (being the only means of access) from the Hometel flats which were within the same management and control. Although the rear land and Hometel flats were held by separate companies, both companies were under the same ownership (the Lee Chang Trust), both companies and both properties were under the same management as was Mrs Chang's own company, Home Management. Thus, Mr Donald, the joiner, carried out work on both properties L
> and made use of the rear land for work on the flats. The wooden fence and the high security fence were partly on the front land. There was evidence, which I accept, that some work was done, as required, to keep down the weeds in the yard, to maintain the fences that were damaged and to board up the window openings. Taking the evidence as a whole, my conclusion is that the undoubted possession of Malory BVI amounted to "actual occupation" within the meaning of section 70(1)(*g*).

[55] The Court of Appeal declined to interfere with this finding. Arden LJ, with whom the other members of the court agreed on this point, expressed herself as follows:

> 80. That leaves the question whether the judge's finding that Malory BVI M
> was in "actual occupation" of the rear land is susceptible to review on appeal. The judge's finding involves questions of primary fact and the application of the correct principles to the facts. What constitutes actual occupation of property depends on the nature and state of the property in question, and the judge adopted that approach. If a site is uninhabitable, as the rear land was, residence is not required, but there must be some physical presence, with some degree of permanence and continuity (*cf Strand Securities Ltd* v *Caswell* [1965] Ch 958). As Lord Oliver of Aylmerton said in *Abbey National Building Society* v *Cann* [1991] 1 AC 56, 93:
>
> "It is, perhaps, dangerous to suggest any test for what is essentially a question of fact, for "occupation" is a concept which may have different connotations according to the nature and purpose of the property which is claimed to be occupied. It does not necessarily, I think, involve the personal

* Editor's note: Reported at [2010] 1 P&CR 16

presence of the person claiming to occupy. A caretaker or the representative of a company can occupy, I should have thought, on behalf of his employer. On the other hand, it does, in my judgment, involve some degree of permanence and continuity which would rule out mere fleeting presence. A prospective tenant or purchaser who is allowed, as a matter of indulgence, to go into property in order to plan decorations or measure for furnishings would not, in ordinary parlance, be said to be occupying it, even though he might be there for hours a time."

81. The requisite physical presence must, as it seems to me, in fairness be such as to put a person inspecting the land on notice that there was some person in occupation: see generally *per* Lord Oliver in *Cann*'s case, at p87). None of the authorities which we have been shown deal with completely derelict land.

82. For my own part, I am not persuaded by Mr Martin's arguments that the judge directed himself otherwise than in accordance with the principles which must be applied when determining "actual occupation". I do not consider that the judge confused actual occupation with possession or occupation of the front land with occupation of the rear land. Nor do I consider that he was wrong in the circumstances to attach significance to the fencing of the rear land. In this particular case, the fencing cannot be regarded as wholly separate from occupation of the rear land. The fencing was one of the factors relevant to be taken into account. The judge was also right in my judgment to attach significance to the access permitted from the front land. Even though there was another gate, the access from the front land supported the notion that some person connected with the front land claimed a right to be on the rear land. On that basis the question of whether applying those principles there was "actual occupation" was essentially a question of fact for the judge. At the relevant time, there were derelict buildings on the rear land which meant that it was not possible to occupy it by living in those buildings or by cultivating the land or by using the land for recreation. The judge had to consider other acts denoting occupation, such as boarding up the windows of the building and fencing the site (in both cases) to keep vandals and trespassers out, and also using the land for storage. In my judgment, the judge was entitled to draw the conclusion that Malory BVI was in occupation from the facts as found by him, and accordingly, his conclusion cannot be disturbed by this court. Moreover, no one visiting the rear land at the time of the sale to Cheshire could have drawn the conclusion that the land and buildings on the rear land had been abandoned; the evidence of activity on the site clearly indicated that someone claimed to be entitled to be on it.

[56] Here, there was no obvious legitimate activity in the drying compound, but the fencing and other points referred to by me ought in my judgment to have made it obvious that the drying compound was being dealt with as part of the common parts of the flats. So far as the remainder of the disputed land is concerned, I have concluded that Roberto Mac is not in possession of it. Accordingly, para 6 of Schedule 4 to the 2002 Act has no application. However, para 5 will also have no application on the facts as found by me, unless any alteration is for the purpose of correcting a mistake. There is also the question of the extent to which the rights of the tenants in the shops and the flats would survive if the register is not altered to restore Knights Construction as freehold owner.

Receipt of rent and profits

[57] The long leases both of the flats and the shops did not involve the payment of rent. Mr Gore sought to contend, however, that one of the shorthold tenancies had been in existence since before the 2002 Act came into force and that Knights Construction was in receipt of the rents and profits of that flat, which were paid not only in respect of the flat but of rights of way and parking rights over the disputed land so that it had an overriding interest by virtue of para 8 of Schedule 12 to the 2002 Act. I am unable to accept that receipt of rent in respect of another property can also be treated as being in respect of the disputed land just because the tenant of that other property has rights over it, but, in any event, no such rights were included in the tenancy. The rent under the tenancy was in respect of the rights granted by it, and it was only subsequently, when he was handing over the keys, that a gratuitous licence was granted to the new tenant by Mr Purser on behalf of Knights Construction.

Mistake and indemnity under 2002 Act

[58] The scope of the provisions of para 5(a) of Schedule 4 to the 2002 Act has recently been the subject of considerable debate. In particular, there are conflicting decisions of two other deputy adjudicators to which I shall refer. The relationship of that paragraph to the indemnity provisions in para 1(1) of Schedule 8 are also of importance, as it would appear that the provisions were drafted to reflect each other.

[59] With regard to the question of whether an alteration of the register in respect of the disputed land, other than the drying compound, so as to transfer title in that land from Roberto Mac to Knights Construction would be for the purpose of correcting a mistake, when the case was heard by me last September, I was referred by Mr Gore to the decision of Mr Owen Rhys, sitting as a deputy adjudicator to the Land Registry, in *Ajibade* v *Bank of Scotland*, REF/2006/0163/0174*.

[60] If that decision is correct, and I shall refer to it in detail later in this decision, then Mr Gore is entitled to contend that the same approach should be adopted in the present case. Unfortunately, neither party was aware of what was then a very recent decision of Mr David Holland, also sitting as a deputy adjudicator to the Land Registry in *Stewart* v *Lancashire Mortgage Corporation Ltd* REF/2009/0086/1556, where Mr Holland reviewed extensively authorities to which Mr Rhys had not been referred, and declined to follow his decision, holding instead that the original owner was not entitled to be restored to the register. Nor had either counsel referred me to any of the potentially relevant authorities identified by Mr Holland in his decision.

[61] In addition, it appeared to me that there was a potential problem as to the relationship between the provisions as to rectification and those as to indemnity, in that, on one construction of the provisions, a party might end up without the land or an indemnity as a result solely of the provisions of the 2002 Act, although wholly without fault. I was concerned as to a possible conflict in this respect both between the losing party in the future and the Land Registry, and, if the result of the 2002 Act was that an innocent party lost its property without qualifying to claim an indemnity under Schedule 8, between the legislation as so interpreted and Article 1 of Protocol 1 of the Human Rights Convention.

[62] As a result, after making findings of fact and of law that were not affected by this issue, I directed that the Chief Land Registrar should be added as second respondent to enable the issues as to the meaning of the relevant provisions to be fully considered between all parties affected, and for a single decision to be given in respect of them. The Chief Land Registrar has been made a party only in respect of those limited issues and nothing I have found on other issues binds him.

[63] Mr Timothy Morshead duly appeared on behalf of the Chief Land Registrar at the further hearing this month, together with counsel for the other parties, to argue the rectification and indemnity issues. I am grateful for the considerable assistance he has provided in comparing the provisions of the 2002 Act with those of the Land Registration Act 1925 (the 1925 Act) that it replaced, and in researching the genesis and purpose of those changes.

[64] In the 1925 Act, the provisions as to rectification were contained in section 82, and the provisions as to indemnity were contained in section 83. Section 82 included provision for rectification pursuant to an order of the court or by the registrar in 8 identified cases. These were as follows:

(*a*) … where a court of competent jurisdiction has decided that any person is entitled to any estate right or interest in or to any registered land or charge, and as a consequence of such decision the court is of opinion that a rectification of the register is required, and makes an order to that effect;

(*b*) … where the court, on the application… of any person who is aggrieved by any entry made in, or by the omission of any entry from, the register, or by any default being made, or unnecessary delay taking place, in the making of any entry in the register, makes an order for the rectification of the register;

(*c*) … with the consent of all persons interested;

(*d*) Where the court or the registrar is satisfied that any entry in the register has been obtained by fraud;

(*e*) Where two or more persons are, by mistake, registered as proprietors of the same registered estate or of the same charge;

(*f*) Where a mortgagee has been registered as proprietor of the land instead of as proprietor of a charge and a right of redemption is subsisting;

(*g*) Where a legal estate has been registered in the name of a person who if the land had not been registered would not have been the estate owner; and

* Editor's note: Unreported 8 April 2008

(*h*) In any other case where, by reason of any error or omission in the register, or by reason of any entry made under a mistake, it may be deemed just to rectify the register.

[65] In addition, the Land Registration Rules 1925 (the 1925 Rules), rr 13 and 14, contain provisions for the correction by the registrar of a clerical error.

[66] Section 83 included the following provisions:

(1) Subject to the provisions of this Act to the contrary, any person suffering loss by reason of any rectification of the register under this Act shall be entitled to be indemnified.

(2) Where any error or omission has occurred in the register, but the register is not rectified, any person suffering loss by reason of such error or omission, shall, subject to the provisions of this Act, be entitled to be indemnified.

(3) …

(4) Subject as hereinafter provided, a proprietor of any registered land or charge claiming in good faith under a forged disposition shall, where the register is rectified, be deemed to have suffered loss by reason of such rectification and shall be entitled to be indemnified under this Act.

[67] In Schedule 4 to the 2002 Act, there was a change of language and ostensibly of content. The 1925 concept of "rectification" of the register became "alteration", and "rectification" was defined for the purpose of Schedule 4, in respect of alteration of the register, as "alteration which: (a) involves the correction of a mistake; and (b) prejudicially affects the title of a registered proprietor"; see para 1 of Schedule 4. Paragraph 2 of Schedule 4 provides for the court to "make an order for alteration of the register for the purpose of: (a) correcting a mistake; (b) bringing the register up to date; or (c) giving effect to any estate, right or interest excepted from the effect of registration." Paragraph 5 of Schedule 4 gives the same power to the registrar together with the additional power to alter the register to remove a superfluous entry.

[68] Both Acts contain provisions to protect to some extent the position of proprietors of a registered estate in respect of land in their possession. The 1925 Act contained no definition of proprietor in possession. The 2002 Act contains such a definition in section 131, which was designed to reverse the decision of the Court of Appeal in *Kingsalton Ltd* v *Thames Water Developments Ltd* [2001] EWCA Civ 20* that the registered proprietor was to be treated as being in physical possession unless he could be shown to have been dispossessed. For the purposes of the 2002 Act, land is in the possession of the proprietor of a registered estate if it is physically in his possession, or in the possession of a person who is entitled to be registered as the proprietor of the registered estate. I have held in the present case that Roberto Mac is not in possession of the disputed land.

[69] The 2002 Act was the result of a review of land registration by the Law Commission and the Land Registry. Where a statute has been enacted as a result of the recommendations of the Law Commission, it is appropriate and permissible to consider those recommendations to help identify both the mischief that the Act is designed to cure and the public policy underlying it, even though the recommendations cannot be conclusive as to the construction of the provisions in the Act: see *Yaxley* v *Gotts* [2000] Ch 162†, at p182, *per* Clarke LJ.

Reports of the Law Commission and the Land Registry

[70] I was referred by Mr Morshead to the two joint reports that they produced. The first was entitled *Land Registration for the Twenty-First Century: A Consultative Document*; and the second, which included the Land Registration Bill, was entitled *Land Registration for the Twenty-First Century: A Conveyancing Revolution*, Law Com no 271. The second of these reports was laid before parliament by the Lord Chancellor pursuant to section 3(2) of the Law Commissions Act 1965.

[71] The consultative document contained the following observations as to rectification and indemnity (I omit the references to the footnotes):

Rectification G

2.37 We explain in detail in Part VIII of this Report the ten grounds on which the register may be rectified. However, they may be broadly summarised as follows —

(1) Where a court makes an order which either gives effect to an established property right or interest, or orders the removal from the register of an entry in respect of a right which has not been established.

(2) Where the court or the registrar decides that the register is incorrect in some way.

(3) Where the registrar decides that a clerical error needs to be corrected.

…

2.39 The present law governing rectification works well in practice and gives rise to few difficulties. Nevertheless, this review of land registration offers H an opportunity to simplify and clarify the legislation, and we make provisional recommendations to this effect in Part VIII.

Indemnity

2.40 The provisions for rectification and for indemnity are directly linked. Where the register is rectified, any person suffering loss by reason of the rectification may be indemnified. However, indemnity also serves a wider purpose in that it protects persons from loss —

(1) caused by errors and omissions on the register, even where the register is not rectified;

…

PART VIII

RECTIFICATION OF THE REGISTER J

INTRODUCTION

8.1 … We conclude that the existing law and practice in relation to rectification are largely sound, and that the main deficiencies lie in the way in which the legislation is drafted. In the light of this, we make recommendations for reform which are primarily clarificatory rather than substantive.

8.2 Although rectification is commonly considered in conjunction with the right to indemnity for loss suffered as a result of errors and omissions in the register, we make no recommendations in this Report for the amendment of the provisions on indemnity. … However, we anticipate that, in the Land Registration Bill that will be drafted to reflect the responses to this Report, there may be a handful of minor amendments to reflect more clearly the current practice in relation to the payment of indemnity. These will not entail any change of substance. K

THE NATURE OF RECTIFICATION

8.3 … There is no definition of rectification in the Act, but it can probably best be described as "any amendment to the register or its plan for the purpose of putting right any substantive error of omission or commission, or any legally recognised grievance". Rectification is, therefore, concerned with the correction of mistakes in the register. This is reflected in the corresponding provisions in the Land Registration Act 1925 for the payment of indemnity. The statutory jurisdiction to award indemnity in cases where the register either is or is not rectified arises only where loss has been suffered as a result of some error or omission in the register. It is of course true that the provisions on rectification and indemnity are not exact correlatives. Nevertheless, it is a pointer to the true nature of rectification and one which we consider should be clarified in any L new legislation.

[72] The report then considered the 10 grounds on which the register could be registered under the then existing law (described in para 8.4 as set out in section 82(1) of the 1925 Act and rr 13 and 14 of the 1925 Rules). In relation to section 82(1)(*g*), the report comments as follows:

GROUND 7: WHERE THE PROPRIETOR WOULD NOT HAVE BEEN THE ESTATE OWNER HAD TITLE BEEN UNREGISTERED

8.18 The seventh ground of rectification is —

[w]here a legal estate has been registered in the name of a person who if the land had not been registered would not have been the estate owner. M

This ground of rectification covers a number of situations where a transfer of *unregistered* land would have been void and would not have vested a legal estate in the transferee. Obvious examples include —

(1) where B was mistakenly registered as proprietor of more land than was actually comprised in the transfer from A;

(2) where A purported to transfer land to B which A no longer owned, either because A had already transferred the land to C, or because C had acquired title to it by adverse possession;

(3) where the transfer from A to B was itself a nullity, because, for example, it was a forgery, or where A could plead *non est factum*, though rectification would commonly be under Ground 1; and

(4) where the transfer by A to B was *ultra vires*, though again, rectification would commonly be under Ground 1.

* Editor's note: Reported at [2002] 1 P&CR 15

† Editor's note: Also reported at [1999] 2 EGLR 181

A 8.19 Because of the reference to the position in unregistered land, this ground of rectification has been much criticised. There is a concern that it may import the law on notice and land charges — the very things which the registered system was meant to obviate. In practice this has not happened, and by far the commonest reason for rectification under this head is the case exemplified in paragraph 8.18(1) above.

[73] Having dealt with the individual grounds, in para 8.36, the report concluded that the then law governing rectification worked without undue difficulty in practice, and that its defects lay primarily in the way it was drafted. It summarised the defects as follows:

B 8.37 No clear distinction is drawn between the *rectification* of an error or omission in the register (which is discretionary) and an *amendment* of the register because the court determines that A is entitled to an estate, right or interest in or over land of which B is the registered proprietor. The latter is not intrinsically different from an amendment to the register to indicate that, say, a lease has terminated. The first two grounds of rectification are situations where the court has determined the substantive rights of the parties and there is in practice no true element of discretion in its order that the register should be rectified to give effect to those rights.

8.38 The yardstick for determining whether there is an error or omission is, in some cases, the position that would have existed had title not been registered. There are a number of objections to this. First, there is the risk that it may bring into play aspects of unregistered land, such as the doctrine of notice
C and the registration of land charges, which the registered system was meant to supersede. The difficulty with this approach is to know just how much of the unregistered system is brought into play. Secondly, as the great majority of titles are now registered, it is inappropriate, particularly as the principles of unregistered conveyancing are becoming unfamiliar to practitioners. Thirdly, it is unnecessary. The existence of an error or omission can be readily defined according to the principles of registered conveyancing.

8.39 There is uncertainty as to the ambit of the restrictions on rectification that exist in favour of a proprietor who is in possession. Although this has not so far been the cause of much difficulty in practice, it could be, and it would be better to address the matter.

D [74] The report went on to make proposals for reform, as follows:

PROPOSALS FOR REFORM

The essentials of an effective scheme of rectification

8.40 Any rectification scheme must overcome the deficiencies outlined above. To achieve this we regard as the essentials of any viable scheme the following features, which should be made clear in the legislation

(1) The only function of the power to rectify is to provide a discretionary mechanism which enables errors and omissions in the register to be remedied. It should remain the case that indemnity is available for any loss that is caused as a result of rectification or where, as a matter of discretion, rectification is not ordered, its refusal.

(2) In the absence of any error or omission in the register, where a court
E makes a determination of substantive rights in or over registered land, it should do so in accordance with the principles of registered land. The principles governing unregistered land should come into play *only* if there is some issue which arises from the time prior to first registration. Following any such determination, the register should be amended to reflect its outcome. That amendment, unlike rectification, should not be a matter of discretion.

(3) Normally, the register should be rectified to give effect to an overriding interest, though as it is an application of the power to rectify, it should remain a matter for discretion.

(4) The existing principle of qualified indefeasibility which applies to a proprietor who is in possession should be retained but its scope should be clarified.

F **Grounds for rectification**

8.41 As we have indicated, rectification is an issue where the register is in some way incorrect. We consider that rectification should be confined to that situation. The consequences of this view are as follows

(1) It would no longer be appropriate to provide for rectification of the register to give effect to a court order that a person is entitled to an estate, right or interest in any registered land or charge. Where the court determines the substantive rights of the parties, the register should be amended to give effect to the order which it makes. There should be no element of discretion about it. Any determination would, of course, be undertaken in accordance with the principles applicable to registered land. This should preclude any possible argument that the principles of unregistered land are in some way relevant.

(2) We have recommended in Part VI of this Report that both the court and the registrar should be given a power to discharge or vary a caution, notice or restriction. The introduction of this general jurisdiction would obviate the need for the present power to rectify the register on the application by a person aggrieved by an entry. The exercise of this power to amend the register should cease to be regarded as rectification.

(3) There would be no need to have a list of grounds upon which rectification might be ordered as there are at present in section 82(1) of the Land Registration Act 1925. They would all be encompassed within the one general principle that the register could be rectified, as a matter of discretion, where there was an error or omission in it.

8.42 In the light of this, **we provisionally recommend that, subject to the qualifications that we explain below, the court and the registrar should have a power to rectify the register whenever there has been an error or omission in it. Do readers agree?**

8.43 *That power would encompass all the cases presently included in section 82(1)(c)-(h) of the Land Registration Act 1925 (Grounds 3-8 above), as well as the ground listed in rule 14 of the Land Registration Rules 1925 (Ground 10 above).*" (Italics provided)

[75] Finally, in para 8.58(10), in summarising its recommendations, the report states that "the register might be rectified even though this might affect any estates, rights, charges, or interests acquired or protected by registration, or by an entry on the register, or any overriding interest".

[76] Following the consultation procedure, the draft bill was laid before parliament with the commentary of the Law Commission and the Land Registry in Law Com no 271. That report contained the following statements as to the fundamental objective of the bill (once again I omit references to footnotes):

1.5 The fundamental objective of the Bill is that, under the system of electronic dealing with land that it seeks to create, the register should be a complete and accurate reflection of the state of the title of the land at any given time, so that it is possible to investigate title to land on line, with the absolute minimum of additional enquiries and inspections.

1.6 Although that ultimate objective may seem an obvious one, its implications are considerable, and virtually all the changes that the Bill makes to the present law flow directly from it.

[77] However, in relation to alteration, rectification and indemnity, it states as follows:

ALTERATION, RECTIFICATION AND INDEMNITY

2.38 Mistakes do occasionally occur in the register and it is also necessary to update it to take account of interests that have determined. The Bill does not make major changes to the present law and practice on the alteration of the register. What it does do is to codify that present practice in a way that makes its working apparent. The present legislation obscures what actually occurs.

2.39 The law governing the payment of indemnity for mistakes that have occurred in the register and other losses for which the Registry is responsible was revised by the Land Registration Act 1997. The Bill does not, therefore, make anything more than minor changes to the present law.

[78] The report goes on, in Part X, to explain the provisions of the draft bill that govern alteration of the register and the payment of indemnity. In para 10.1, it again refers to the 10 grounds on which rectification may be ordered under the 1925 Act and Rules. It repeats the summary of those grounds contained in para 2.37 of the consultative document. In paras 10.3 to 10.4, it continues:

10.3 Indemnity is, in some senses, the correlative of rectification. It is payable where a person suffers loss as a result of an error or omission in the register, whether or not the register is rectified, and in certain other circumstances. The availability of indemnity is of great importance to the system of land registration. It is the basis of the so-called "State guarantee of title" which registration confers.

10.4 In the Consultative Document, we explained that the practice that has developed both in the courts and in the Land Registry in relation to the rectification of the register was largely sound and that the deficiencies lay in the way in which the legislation was drafted. The provisions of the Land Registration Act 1925 tend to obscure the real nature of rectification and the manner in which it operates. We considered that the main task of any new legislation should be to recast the legislation in a much more transparent form. However, we did make a number of recommendations to change the law where it was either unsatisfactory or could be improved. We did not propose any substantive changes to the law governing indemnity because the subject had been considered recently by the Joint Working Group and its recommendations had been implemented by the Land Registration Act 1997. The recommendations in the Consultative Document on rectification, amendment and indemnity were strongly supported on consultation.

[79] The report then explains the new definitions of alteration and rectification (paras 10.6 to 10.7) and in 10.8 states that the bill makes it clear that "rectification of the register… can (as now) affect derivative interests", but any such changes will be prospective only, which accords with the way in which the analogous provisions of the 1925 Act have been interpreted. In para 10.10, it describes three cases in which the court may order the rectification of the register, the first being for the purpose of correcting a mistake, the second being for the purpose of bringing the register up to date and the third being to give effect to any estate, right or interest excepted from the effect of registration. In para 10.29, the report states that although the provisions have been recast in the bill, the substance of them has not been altered in any significant way.

[80] It is clear that there is nothing in these reports to suggest that it was the intention of the Law Commission or the Land Registry that the law should be altered so as to remove from a former owner of land the right to have the register rectified just because, following a mistake, a third party had acquired an interest in it for valuable consideration. Rather, it was clearly the intention that the law should not be changed in that respect.

[81] In this case, the Salvation Army became registered with land to which it had no title, land that belonged to Knights Construction. It then transferred its registered title to Roberto Mac for a valuable consideration. There is no doubt that the registration of the Salvation Army as proprietor of the disputed land was a mistake. The position in that respect, for present purposes, is the same as if a fraudster had forged the name of the freehold owner of unregistered land on a transfer to himself, got himself registered as proprietor and then transferred the title to Roberto Mac. In addition, there is a further mistake. This was a first registration, and it may be said to be a mistake that the land is included in the registered title at all. In this respect, it is different from the typical case of forgery because there the land has remained throughout in the correct title.

Law under the 1925 Act

[82] In the light of the Law Commission's reports, it is necessary first to consider what the law was in such a case before the 2002 Act came into force. That question was considered by the Court of Appeal in *Argyle Building Society* v *Hammond* (1984) 49 P&CR 148. In that case, A was the registered proprietor of a property. That property was transferred to B by a transfer for value that A alleged was a forgery. B was registered as owner in the proprietorship register in place of A. On the same date as the transfer, B charged the property to C as security for an advance of £15,000 and the charge was also registered in the charges register of the property. A preliminary issue was directed as to whether, if C took the charge in good faith and for value, without actual knowledge of the forgery, there was any possibility of A obtaining rectification of the register. I note that if the charge was granted on the same day as the transfer, B could not yet have been the registered proprietor.

[83] The Court of Appeal held that the effect of registration was to vest in B the legal fee simple, regardless of any forgery, but that, by section 82(2) of the 1925 Act, the court had the power to rectify the charges register in a proper case even against a bona fide charge, and that on the basis of the assumed facts the court would have a discretion as to whether to order rectification. In delivering the judgment of the court, Slade LJ pointed out, at the foot of p152, that section 69(1) of the 1925 Act operated to confer protection even on a forger and those claiming through him, a result described, on p153, as being achieved by a kind of statutory magic. Slade LJ went on, at p154, to refer to a passage in *Ruoff & Roper: The Law and Practice of Registered Conveyancing*, where it was stated that although rectification could be granted against the proprietor of the land, a charge created by the transferee in favour of a bona fide mortgagee would not be disturbed.

[84] At p156 of the report, the judgment concluded that once a person had been registered as the proprietor of freehold land, he had, so long as he remained so registered, the statutory power to create valid charges even though the purported transfer of the freehold under which he claimed was a forged instrument. Likewise, so long as he remained so registered, he had the power to transfer the freehold title itself. The court took account of the fact that the charge had been executed before the title had been registered, but considered the legal issues:

> on the assumptions that (a) the protection afforded by the 1925 Act to the respondent as mortgagee is no less effective than it would have been if the charge in its favour had been executed *after* the title of [B] as proprietors had been registered; and (b) that the protection in law afforded to mortgagees whose charges have been registered under the 1925 Act is no less effective than the protection afforded to freeholders whose title has been registered under that Act, even though it is common ground that the 1925 Act contains no explicit provisions relating to mortgagees precisely corresponding with section 69(1).

[85] Nevertheless, the court continued, at the foot of p156: "it is clear that any protection enjoyed by a proprietor or mortgagee under the 1925 Act is conferred subject to the provisions for rectification of the Register contained in section 82". After setting out section 82(1), so far as material, the court concluded firstly that the jurisdiction to rectify plainly extended to the property register, the proprietorship register and the charges register. Second, it concluded, at p157, that on the assumed facts: "the court would, in our judgment, have clear jurisdiction to rectify the proprietorship register of the house by substituting the name of [A] for that of [B]". It went on to consider the position of C, and drew attention to the provisions of section 82(2) that the register could be rectified under that section "notwithstanding that the rectification may affect any estates, rights, charges, or interests acquired or protected by registration, or by any entry on the register, or otherwise". The court concluded, at the foot of p157, that this made it clear that the court had jurisdiction in a proper case to rectify the charges register even though the rectification may affect a charge acquired or protected by registration. It went on, at p158, to draw attention to the provisions for an indemnity for innocent parties who were caused loss if the court directed that the register should be rectified or if it exercised a discretion not to rectify it. The court concluded, at the foot of p162, that a court always had at least theoretical discretion to rectify any part of the register, even as against innocent third parties, subject only to the limiting provisions of section 82(3).

[86] The case then proceeded to trial, following which there was a further appeal to the Court of Appeal under the name of *Norwich & Peterborough Building Society* v *Steed (No 2)* [1993] Ch 116. By that time, the forgery allegation had been abandoned, and a claim of *non est factum* also failed. The appeal against the latter finding was unsuccessful, with the result that the transfer was voidable only. It had not been avoided when C had advanced the money. The Court of Appeal declined to follow *dicta* in *Argyle Building Society* to the effect that the court also had a discretionary power to rectify when the original transaction was voidable and not void, and as a result A was unsuccessful as against C. The leading judgment, with which the other members of the court agreed, was given by Scott LJ. At p125A-C, he defined the issues as being whether the transfer was void or voidable. He continued by stating that if the transfer was held to be void, "there would, strictly, be an issue as to whether or not an order for rectification of the register, as against [C], ought to be made. But for reasons which I will later mention, it seemed to us that there would be no serious answer to the defendant's rectification claim and [counsel for C] did not contend otherwise."

[87] After considering sections 82 and 83 of the 1925 Act, Scott LJ stated, at p132A-B, that there was no doubt that if the forgery or *non est factum* plea had succeeded the case would have fallen squarely within para (*g*) of section 82(1). He went on:

> In neither case, if the land had been unregistered, would [B or C] have obtained a legal estate. I cannot see any reasonable basis on which an order of rectification could have been withheld.

A [88] Where the transfer was voidable only, however, with unregistered land title would have been lost and have become irrecoverable once it had passed to an innocent third party for valuable consideration. Scott LJ was unwilling to spell out of the provisions of section 82 an intention to allow even a discretionary power to rectify in such a case. Having analysed the provisions of section 82(1), he continued, at p134G-135C:

> In my opinion the scheme is reasonably clear. Paragraphs (*a*) and (*b*) give power to the court to make orders of rectification in order to give effect to property rights which have been established in an action or which are clear. Paragraph (*c*) enables orders to be made by consent. The remaining
> B paragraphs, (*d*) to (*h*), are intended to enable errors to be corrected. Paragraph (*d*), paragraph (*e*), paragraph (*f*) and Paragraph (*g*) each deals with an error of a particular character. But, since these paragraphs might not cover comprehensively all errors, paragraph (*h*) was added as a catch-all provision to cover any other errors. …
>
> Paragraph (*h*) is relied on by [counsel for A]. But in order for the paragraph to be applicable some "error or omission in the register" or some "entry made under a mistake" must be shown. The entry in the charges register of [C's] legal charge was not an error and was not made under a mistake. The legal charge was executed by [B], who were at the time transferees under a transfer executed by [the attorney for A]. The voidable transfer had not been set aside. The registration of B as proprietors took place at the same time as
> C the registration of the legal charge. Neither registration was an error. Neither entry was made under a mistake. So the case for registration cannot be brought under paragraph (*h*).

[89] The Law Commission and the Land Registry were well aware of these decisions, and they are referred to in the first of the above reports. It is plain that they did not intend to alter the law as set out in them except in the relatively minor respects identified by them. It is equally plain that their efforts to achieve clarity by the redrafting of the provisions has had the opposite effect. There have been a number of cases where conflicting views as to the effect of the changes have been expressed.

D *Decisions as to the law under the 2002 Act*

[90] I turn first to the decision of Blackburne J in *Pinto* v *Lim* [2005] EWHC 630 (Ch), another case of a forged disposition. In this case, in September 1999, one of two joint proprietors of a property forged the signature of the other to a transfer by which the property was purportedly transferred into the forger's sole name. The forger was duly registered as sole proprietor. Some months later she transferred the property to her brother for £119,000, part of which was provided by a building society on mortgage. The brother was then duly registered as proprietor and the building society as chargee. It was common ground that the brother purchased in good faith and for a proper price. The
E original proprietor whose signature was forged claimed rectification of the register. As the proceedings commenced before 13 October 2003, it was common ground that the question of rectification was to be governed by the 1925 Act, although, in due course, any question of indemnity would be governed by Schedule 8 to the 2002 Act, and not by section 83 of the 1925 Act.

[91] It was common ground that the application fell within at least one of the grounds in section 82(1) of the 1925 Act, the most obvious case being para (*g*). As the person against whom rectification was sought (the brother) was in possession of the property, the discretion had to be exercised giving special weight to the fact of possession: see paras 84
F to 87. There was, however, no suggestion that the fact that the original fraudster was no longer on the register made any difference at all.

[92] In exercising his discretion, Blackburne J also had regard to the provisions for indemnity in Schedule 8 to the 2002 Act. It is sufficient for present purposes to note that Blackburne J clearly proceeded on the basis that the unsuccessful party would be entitled to an indemnity. In deciding to exercise his discretion to refuse rectification, he pointed out, in [100], that para 1(1)(a) and 1(1)(b) of Schedule 8 to the 2002 Act were the converse of each other. He continued:

> 100. … If a mistake (ie an error or omission) has been made which entitles the court, in the exercise of its discretion, to order rectification of the [brother's] registration as proprietor of the property… it is difficult to see why, if the court declines in the exercise of that discretion to make the order sought, the claimant who can show that, as a result of the mistake, he has suffered a loss should not be entitled to an indemnity for that loss (provided only that the loss has not resulted, wholly or partly, from his own fraud or… lack of proper care). In the instant case, [the original owner] has, by the registration of the [brother] as sole proprietor of the property… been deprived of his beneficial interest in the property. This occurred without fault on his part. It is not relevant that, had he made his claim at the time when [the forger] was still registered as proprietor (ie prior to the [brother] purchasing the property from her and becoming proprietor in her place) he would have been seeking rectification of a different mistake (namely the registration of [the forger] as proprietor).

[93] It is plain that Blackburne J regarded there as being two mistakes: the initial registration of the forger and the subsequent registration of the brother. Immediately thereafter, he cited a passage from the 1986 edition of *Ruoff & Roper*, setting out the legal position following *Argyle Building Society* as an accurate statement of the 1925 Act, overlooking that in relation to voidable titles the Court of Appeal in *Norwich & Peterborough Building Society* had held that the *dicta* in that case were incorrect.

[94] The passage in *Ruoff & Roper* approved by Blackburne J, in [100], did, however, contain the following proposition in relation to a void transaction:

> Thus for example, if a forger succeeds in getting himself registered as the proprietor and thereafter executes a disposition for value in favour of a bona fide purchaser or mortgagee who completes the disposition by registration, then without a judicial hearing, that purchaser or mortgagee will not be deprived of his registered estate or interest in favour of the true original owner… However, there will, in such circumstances, be grounds on which the register may be rectified and the Court will have a discretion, subject to the limitation on its powers, to order rectification where this is considered proper having regard to the balance of equities.

[95] Blackburne J concluded, in [101], by holding that that passage: "illustrates that where … the court could, as a matter of discretion, have ordered rectification but declines to do so, the Claimant will be compensated notwithstanding that the forger has subsequently disposed of the property for valuable consideration to a wholly innocent purchaser who has since become registered as its proprietor". He clearly proceeded on the basis that this was a case where the court had a discretion to order rectification and that, in the event of rectification being refused, the entry of the bona fide purchaser was a mistake for the purposes of Schedule 8 to the 2002 Act whose correction would involve rectification of the register.

[96] Paragraph 19 of Schedule 12 to the 2002 Act provides that Schedule 8 "applies in relation to claims made before the commencement of that Schedule which have not been settled by agreement or finally determined by that time (as well as to claims for indemnity made after the commencement of that Schedule)". It would to my mind be an extraordinary lacuna if the references to a mistake whose correction would involve rectification of the register would exclude from compensation cases that had arisen before the Act came into force that did not fall within section 82(1)(*h*). It would also be wrong to construe Schedule 8, unless compelled to do so, in a way that would deprive the original owner of an existing, pre-October 2003, statutory property right, the right to an indemnity, in breach of Article 1 of Protocol 1 to the European Convention on Human Rights.

[97] Although the decision in *Pinto* was given after the 2002 Act came into effect, the passage in *Ruoff & Roper*, corrected in a later edition following the Court of Appeal decision in *Norwich & Peterborough Building Society*, must have been known to the Law Commission and the Law Society when putting forward proposals for reform, and had they wanted to propose the removal of the right to rectification and indemnity once the original invalid transaction was no longer recorded on the register, they could have done so.

[98] The next decision was that of Mr Terence Mowschenson QC, sitting as a judge of the Chancery Division of the High Court in *Barclays Bank plc* v *Guy* [2008] EWHC 893 (Ch). In that case, on an application by the bank for summary judgment, there was an issue whether a transfer was void or voidable. The judge concluded that it

was voidable only, and on that basis the judge followed the decision of the Court of Appeal in *Norwich Building Society*. He set out the position, in paras 16 to 27, as follows:

15. The Claimant contends that assuming in the Defendant's favour that the transfer was fraudulently procured by Ten Acre …, that cannot affect the Claimant's right as registered chargee. It does not provide the Defendant with any basis for rectification of the register as against the Claimant.

16. The grounds for rectification of the register pursuant to [s]65 of the Land Registration Act 2002, which I shall refer to as the 2002 Act, are set out in s65 and para 2 of Sch 4 and include the alteration of the register for the purposes of correcting a mistake. By para 1 of Sch 4, references to rectification of the register, in relation to alteration of the register are to alterations which involve the correction of a mistake and which prejudicially affects the title of a registered proprietor.

17. The starting point is that at the date of the charge and its registration against the property Ten Acre was the registered proprietor of the property. It was entitled to exercise its power to charge the property. By s23 sub-s1 and s24 of the 2002 Act the registered proprietor of a registered estate is entitled to make a disposition of any kind permitted by the general law, subject to various exceptions. This includes under s23(1)(b) a power to charge the estate at law with the payment of money, s58 sub-s1 of the 2002 [Act] provides under the heading "Conclusiveness": "If on the entry of a person on the register as the proprietor of a legal estate the legal estate would not otherwise be vested in him, it shall be deemed to have been vested in him as a result of the registration."

18. It follows that at the time the property was charged to the Claimant Ten Acre was the registered proprietor. The legal estate was deemed to be vested in it, and it was entitled to exercise its owner's powers to charge the property. It follows that there can be no mistake in the Claimant's registration as chargee. I refer to the decision of Scott LJ in *Norwich Building Society* v *Steed* [1993] Ch 116 at p135 between letters A to H.

"Paragraph (*h*) is relied on by Mr Lloyd. But in order for the paragraph to be applicable some "error or omission in the register" or some "entry made under a mistake" must be shown. The entry in the charges register of the building society's legal charge was not an error and was not made under a mistake. The legal charge was executed by the Hammonds, who were at the time transferees under a transfer executed by Mrs Steed as attorney for the registered proprietor. The voidable transfer had not been set aside. The registration of the Hammonds as proprietors took place at the same time as the registration of the legal charge. Neither registration was an error. Neither entry was made under a mistake. So the case for rectification cannot be brought under paragraph (*h*)."

19. Until the chargor's title is avoided, his title as the person on the register is a real registration carrying with it all the incidents of the proprietor of the property which registration carries with it.

20. The Claimant contends that for the purpose of a rectification claim no further investigation is required or is relevant. It matters not whether the transfer is ultimately found to be fraudulently procured or whether the transfer is void or voidable, and I was referred to Ruoff & Roper on the *Law and Practice of Registered Conveyancing* at paras 46.028 to 46.030 on the law and which conveniently consider the position where the transfer to the chargor is void, in example 1.2 at 46.028:

"*Example 1.2.* R1 is the proprietor of a registered freehold estate in land. R2 forges R1's signature on a form of transfer and is registered as the new proprietor of the estate. R2 then grants a legal mortgage charge over the land to M1, which is duly registered. R1 applies for rectification so that R2's may be removed from the proprietorship register and his own name replaced, and the charge to M1 deleted from the charges register.

The registrar will rectify the proprietorship register against R2, as in Example 1.1 above. He will not rectify the charges register against M1. R1 will hold the freehold estate subject to M1's charge.

The legal estate is "deemed to be vested" in R2 when he was registered as its proprietor, even though it would not otherwise have vested in him owing to invalidity of the purported disposition from R1 to R2. No special significance attaches to the use of the expression "deemed to be vested", which in this context simply means that the estate is actually vested in R2. The consequence is that R2 had full owner's powers of disposition over the estate, including the power to grant the mortgage charge to M1, since he was the registered proprietor. His powers of disposition remained unqualified so long as his registration as proprietor remained effective. It cannot be said therefore that the disposition by way of legal mortgage to M1 was in any way invalid, nor that the registration of that mortgage was a mistake. Accordingly, R1 has no right to have the register rectified as against M1. R1 could have prevented this result by applying to the registrar to enter a restriction on R2's powers of disposition as soon as he discovered that a forgery had taken place. R2 could be restricted from making a registered disposition of the estate without the consent of the court or registrar.

This result is different from what would have been happened with G
unregistered conveyances, where the title of R2 and M1 would have depended on the rule nemo dat quod non habet. It also differs from what might have been the result under the rectification provisions of the former Land Registration Act 1925. Under that Act, there was equivocal authority that rectification would have been ordered against a third party transferee or chargee of an registered estate, where the original disposition which transferred the estate was void. The power to rectify would exist, though the court, in its discretion, might refuse to exercise it. One of the reasons for this result was the courts' wish to achieve by rectification the same result as would have followed according to the principles of unregistered conveyancing.

It is submitted, however, that this approach is not justifiable under the 2002 Act. First, it was the explicit policy of HM Land Registry and the Law H
Commission in the consultation paper that preceded [the] 2002 Act that, in the absence of some error on the register, the principles of unregistered land should not determine whether the register should be rectified. Any other result would undermine the general aim of the Act that the register should indicate accurately and comprehensively the state of the registered proprietor's title. A registered proprietor cannot be deemed to lack powers of disposition which the fact of registration indicates that he actually has. Secondly, the 2002 Act has narrowed the grounds of rectification provided in the 1925 Act so that it no longer allows rectification where the proprietor of a legal estate would not have been the estate owner if the land had been unregistered. The applicant for rectification must prove some error in registration."

21. Obviously, where the title of the proprietor is voidable then the example J
is even stronger, and I was referred in the Claimant's helpful skeleton argument to a passage by Charles Harpum and Janet Bignell in *Registered Land: Law and practice under the Land Registration Act 2002* [2004] at 15.1 where they comment on s58 under the Land Registration Act 2002 at para 15.1:

"It is a fundamental principle of registered conveyancing that it is registration that vests the legal estate in the registered proprietor (cf LRA 1925, s69(1)). In accordance with this principle, the LRA 2002 provides that if on the entry of a person in the register as the proprietor of the legal estate, the legal estate would not otherwise be vested in that person, the legal estate will be deemed to be vested in him as a result of the registration (LRA 2002, s 58(1)). What this means is that the legal estate will vest in a person even if he is registered as proprietor on the basis of a forged transfer which is itself a nullity (See (2001) Law Com No 271, para 9.4)." K

22. I am satisfied that the transfer to Ten Acre was voidable and not void. … At the date that the charge was taken and registered the transfer had not been avoided. Indeed, as I understand it from the argument that was put forward by the Defendant, it is not suggested that any attempt has been made to avoid it.

23. In *Norwich & Peterborough Building Society* v *Steed* the Court of Appeal held that where a power of attorney had been validly granted, and a transaction entered into in excess of that authority by fraud, the transaction was voidable and not void. Scott LJ agreed with the judge (Knox J):

"The judge dealt very shortly with the ultra vires point. He said: 'the transfer on its face was a perfectly regular sale and as against third parties taking a legal estate for value and in good faith it was within the ostensible authority of Mrs Steed and cannot now be repudiated against such third parties. Nothing short L
of non est factum rendering the transfer void, or forgery which has the same legal effect but is not now alleged, will suffice to defeat the building society's innocent reliance upon the transfer.'

"Save that I would not describe Mrs Steed's authority to execute the transfer as 'ostensible" authority, I am in complete and respectful agreement with the judge."

24. Accordingly, this is not a case of forgery or non est factum and there has been no suggestion by Mr Guy that he lacked capacity or title. …

25. … In effect, regrettably, Mr Guy knowingly signed a document which he knew could have the effect of transferring the property to Ten Acre. As at the time the charge was granted the voidable transfer had not been set aside and the register had not been rectified as against Ten Acre. No such rectification could M
affect the Claimant in this case.

26. …

27. In the light of my holding that the transfer was voidable and not void *I do not need to address the question of whether the transfer, the charge, would have been effective if the transfer was void.* [emphasis provided]

[99] It is clear, in particular from [27] of his decision, that the deputy judge decided this case purely on the basis that the transaction was voidable, and did not consider the position if the transaction was void. It is also clear that the editors of *Ruoff & Roper* were wrong to describe the two Court of Appeal cases to which I have referred as equivocal authority as to the rectification of void transactions, and were wrong as to the intentions of the Law Commission with regard to the scope of rectification under

the 2002 Act. When, however, the Court of Appeal refused Mr Guy permission to appeal (*Barclays Bank plc* v *Guy* [2008] EWCA Civ 452*), Lloyd LJ, with whom Carnwath LJ agreed, said, in [23]:

> while the scope of the phrase "correcting a mistake" is no doubt something that requires to be explored and discussed and developed in the course of future litigation, which will be decided upon the facts and upon the merits of each case, I cannot see that it is arguable that the registration of the charge can be said to have been a mistake, or the result of a mistake, unless at the least Mr Guy can go so far as to show that the bank, the mortgagee, had either actual notice, or what amounts to the same, what is referred to as "Nelsonian" or "blind eye notice", of the defect in the title of the mortgagor, Ten Acre Limited in the present case. I simply cannot see how it could be argued that if the purchaser or chargee knows nothing of the problem underlying the intermediate owner's title, that the registration of the charge or sale to the ultimate purchaser or chargee can be said to be a mistake. That seems to me inconsistent with the structure and terms of the 2002 Act. So the question is whether Mr Guy can show an arguable case, on the evidence, for saying that Barclays Bank had actual notice or was turning a blind eye to matters that it knew, which would if it addressed them properly, have shown it that Ten Acre Limited did not have a good title to the property.

[101] Mr Guy was unrepresented at that hearing, and the decision refusing leave does not constitute a binding decision of the Court of Appeal by way of precedent, as was pointed out by Lloyd LJ himself in *Guy* v *Pannone LLP* [2009] EWCA Civ 30†, in [28].

[102] Following criticisms of the decision of Mr Mowschenson and of the *dicta* in the Court of Appeal, Mr Guy attempted to renew his application for permission to appeal, and that attempt was heard by a three-judge Court of Appeal last December, when Mr Guy was represented by leading and junior counsel. The attempt failed as being too late, but in the course of his judgment ([2010] EWCA Civ 1396‡), with whom the other two members of the court agreed, Lord Neuberger MR observed as follows:

> 14. In order to understand the arguments advanced on behalf of Mr Guy, it is necessary to explain briefly the relevant law relating to the substantive issue which Mr Guy wishes to raise on his projected appeal. As the Bank's application for judgment was under CPR 24, the facts were assumed to be as Mr Guy contended, and, at least before Carnwath and Lloyd LJJ, it was accepted that, on this basis, the purported transfer to TAL was void (although Mr Nugee has made it clear on behalf of the Bank that, if this application succeeds, and the judgment of Mr Mowschenson is set aside, Mr Guy's case on the facts is likely to be challenged).
>
> 15. Even assuming that the Transfer was void, the Bank's case is and was that, as the Charge was nonetheless a valid and genuine charge executed by the registered proprietor of the land, TAL, to secure monies owing to the Bank, and as the Charge was duly registered by the Bank, the Bank's rights as a registered chargee are unassailable by Mr Guy.
>
> 16. There is no doubt that, under the provisions of the Land Registration Act 2002, registration is normally conclusive in the sense of irrebuttably establishing title (see eg sections 23 to 26 and 58). However, paragraph 2(1)(a) of schedule 4 to the Act enables the court to "make an order for alteration of the register for the purpose of… correcting a mistake." The central issue is, therefore, whether Mr Guy could get the registration of the Bank as chargee in the charges register relating to the Land removed under the provisions of this paragraph.
>
> 17. This issue can fairly be said to turn on whether the concept of "correcting a mistake" is given a wide or a narrow meaning. [Lord Neuberger then cited passages from the judgment of Lloyd LJ in [2008] EWCA Civ 452].
>
> …
>
> 20. Mr Stewart argued that the question whether Mr Guy could invoke paragraph 2(1)(a) of schedule 4 to the 2002 Act for the purpose of removing the registration of the Charge raised a difficult point of law, for which permission to appeal should have been given.
>
> …
>
> 22. … this is not a case where the court proceeded on 9 April 2008 in ignorance of any relevant cases, articles or textbooks. As Lloyd LJ said at [2008] EWCA Civ 452, para 22, there were differing views expressed on the point at issue in the then-current editions of two leading textbooks, *Ruoff & Roper on Registered Conveyancing* (which, as Lloyd LJ said, "took the view that the reading of 'a mistake' is narrow and the concept of mistake is restricted to mistake in registration"), and *Emmet and Farrand on Title*, (which, he said, "expresses a different view"). Lloyd LJ also considered an interlocutory decision of His Honour Judge Peter Langan QC, *Fretwell* v *Graves*, handed down in the Leeds District Registry of the High Court on 16 March 2005, in which the point was held to be arguable either way.
>
> 23. Nor is this a case where the applicant is able to refer to any subsequent decision of this court or of the Supreme Court in which it has been suggested, let alone held, that the conclusion reached by Lloyd LJ was wrong. The material we were referred to by Mr Stewart which post-dates the 9 April 2008 hearing consisted of a more recent edition of *Emmet and Farrand* (in which the editors adhere to their view and criticise the decision of 9 April 2008), a decision of Mr Owen Rhys, a Deputy Adjudicator of the Registry, in *Ajibade* v *Bank of Scotland PLC* (REF/2006/0163/0174, 8 April 2008), and a concession made by the Chief Land Registrar in the High Court Registry, *Odogwu* v *Vastguide*. Not only do those documents not contain anything which is binding on this court, but they do not include material which seems likely to have taken matters further than they were on 9 April 2008.

[103] After explaining why the appeal could not be reopened, Lord Neuberger continued, in [33] to [35]:

> 33. It is fair to say that Mr Stewart has advanced Mr Guy's case, in so far as it relies on para 2(1) of schedule 4 to the 2002 Act, much more strongly than it appears to have been argued on 9 April 2008, and on at least one basis which does not seem to have been raised on that date.
>
> 34. The skeleton argument produced on behalf of Mr Guy for the 9 April 2008 hearing seems, albeit impliedly, to concede the point which he is now raising, and really puts his case on the alleged blind eye knowledge which the Bank had of the wrongful registration of TAL as proprietor. On that basis, it appears to me that, on 9 April 2008, Lloyd LJ could have been forgiven for assuming that the point now raised was not being strongly advanced.
>
> 35. It also seems clear that Lloyd LJ's analysis proceeded on the basis that the alleged "mistake" for the purposes of para 2(1) of schedule 4 to the 2002 Act was the registration of the Charge in the charges registers. However, there are other ways of putting Mr Guy's case, namely (a) that the removal of his name from the proprietorship register was a mistake, and, in order to correct that mistake, the Charge would have to be removed from the charges register, or (b) that the registration of the Charge flowed from the mistake of registering the Transfer, and therefore should be treated as part and parcel of that mistake.

[104] A few weeks after the decision of Mr Mowschenson in *Barclays Bank* v *Guy,* and well before the hearing of the initial application for permission to appeal, a similar issue arose in *Ajibade*, to which I have already referred, and to which reference was made by Lord Neuberger in the passage quoted above. In that case, the applicant had been the registered proprietor of a property. By means of a forged power of attorney, the applicant's sister had transferred that property to her own husband, who was registered as proprietor. The transfer was a purported sale, partly funded by an advance from a bank, which was duly registered in the charges register. The husband then borrowed further money from the second respondent, Endeavour Personal Finance Ltd, and that advance was secured by a second charge that was also registered in the charges register. The applicant sought that she should be reinstated as proprietor and that both charges should be removed from the register. Once it was established that the transfer to the forger's husband was a nullity, there appears to have been no question but that the applicant should be restored as proprietor. In addition, the Bank of Scotland, the first chargee, conceded that the entry relating to its charge should be cancelled, that charge having been granted before the transfer to the husband was registered. The second chargee, however, contended that its charge should remain on the register because there was no mistake in its registration.

[105] In [4] of his decision, Mr Rhys set out, and dealt with, the contentions of counsel for the second respondent as follows:

> Endeavour's argument is as follows. It contends that at the time its charge was created and registered, Mr Abiola was in fact the registered proprietor, with all that that entails. In particular:
>
> 4.1 he was at that time deemed conclusively to be the proprietor of the legal estate in the property under section 58 of the 2002 Act
>
> 4.2 in that capacity he had the power to charge the estate at law with the payment of money — see sections 23(1) and 24 of the 2002 Act

* Editor's note: Reported at [2008] 2 EGLR 74; [2008] 25 EG 174

† Editor's note: Reported at [2009] 07 EG 90 (CS)

‡ Editor's note: Reported at [2011] 1 WLR 68

4.3 this was so irrespective of whether he was registered as proprietor by virtue of a mistake — see section 26 of the 2002 Act.

For my part, I entirely accept the propositions at 4.1 and 4.2. However, I must question whether section 26 of the 2002 Act does in fact support the proposition at 4.3. Section 26 is entitled "Protection of disponees" and is in these terms:

"(1) Subject to subsection (2), a person's right to exercise owner's powers in relation to a registered estate or charge is to be taken free from any limitation affecting the validity of a disposition.

(2) Subsection (1) does not apply to a limitation —

(a) reflected by an entry in the register, or

(b) imposed by, or under, the Act.

(3) This section has effect only for the purpose of preventing the title of a disponee being questioned (and so does not affect the lawfulness of a disposition)."

5. Clearly, section 26 does not in terms refer to registration as a result of a mistake: it refers to "limitations affecting the validity of a disposition". In my view, this expression is not apt to cover a situation where the owner has been registered as a result of a mistake, because "limitations affecting the validity of a disposition" more naturally refers to express or implied limitations which fetter an owner's powers — typically, the requirement to obtain the consent of a beneficiary or other person to a trustees' sale. Furthermore, the power to alter or rectify the register under Section 65 expressly applies in cases of mistake — see Schedule 4 paragraph 5(1)(a). Accordingly, section 26 does not in my view assist at all in determining whether Endeavour's charge should be removed. All it purports to do is to protect a disponee where he or she has taken an interest in land as a result of a disposition which is beyond the powers of the "disponors" because their powers are subject to specified limitations. In effect, it supplements the provisions of sections 23 and 24, but has no bearing, it seems to me, on the power to correct a mistake in the register.

6. Mr Greville Healey, who appeared for Endeavour, supported the arguments outlined above by reference both to the Law Commission Report Number 271 ("the Report") which gave rise to the repeal of the Land Registration Act 1925 ("the 1925 Act") and the enactment of the 2002 Act, and to a passage in *Ruoff & Roper's Registered Conveyancing*. With regard to the Report, he cited two particular passages, at 9.4 and 4.10. Paragraph 9.4 reads as follows:

"One of the most fundamental principles of registered conveyancing is that it is registration that vests the legal estate in the registered proprietor. Clause 58(1) of the Bill provides accordingly that if, on the entry of a person in the register as the proprietor of a legal estate, the legal estate would not otherwise be vested in him or her, it shall be deemed to be vested in him or her as a result of registration. Thus, for example, if a person is registered as proprietor on the strength of a forged transfer, the legal estate will vest in that transferee even though the transfer was a nullity".

This passage is incontrovertible and uncontroversial — it expresses the underlying principle of land registration as embodied both in the 1925 and 2002 Acts, that which has been described as "the statutory magic". However, this passage in itself does not take the matter any further, since although a person may become the legal estate owner through registration where he would not have done so in unregistered conveyancing, this registration is still subject to the Court's and the registrar's curative powers under Section 65 and Schedule 4. This passage does not — and could not — claim that a person once registered as a proprietor is safe from *any* challenge to his title, since all registrations are subject to Section 65 and Schedule 4.

7. Mr Healey also referred me to paragraph 4.10 of the Report, which is to be found in the section which explains the rationale for Clause 26 — which became section 26 of the 2002 Act — which deals with "Protection for disponees" and which is set out in full above. Interestingly, the specific example given by the Law Commission to illustrate the operation of Clause 26 is the case where:

"(1) W and X held land on a bare trust as nominee for Y, on terms that they could not make any disposition of land without Y's written consent;

(2) Y, who was in actual occupation of the land held in trust, did not protect her interest by the entry of a restriction; and

(3) W and X fraudulently charged the land to Z without Y's consent in breach of trust."

The Law Commission states that in these circumstances "Z's charge would be valid and could not be called into question by Y". This is entirely right, and consistent with my understanding of the operation of section 26 of the 2002 Act, namely that it is directed to a disposition that would otherwise be invalidated by a failure to comply with the terms of the power of disposal. Again, however, it does not assist in understanding the scope of the power to correct a mistake.

[106] Mr Rhys then set out, in [8], the arguments for the second respondent that "the original mistake the registration of Mr Abiola pursuant to a forged disposition is no longer operative once he has become registered as proprietor. Although he was first registered owing to a mistake, once he is registered he is entitled to deal with the property in the same way as any registered proprietor. Since he was entitled to charge the property… there is no operative mistake in respect of which the registrar may exercise his powers." Counsel for Ms Ajibade, on the other hand, contended that fraud unravelled all and that the power to alter the register for the purpose of correcting a mistake was not to be limited so as to exclude the removal of the second charge.

[107] Mr Rhys drew attention to the scope of the right of the unsuccessful party to apply for an indemnity under Schedule 8 to the 2002 Act, but held that this could not apply to Ms Ajibade if, as the second respondent contended, the registration of the second charge was not a mistake and there was not even the possibility of rectification. He continued:

11. The case turns, quite simply, on the meaning of the phrase "correcting a mistake…" as it appears in Schedule 4 paragraph 5. … In reaching my conclusion I have regard to the following points (among others):

11.1 Although obviously the underlying purpose and intention of our system of land registration is to make a title, once registered, as secure as possible, it has always been recognised that there must be a power in the Court and the registrar to rectify the register where a mistake has been made. A registered title is not, therefore, absolutely guaranteed by the State — it is always subject to rectification on proper grounds.

11.2 However, where rectification has taken place, or where it has been refused notwithstanding the existence of a mistake, indemnity may be payable by the Land Registry to compensate for any loss. In the present case, however, if the arguments of Endeavour are accepted, it would appear that indemnity is not payable to the Applicant if rectification is refused.

11.3 The phrase "correcting a mistake" in Schedule 4 is not defined.

11.4 I have found that Mr Abiola was registered as proprietor due to a fraud on the Applicant. Mr Abiola did not object to the rectification, has taken no part in this adjudication and it is entirely possible that he was party to the fraud carried out by his wife.

12. In one sense, the arguments addressed to me by both the Applicant and Endeavour have been focused on the wrong word — namely "mistake". The Applicant contends that the registration of Endeavour is, in effect, a mistake. Endeavour contends that is not. However, it is manifest, and not disputed, that there was an operative mistake when the Land Registry registered Mr Abiola as proprietor of the Property. The real question, it seems to me, is to decide how far the registrar can go in "correcting" that mistake. It is accepted that Mr Abiola should be removed from the register and the Applicant replaced as proprietor in the Proprietorship Register. It is also accepted that Halifax Plc should be removed from the Charges Register. Would these steps entirely correct the original mistake? The Property would still remain subject to the second charge created by the fraudulently-registered Mr Abiola, and the Applicant would regain her title but encumbered by this charge. It does not seem to me that, in such circumstances, the mistake has been fully corrected. Certainly, all the consequences of the mistake would not have been corrected, even though it would be open to the registrar to [do] so by rectifying the Charges Register by removing Endeavour's charge. In my judgment, there is no basis for construing the expression "correcting a mistake" in such a restrictive way. It seems to me that this formulation is intended to give the registrar wide scope to exercise the discretion to rectify in a fair and just way. It would have been possible to have imposed express limitations on the power in the sense contended for by Endeavour if this had been the intention — as indeed limitations have been imposed in regard to proprietors in possession (see Schedule 4 Paragraph 6). It seems to me that it would be perverse to limit the registrar's power to rectify the register to the correction of only one consequence of the mistake, leaving uncorrected the other direct consequences of the original mistake. Protection for parties in the position of Endeavour is to be found both in the undoubted requirement that the registrar should exercise his power in a fair and just way, and in the existence of an indemnity from the Land Registry.

13. It may be argued that the ability of the registrar to correct not just the original mistake but also the consequences of a mistake could lead to an undermining of the "sanctity" of a registered title. It might be said that this power could be used to rectify the register against purchasers at several removes from the original fraudster. However, in practice this would not be a problem in my view. In particular, the restrictions contained in Paragraph 6 in respect of proprietors in possession operate to limit very substantially the ability to rectify against successors in title. Furthermore, there is no suggestion that the more numerous grounds for rectification of the register as set out in section 82 of the 1925 Act undermined confidence in the operation of the land registration system under the previous regime. The power to rectify will necessarily be used

A sparingly, but the exercise of the power is a matter for the Court or the registrar. It may be noted that the Court of Appeal in *Argyle Building Society* v *Hammond* (1985) 45 P&CR 148 made it clear that the power to rectify under the 1925 Act was not to be construed restrictively, but was a discretion to be exercised in accordance with all relevant circumstances. In doing so the court expressly disagreed with the analysis in the then current edition of *Ruoff & Roper* which — as is the case today — sought to construe the power to rectify under section 82 in a restrictive way.

[108] Despite the fact that counsel appears not to have drawn Mr Rhys' attention to the passages in the two Law Commission reports that I have had cited to me, Mr Rhys' approach does, in my
B view, accurately reflect the intended effect of para 5 of Schedule 4 as explained in those passages. It also draws attention to the potential result of the construction put forward on behalf of the second chargee that the original owner, deprived of his title by the effect of the legislation, could get neither rectification nor indemnity. Mr Rhys did not have Mr Mowschenson's decision in *Barclays Bank* v *Guy* cited to him, but the issue before him was not addressed in that case, and it is difficult to see how it could have assisted him beyond drawing his attention to the passage cited from *Ruoff & Roper* on which I have already commented.

[109] The next decision is that of Sir Donald Rattee in *Odogwu*
C v *Vastguide Ltd* [2008] EWHC 3565 (Ch), a case where the Chief Land Registrar was added as a defendant. In that case, the claimant, who lived in Nigeria, was the registered proprietor of a property. It was alleged that a fraudster had forged the claimant's signature on a charge to secure a loan to the fraudster of some £750,000. The charge was registered. In due course, the chargee appointed a receiver and sold the property to Vastguide in the exercise of its power of sale. Following the sale, the charge was removed from the register. Vastguide was registered as proprietor and there was also registered a charge granted by Vastguide to the second defendant, Investec Bank UK Ltd, to secure the repayment of sums loaned to it by that bank. The claimant applied to rectify the register pursuant to Schedule 4 to the 2002 Act to remove
D Vastguide as proprietor and the bank as chargee. A preliminary issue was directed as to whether the court had jurisdiction to remove the registration of Vastguide and its chargee on the assumption that the original charge was a forgery.

[110] The preliminary issue was never heard because all the defendants conceded that the court had jurisdiction to rectify the charges register. At the trial, Vastguide sought to resile from this concession, but was held to be estopped from doing so. Investec was not represented at the trial following a concession by the claimant that if the claim against Vastguide failed, the claim against Investec would also fail. Investec did not seek to resile from the concession as to jurisdiction, and was
E not represented at the trial.

[111] Vastguide accepted that it was estopped from contending that there was no jurisdiction to rectify, but sought to contend that the jurisdiction arose not because the entry of Vastguide on the register was a mistake but because the jurisdiction under para 2 of Schedule 4 to the 2002 Act included jurisdiction to make an order for the alteration of the register for the purpose of correcting not only a mistake on the register, but the consequences of such a mistake: see [50] of the decision. The judge held that the concession was that the registration of Vastguide was a mistake, and that it could no longer, therefore, contend that the jurisdiction arose on another basis: see [54].

F [112] The judge went on, however, to consider the new argument in case he was wrong as to the estoppel point. He continued as follows:

> 56. The relevant jurisdiction of the court is that conferred by paragraph 2(1)(a) of Schedule 4 to the Act, namely "to make an order for alteration of the register for the purpose of correcting a mistake." In my judgment the reference in that quotation from paragraph 2(1) to a mistake must be to a mistake on the register, so that it can be corrected by an alteration of the register. I accept the submission of counsel for the Claimant that a fundamental objection to Vastguide's argument is that the Credit & Mercantile charge, which Vastguide argues is the relevant mistake, is no longer on the register. It was deleted on completion of the sale to Vastguide when Credit & Mercantile was paid its debt.
>
> 57. Counsel for Vastguide sought to overcome this initial obstacle by submitting that, for this purpose, the register is not limited to the current edition of the register showing currently effective entries, which admittedly does not show the charge, but extends to the records still held by the Land Registry showing all previous entries, including those since deleted. I find this argument misconceived. In my judgment the reference in paragraph 2 of schedule 4 to the Act to a mistake is to a mistake appearing on the current edition of the register which shows subsisting entries. Only such a conclusion makes sense in the context of the jurisdiction in paragraph 2 to alter [the] register for the purpose of correcting a mistake. I cannot see what purpose would be served by making an alteration to a previous edition of the register to remove an entry such as the Credit & Mercantile charge which has already been deleted.
>
> 58. In my judgment Vastguide's new argument is bad for this reason alone. However, quite apart from this point it does not seem to me obvious that the reference to "correcting a mistake" in paragraph 2(1)(a) of Schedule 4 to the Act can properly be construed as including a reference to correcting the consequences of the mistake, which is the next step in Vastguide's argument. However, counsel for Vastguide relied in this context on a recent decision of a Land Registry Adjudicator in *Ajibade* v *Bank of Scotland plc,* in the course of which the Adjudicator seems to have taken the view that the correction of a mistake by rectification of the register could extend beyond the correction of an entry on the register to include the correction of the consequences of such a mistake. As it is not necessary for me to decide this point in this case in the light of my rejection of Vastguide's argument on the ground that the Credit & Mercantile charge has already been deleted from the register (a point that did not arise in the *Ajibade* case) I do not think it appropriate that I express any conclusion on it.
>
> 59. Thus, even if I had not concluded that Vastguide is estopped from relying on the argument I have just been considering, I would have rejected the argument as misconceived. Counsel for Vastguide accepted that, if he failed on that argument, Vastguide would be estopped by the concession recited in the consent order from denying that the entry in the register of Vastguide as proprietor was a mistake within the meaning of Schedule 4 to the Act, since that would then be the only ground on which the court would have the jurisdiction which Vastguide had accepted by its concession. It is accordingly unnecessary, and therefore inappropriate, for me to express any conclusion on the question whether, as a matter of law, the entry of Vastguide as proprietor is capable of being such a mistake. I proceed to consider the parties' cases on the footing that it is such a mistake which the court has jurisdiction in these proceedings to rectify by removing that entry from the register, subject to the provisions of paragraph 3 of schedule 4.

[113] While not making any findings as to the correctness of the decision of Mr Rhys in *Ajibade*, therefore, the judge concluded that if the registration of Vastguide was not itself a mistake, he would not have held that he had jurisdiction to correct it at a time when the original mistake, the registration of the forged charge, had already been removed from the register. In *Ajibade*, the original mistake was still on the register. The judge's reasoning was that only this made sense in the context of a provision permitting and alteration for the purpose of correcting a mistake, which he construed as meaning a mistake on the current edition of the register. If this is correct, it would involve either a substantial change from the law under the 1925 Act or disagreement with the decision of Blackburne J in *Pinto* that the subsequent registration of the purchaser was also a mistake for this purpose. It would also mean that, if the original mistake had been removed from the register already, and the new entry was not a mistake itself (the judge carefully refrained from expressing any view on this point), there would be difficulties in seeing how the provisions of Schedule 8 to the 2002 Act would entitle the original owner to indemnity, a point that was not considered by the judge. So far as appears from the decision, none of the relevant provisions of the Law Commission's reports, the human rights issues and *Pinto* appear to have been drawn to the judge's attention.

[114] Finally, I turn to a recent decision by another deputy adjudicator, Mr David Holland, in *Stewart*. In that case, the applicant, Ms Stewart, lived abroad. The dispute concerned industrial premises that had been purchased in the name of Ms Stewart, but were occupied by her half-brother under a family arrangement. The brother claimed that Ms Stewart held the property as bare trustee for him. A solicitor drafted a form TR1 by which, on its face, Ms Stewart appointed the brother as joint trustee and Ms Stewart and her brother then transferred the property to the brother alone. Although what purported to be Ms Stewart's signature appeared on the TR1, it was common ground before Mr Holland that it was a forgery. The solicitor, in ignorance

of the forgery, sent the TR1 to the Land Registry and the brother was registered as sole proprietor of the property on 21 November 2005, on which date a charge was also registered in favour of Lloyds TSB to secure repayment of £18,000.

[115] In May 2008, the brother redeemed the original charge, which was removed from the register, and charged the property to the respondent to secure repayment of a loan of over £57,000. He also executed a further charge on 6 June 2008 to secure repayment of further advances. The respondent applied to register the first of these charges on 19 May 2008. On 7 June 2008, the brother was killed in a road accident. The application to register the second charge was made on 10 June 2008. Objection was made to the registration of the charges, initially by the father of Ms Stewart and her brother, and in due course Ms Stewart applied to alter the register to remove her brother as registered proprietor and to restore her name as such. The matter was referred to the adjudicator and came on for hearing, when Ms Stewart was represented by one of the administrators of the estate of her deceased brother (a non-legal representative) and the respondent was represented by counsel. There was no dispute that Ms Stewart was entitled to be restored to the register. The issue was whether the respondent's charges ought to be registered and to remain on the register.

[116] Mr Holland found that the brother was intended to have a substantial beneficial interest in the property but that it had been put into Ms Stewart's name to prevent him disposing of it without the consent of Ms Stewart and their father. He did not decide if she was a bare trustee, but did find that if she had an interest, then she and the brother were tenants in common, so that no question of survivorship arose.

[117] Mr Holland went on to find that he should at least notionally give effect to the respondent's application to register the charges, with the result that Ms Stewart, in order to remove the charges, had the burden of bringing herself within the terms of Schedule 4 to the 2002 Act. He continued, in [51], by setting out the submissions of counsel for the respondent:

> Miss Tozer's principle submission is that, by paragraph 5 to Schedule 4 to the act, the Registrar only has (relevantly) power to alter the register to correct a "mistake". The registration of the Respondent's charges was not a mistake therefore the entries cannot be removed. She submits that the effect of the provisions of the act set out above, particularly section 58, is to vest in the son both the title and the power to create a valid charge. He was the registered proprietor at the date the charges were created. He remained and remains the registered proprietor until a successful application is made to alter the register. Thus at the date they were created and (at least notionally) registered the charges were lawfully created. Indeed the Registrar could not have refused to register them. Thus, she submits, the effect of the act is that, even though the registration of the son as freehold proprietor could be said to be a mistake, any subsequent dispositions by him are deemed to be lawful and the registration of them cannot be a mistake.

[118] Mr Holland then referred to passages from *Ruoff & Roper* and from another textbook that supported counsel's contentions and continued, in [53]:

> More importantly this view derives support from the decision of Mr Terence Mowschenson QC sitting as a Deputy Judge of the Chancery Division in *BARCLAYS BANK* V *GUY* [2008] EWHC 893 (Ch). The relevant facts of that case are very similar to those of this case save that it was alleged that the transfer which resulted in the registration of the person who then charged the freehold was said to have been procured by some form of fraudulent misrepresentation or undue influence such that it was at law voidable rather than void (the latter of which is the effect at law of a forged transfer such as one has in this case).

[119] Mr Holland then continued by quoting from [17] to [20] of that decision. I have already cited the relevant passages. However, he ends with only a portion of [20] reading: "It matters not whether the transfer is ultimately found to be fraudulently procured or whether the transfer is void or voidable …". This omits the initial sentence of that paragraph: "The claimant contends that for the purpose of a rectification claim no further investigation is required or is relevant." If there is any doubt as to whether the sentence quoted is more than a recital of counsel's submissions and not a finding by the judge, that doubt is dispelled by [27], where the judge says that in the light of his holding that the transfer was voidable and not void, he did not need to address the question of whether the transfer, the charge, would have been effective if the transfer was void. It is abundantly plain that the judge, who essentially based his decision on *Norwich Building Society*, was not making any findings as to the position if the original transfer had been void and not voidable. Mr Holland is plainly wrong in finding that this decision assisted in any way in relation to a forged transfer.

[120] Mr Holland then went on to consider the views of the Court of Appeal when permission to appeal was first sought by Mr Guy, while acknowledging that, as pointed out by Lloyd LJ in *Guy* v *Pannone*, it could not be cited and was not authority, a proposition that, it is apparent from [57] of his decision, caused him some difficulty. In the end, however, he concluded that the most that could be said was that there was nothing in either judgment of Lloyd LJ that would indicate that he thought that the approach and conclusions of Mr Mowschenson at first instance were incorrect.

[121] After finding that other cases including *Odogwu* were not relevant, he went on to consider *Norwich Building Society* and *Pinto*, but only in relation to the findings in both cases that under the 1925 Act: (1) the registration of the voidable transaction and the accompanying charge in *Norwich Building Society* were not mistakes and could not be brought under section 82(1)(*h*) of that Act; and (2) the registration of the purchaser from the fraudster in *Pinto* also did not fall under that subsection. He continues, in [65]: "Thus both of these cases support the view that a narrow interpretation equivalent to that favoured by the Respondent was placed on the equivalent provision in the 1925 legislation."

[122] This ignores the analysis of Scott LJ, at p134G of the report of *Norwich Building Society*, which I have cited above, which points out that all the provisions of section 82(1)(*d*) to (*h*) deal with the correction of errors. It also ignores the provisions in the Law Commission and Land Registry reports that make it clear that para 2(1)(a) of Schedule 4 to the 2002 Act was intended to cover all the cases of mistake in section 82(1). It was not an equivalent provision to section 82(1)(*h*) alone. In this respect also, I am satisfied that Mr Holland was wrong in his approach.

[123] In [72], Mr Holland concluded that he had two conflicting first instance decisions, *Guy* and *Ajibade*, and that he had to decide which to follow. This is also plainly wrong for two reasons: the decisions do not conflict, for the reasons I have given; and if they had conflicted, then, subject to the usual exceptions under the doctrine of precedent, he would have been obliged to follow the decision of the High Court, the court to which appeals lie from the adjudicator. In my judgment, a decision of the High Court at first instance is as much binding on the adjudicator as a decision on appeal from the adjudicator.

[124] Mr Holland then goes on to give his reasons for not following the decision of Mr Rhys in *Ajibade*. In [73], he states:

> It seems clear to me that the effect of sections 58, 23, 24 and 132 of the act is that the son, as the person entered on the register as the registered freehold proprietor of the unit, could lawfully charge the freehold interest. The clear intention of Parliament was to give owner's powers to the person who happened to be registered as freehold proprietor even if this registration was as a result of fraud. The son remained the registered proprietor until a successful application was made to alter the register. Thus at the date they were created and (at least notionally) registered the charges were lawfully created. Indeed the Registrar could not have refused to register them. As such it is difficult to see how the registration of the charges could be described as a "mistake". They were lawfully created and lawfully registered. As is clear from paragraph 1.5 of the Law Commission report (cited above), the fundamental objective of the act was that the register should be a complete and accurate reflection of the state of the title to land at any given time. This was intended to facilitate the move to electronic conveyancing. It was intended to allow investigation of title on line "with the absolute minimum of additional enquiries and inspections". To allow ex post facto removal of entries which were at the time lawfully made to reflect lawful dispositions would, it seems to me, undermine this intention. Section 58 reflects a balance struck by Parliament which preferred certainty of title over the property rights of those who had been the victims of fraud. They are to have their remedy by way of indemnity under Schedule 8.

[125] The principal objection to this approach is that there were equivalent provisions to sections 58, 23, 24 and 132 in the 1925 Act, and much had been said in relation to that Act about the sanctity of the register. Nevertheless, the statutory magic had been held in *Argyle* to be subject to the provisions as to rectification. Further, para 1.5 of the Law Commission report should not be read in isolation. The provisions of the report, and of the consultative report, to which I have already referred in detail, clearly show that no substantial changes were intended in this respect by the Law Commission.

[126] Mr Holland suggested that the wronged original owner would have his remedy by way indemnity under Schedule 8. That suggestion was also reflected in the submissions made to me by Mr Jones for Roberto Mac. One problem with that is that it would not accord with the intentions of the Law Commission (see, in particular, the consultative report, para.8.40(1) to which I have already referred, as to when indemnity was to be available). Also, as Mr Gore pointed out for Knights Construction, para 8(3) provides that "No indemnity under sub-paragraph (1)(b) is payable until a decision has been made about whether to alter the register for the purpose of correcting the mistake; and the loss suffered by reason of the mistake is to be determined in the light of that decision." This would presumably involve an application to the registrar to decide whether to alter the register, even though it was clear to all that there was no power for him to do so. There would have to be, for the purposes of para 1(1)(b), a loss suffered by the applicant by reason of a mistake whose correction would involve rectification of the register. But if the only mistake was the original registration of the fraudster, and the correction of the mistake would involve rectification of the register for the purposes of Schedule 8, it becomes difficult to see why, under Schedule 4, there should be no power to rectify the register for the purpose of correcting that mistake.

[127] Although such an artificial construction may be possible, if there were no jurisdiction to rectify under Schedule 4, to ensure that an innocent original owner did not lose his property as a result of the provisions of the legislation concerning registered land without compensation, it seems to me that such a construction could be justified only under the extended rules of construction for the purposes of ensuring compliance with the Human Rights legislation as stated by the House of Lords in *Ghaidan* v *Godin-Mendoza* [2004] UKHL 30; [2004] 2 AC 557*.

[128] Further, as was pointed out by Mr Morshead, if the approach favoured by Lloyd LJ, Sir Donald Rattee and Mr Holland were correct, it would mean that if a fraudster forged a transfer to himself from the true owner, became registered as proprietor and then sold on, and the purchaser was then registered in place of the fraudster, rectification would not be possible because there was no longer the original mistake on the register to correct. Yet if the forger forged a transfer from the original owner to the same purchaser, who then became registered, that would be the original mistake on the register and rectification would be possible. No rational basis for such a distinction has been suggested to me. Further, that would seem to be the case on that interpretation even if the subsequent purchaser were on notice of the original forgery, at least if he were not a party to it and there were no notice or restriction on the register, since he would be entitled to rely on the integrity of the register. Mr Morshead also suggested that the same would be the position if the purchaser were the fraudster's accomplice, but, in those circumstances, it appears to me that the second fraudster would hold the property on trust for the original owner absolutely.

[129] So too, in the present case, there is still an error on the register in that property has been registered that ought not to have been registered. However, if, instead of a new title being created, either because, as here, the land was previously unregistered, or because it was being transferred out of a larger plot of land registered under a different title, the whole of an existing title was being transferred, there would, on those arguments, be no remaining mistake to be corrected. I can see no logic in such a distinction and no reason to attribute to parliament an intention to limit the right to rectification in this way, rather than in accordance with the proposals in the report on which the legislation is based.

[130] By contrast, there is nothing irrational in the system that operated under the 1925 Act and that the Law Commission intended should continue to operate under the 2002 Act.

Conclusion as to the construction of Schedule 4 to the 2002 Act

[131] As Lord Neuberger pointed out, in [35] of his judgment in round 3 of *Barclays Bank* v *Guy*, there are various ways of approaching the construction of para 2(1)(a) of Schedule 4 to the 2002 Act, and various approaches, none of them binding on me, have been put forward in the cases to which I have referred. I am satisfied that I can and should construe the provision in a manner that gives effect to the intention of the Law Commission in its bill and that ensures that where a person is deprived by legislation of property to which they would otherwise have title, then they should be compensated appropriately, at least if they are not at fault. That is so whether the person deprived is the original owner or the subsequent purchaser of registered land who finds that his title has been removed by rectification, and whether there is an error as to the amount of land included in the title. Any other result would fall foul of Article 1 of Protocol 1 of the Human Rights Convention unless the provisions of Schedule 8 were construed in a manner that would stretch them almost to breaking point. Schedule 4 and Schedule 8 need to be read together for this purpose.

[132] I am therefore satisfied that the remedy of rectification is available in the present case to Knights Construction. It would be so available whether, adapting the two possible interpretations suggested by Lord Neuberger: (a) the original registration of the Salvation Army was a mistake, and, in order to correct that mistake, which here persists, the register should be corrected by removing this part of the land, which should never have been registered at all, from the title; or (b) the registration of Roberto Mac as proprietor of the land flowed from the mistake of including the land in the original title, and therefore should be treated as part and parcel of that mistake. It would also be available if Blackburne J is correct in *Pinto* in treating the registration of the second transfer as a mistake. It is unnecessary for me to decide which of these approaches is the correct one, as they produce the same result. Even if there had to be an existing mistake in the narrow sense suggested by Lloyd LJ, it would exist in the present case by reason of there being more land included in the title than ought to have been included. I would not, however, wish to base my decision on such a distinction but rather on a proper construction of para 5(a) of Schedule 4 in accordance with the intentions of the Law Commission and the Land Registry.

Are there exceptional circumstances that militate against rectification?

[133] On the basis that rectification is available to Knights Construction, Mr Jones contended that there are exceptional circumstances in the present case that mean that I should refuse to order rectification. He said that the land has practical utility to Roberto Mac as it needs it in conjunction with the old chapel and that that practical utility is such that an indemnity under Schedule 8 would not provide adequate compensation. He also pointed out that Knights Construction has neglected the land over the years. It has not actually occupied the land or received rent and profits in respect of it. User of the land is by the lessees of the flats and shops and those rights are protected as overriding interests by Schedule 3 of the 2002 Act.

[134] Mr Gore responded that there are no exceptional circumstances. The car park had not been maintained, but the tenants had no complaint about this. They could still use it to park their vehicles, and had had no maintenance charges. It had not been abandoned by Knights Construction. Its officers had used it to park when they wished, and also to access the garage/store under flat 1. It had responded promptly when it found out that Roberto Mac was seeking to take it over and had safeguarded its tenants' rights. Roberto Mac was in no way misled by the lack of maintenance. If it had been better maintained, Roberto Mac would simply have assumed that the maintenance had been carried out by the Salvation Army. Mr Gore also refers to the failure of Roberto Mac to make enquiries locally or to query the boundary of the car park

---

* Editor's note: Also reported at [2004] 2 EGLR 132

and the highway, the difficulty of assessing compensation payable to Knights Construction, and the difficult position of the tenants in ascertaining and enforcing their rights.

[135] As to the rights of the tenants of the flats and shops over the land, if it remains in the ownership of Roberto Mac, Mr Gore referred me to the decision of the Court of Appeal in *Nevill Long & Co (Boards)* v *Firmenich & Co* (1984) 47 P&CR 59*, as establishing that Roberto Mac would take the disputed land subject to the tenants' rights. That was a case in which the landlord had sold land over which he had granted tenants a right of way. The Court of Appeal concluded that, notwithstanding the severance of the reversions, there continued in existence under each of the leases a single tenancy of the property thereby demised including the right of way.

[136] It does not appear to me that that addresses the problem in the present case, which is that section 29 of the 2002 Act provides that the disposition, when completed by registration, has the effect of postponing to the interest under the disposition any interest affecting the estate immediately before the disposition whose priority is not protected at the time of registration. So far as the various tenants are concerned, their rights would be protected only in accordance with Schedule 3 and, where applicable, para 9 of Schedule 12 to the 2002 Act. If there are any services running under the ground for the benefit of the demised flats, or the demised shops, it is highly probable that they will be protected by para 3(2) of Schedule 3. That protection will also apply to parking rights and rights of way in each case where the right has been exercised in the period of one year ending with the day of the disposition. If there is any legal or prescriptive easement to keep the bins on the disputed land for the benefit of any of the properties, it appears to me that that
also would be protected, as the presence of the bins used by the tenants D
was apparent on an inspection of the land. It is not appropriate for me to make any findings as to the extent to which the tenants' rights survive, but it would appear that most of them would survive.

[137] It is normal for the purchaser of land to have a use for it, and particularly where it is only part of the land that may be removed from the title, it is common for that land to have a value to the purchaser in conjunction with the remainder of the land acquired that can be difficult to put a monetary value on, and that may exceed, for that purchaser, any compensation payable. In addition, while Knights Construction do not receive any rent for the land, it is being used as a car park for the rest of
their property. Their tenants have rights over it under their leases, and E
some of them may lose those rights if the property is not restored to Knights Construction. Others may have greater difficulty in benefiting from the rights given to them. There will also be problems in dealing with rights under the leases where the rights are over land that Knights Construction has ceased to own. Informal assurances as to parking have also been given to the lessees of flats 1 and 15 by Knights Construction, which could be prejudiced, and the ability of Knights Construction in future to let the shops will also be affected if parking rights cannot be granted at the same time. Finally, there was some evidence of limited user by Knights Construction itself, and if there is spare parking space,
that could be commercially let by it. F

[138] I am therefore satisfied that there are no exceptional circumstances and that the register should be rectified to exclude from it the disputed land.

Application allowed.

---

* Editor's note: Also reported at (1983) 268 EG 572

TOWN AND COUNTRY PLANNING

Milebush Properties Ltd *v* Tameside Metropolitan Borough Council and another

COURT OF APPEAL

[2011] EWCA Civ 270

17 March 2011

MUMMERY, MOORE-BICK AND JACKSON LJJ

Town and country planning — Planning obligation — Enforcement — Declaratory relief — CPR 40.20 — Planning permission for office development — Planning agreement under section 106 of Town and Country Planning Act 1990 requiring grant of rights of way to owners of neighbouring properties — One owner seeking declaration as to obligation to grant right of way and proper construction of section 106 agreement — Whether appropriate to grant declaratory relief to non-party to section 106 agreement

In 2000, the respondents acquired, as an investment, a site in Uxbridge with planning permission for an office development and associated car parking. The appellant owned properties on the High Street nearby. The respondents' planning permission contained a condition requiring them to construct a service road to provide rear servicing to properties on the appellant's side of the High Street, so as not to prejudice pedestrianisation plans. Under a planning agreement, entered into by their predecessors in title under section 106 of the Town and Country Planning Act 1990, the respondents were to grant to "authorised properties", including the appellant's, a right of way over the service road for rear servicing, exercisable at all reasonable times on such reasonable terms as the respondents might impose.

In 2009, the respondents wrote to the appellant proposing to grant a right of way limited by certain conditions; these included a prohibition on use as a pedestrian emergency exit. The appellant objected to the proposed terms and applied to the court for declarations that the respondents were obliged to grant a right of way in accordance with the section 106 agreement and as to the proper construction of that agreement.

In the High Court, the construction point was decided against the appellant. The judge further held that the case was not an appropriate one for declaratory relief under CPR 40.20 since a planning obligation in a section 106 agreement was enforceable only by the local planning authority and parties deriving title under it, as a public function and subject to judicial review, not in private law proceedings by a party that was not a party to the agreement: see [2010] EWHC 1022 (Ch); [2010] 2 EGLR 93; [2010] 30 EG 64. The appellant appealed.

Held **(Moore-Bick LJ dissenting in part): The appeal was dismissed. (1) Although declarations under CPR 40.20 are limited to the resolution of justiciable issues about legal rights that are the subject of a real and present dispute between the parties, it is not necessary that the legal right should be vested in one of the parties. A declaration can be granted in private law proceedings as to the disputed construction of a document that affects a claimant, even though that claimant is not a party to the document in question. When considering whether to grant a declaration, the court should take into account what is just for both parties, whether the declaration would serve a useful purpose and whether there are special reasons why a declaration should not be granted. (2) It was not appropriate to grant a declaration to the appellant. The dispute between the appellant and the respondents was not a private law dispute concerning the construction of a deed for the grant of a private right of way, or concerning the legal enforcement of an agreement, or even concerning the nature and extent of an easement, Instead, it concerned the planning objectives of a local planning authority (LPA) and the performance of planning obligations on which the decisions of the LPA were important or even paramount. The LPA were the only party entitled to enforce the planning agreement and their decisions and actions in that respect were challengeable by judicial review. The appropriate forum for determining the public law planning matters involved in the case, and in future similar cases, was a judicial review to which the LPA were a party. (3) The section 106 agreement had been made in circumstances where the pedestrianisation of the High Street had created a need for alternative vehicular access to service the rear of the High Street properties. The provision of an emergency pedestrian exit at the rear of the authorised properties had nothing to do with servicing the rear of those properties and fell outside the agreement. Accordingly, it was not included in the required right of way.**

F G H

The following cases are referred to in this report.

Attorney-General ex rel Scotland v *Barratt Manchester Ltd* (1992) 63 P&CR 179; [1992] JPL 148, CA

Bank of Scotland v *A Ltd* [2001] EWCA Civ 52; [2001] 1 WLR 751; [2001] 3 All ER 58

Feetum v *Levy*; *sub nom Cabvision Ltd* v *Feetum* [2005] EWCA Civ 1601; [2006] Ch 585; [2006] 3 WLR 427

Financial Services Authority v *Rourke (t/a JE Rourke & Co)* [2002] CP Rep 14, Ch

Gouriet v *Union of Post Office Workers*; *sub nom Attorney General* v *Gouriet* [1978] AC 435; [1977] 3 WLR 300; [1977] 3 All ER 70, HL J

Meadows Indemnity Co Ltd v *Insurance Corporation of Ireland plc* [1989] 2 Lloyd's Rep 298, CA

Mercury Communications Ltd v *Director-General of Telecommunications* [1996] 1 WLR 48; [1996] 1 All ER 575, HL, [1994] CLC 1125, CA

Milebush Properties Ltd v *Tameside Metropolitan Borough Council* [2010] EWHC 1022 (Ch); [2010] 2 EGLR 93; [2010] 30 EG 64; [2010] JPL 1303

Rolls-Royce plc v *Unite the Union* [2009] EWCA Civ 387; [2010] 1 WLR 318

S (Hospital Patient: Court's Jurisdiction) (No 1), Re [1995] 3 WLR 78; [1995] 3 All ER 290; [1996] Fam 1, CA

This was an appeal by the appellant, Milebush Properties Ltd, from a decision of Arnold J, sitting in the Chancery Division of the High Court, dismissing a claim against the respondents, Tameside Metropolitan Borough Council, for declaratory relief as to the meaning and effect of a planning agreement, entered into by the respondents' predecessor in title with the local planning authority under section 106 of the Town and Country Planning Act 1990, for the provision of rights of way over a service road. K

Mark Warwick (instructed by JE Kennedy & Co Solicitors) appeared for the appellant; Judith Jackson QC and Alan Johns (instructed by Eversheds LLP, of Manchester) represented the respondents.

A Giving judgment, **Mummery LJ** said:

Introduction and background

[1] This appeal raises a point of some interest on the availability, in private law proceedings between non-contracting parties, of a declaration on the meaning and effect of a planning obligation in a deed made pursuant to provisions in the planning legislation.

[2] Section 106 of the Town and Country Planning Act 1990 (the 1990 Act) governs the enforcement of "planning obligations" entered into "by agreement or otherwise" by any party interested in land in the area of a local planning authority. The obligations, in what are commonly called "section 106 agreements", are "enforceable to the B extent mentioned in subsection (3)", which provides that:

> (3) Subject to subsection (4) a planning obligation is enforceable by the authority identified in accordance with subsection (9)(d) —
> (a) against the person entering into the obligation; and
> (b) against any person deriving title from that person.

[3] I underline two features of the statutory framework governing the extent of enforceability of a section 106 obligation: first, the subject matter of enforcement is described as "a planning obligation"; and, second, the planning authority is identified as the person by which that obligation is enforceable against the person entering into the obligation, C or that person's successor.

[4] Subsection (9) provides that a planning obligation may not be entered into except by an instrument executed as a deed conforming to certain specified requirements. Those requirements are satisfied in this case.

[5] This section 106 agreement is dated 15 February 1991. It was entered into between the local planning authority, Hillingdon Borough Council (Hillingdon), which are also the highway authority, with developers of a freehold office development on the site, London & Metropolitan Estates Ltd and Traffichold Ltd. A condition of the planning permission required the construction of a service road. The D objective was to provide access for the rear servicing of the High Street properties as part of Hillingdon's pedestrianisation proposals for Uxbridge High Street.

[6] The development, including the service road subject to the section 106 agreement, was acquired by Tameside Metropolitan Borough Council (Tameside) on 18 August 2000 as an investment for their pension fund. Tameside's involvement in the development is as an investor, not as a planning or highway authority. It is common ground that, having derived title from the developers and to the extent provided for in section 106, Tameside can be obliged by Hillingdon to perform the planning obligation in the material clause in the section 106 E agreement, as amended (clause 3.5), for the grant of a right of way over a service road to be constructed on the freehold land now in Tameside's ownership.

[7] The land in question is situated at the rear of properties in High Street, Uxbridge. The properties are referred to in the agreement and in this judgment as "the authorised properties". They include 276-280 High Street, owned by Milebush Properties Ltd (Milebush), the appellant. They are used for commercial and residential purposes and include a covered shopping arcade.

[8] The section 106 agreement was amended by a supplemental agreement dated 29 October 1997 inserting the new clause 3.5, which F is at the heart of the case. The amended agreement will be called "the principal agreement" in this judgment. Clause 8.7 of the principal agreement provided that, in the event of any dispute arising between the parties in respect of any matter the subject of the agreement, the same should be referred to a chartered surveyor agreed by the parties (with a procedure, in default of agreement, for the outside appointment of a surveyor), who was to act as an expert and not as an arbitrator.

[9] The dispute between the parties to the appeal centres on the construction of clause 3.5 of the principal agreement. The parties have been unable to reach agreement on the nature and extent of the right of way to be granted over the service road to provide access for the rear servicing of the authorised properties. It would be a mistake to think of this as a run-of-the-mill dispute about a right of way. (*Gale on Easements* was cited in argument, but does not help to resolve this case).The complication in the legal landscape lies both in section 106 and in the fact that these proceedings have not been brought, as contemplated in section 106, by Hillingdon as the planning authority. Quite the contrary: the proceedings have been brought *against* Hillingdon by Milebush, which was not a party to the principal agreement, and also against Tameside, which were not a party to the principal agreement. However, no separate relief was claimed against Hillingdon, which objected that they ought not to have been joined in that action at all. The judge concluded (see [2010] EWHC 1022 (Ch)*) that the reason for the joinder of Hillingdon was an attempt to ensure that they were bound by the judgment on the construction of the planning obligation that Milebush hoped to obtain against Tameside, but, as the judge held, that course was not open to Milebush. He made an order that Milebush pay Hillingdon's costs on an indemnity basis.

[10] For their part, Tameside do not dispute, as such, the binding nature of the planning obligation relating to the service road. Their case is that: (a) by virtue of section 106(3) of the 1990 Act, Milebush is not entitled to take private law proceedings against it for a declaration on the construction of a planning obligation unenforceable by it; and (b) as a matter of construction, the rights over the service road under clause 3.5 are not of the kind claimed by Milebush.

[11] On the construction issue, the essential question is whether Milebush is correct in its contention that the right of way to be granted under clause 3.5 covers a pedestrian emergency exit from the rear of the authorised properties. Tameside are willing to grant a right of way over the service road to Milebush in more limited terms. In correspondence and in this action, Milebush says that the proposed terms are "too restrictive." It asserts that the terms set by Tameside should be amended to confer a wider form of right of way than Tameside are prepared to grant.

[12] The dispute was brought before Arnold J last year in the private law proceedings by Milebush against Tameside and Hillingdon for declarations that Tameside "is obliged to grant to the Claimant [Milebush] the rights of way referred to in clause 3.5 of the Principal Agreement…" and "As to the extent of the rights of way." Tameside applied for summary judgement dismissing the claim. In his judgment, the judge held that it was unnecessary in the circumstances to consider that application. By his order dated 13 May 2010, he simply dismissed the action. He did so for two main reasons, both of which are covered by his grant of permission to appeal. On both points, Milebush contends that the judge's decision was wrong in law.

[13] First, Arnold J rejected the construction of clause 3.5 advanced by Milebush in respect of the pedestrian emergency exit from the rear of its property via the service road. He held that other points in dispute, concerning the locking of a security gate and the hours during which the service road could be used, were not matters for the determination of the court, because they are covered by the provision in the principal agreement for determination by an expert in the event of a dispute between Tameside and Hillingdon and there was no such dispute.

[14] Second, he accepted Tameside's submission that it was not, in any event, an appropriate case for declaratory relief under CPR 40.20. Having referred to section 106 and a line of authorities on the court's discretion to grant a declaration, the judge accepted Tameside's submission that it would be an inappropriate and pointless exercise for the court to make the declaration sought by Milebush.

[15] In brief, Tameside's point is that section 106 provides for the enforcement of planning obligations at the instance of the local planning authority: it does not provide for enforcement by a non-party beneficiary of the planning obligation, which is what Milebush claims to be.

[16] Further, subject to judicial review, Hillingdon have the power, in order to further their planning objectives, to release Tameside from the planning obligation to them under clause 3.5, or not to enforce that clause in whole or in part, or to enter into an arrangement with Tameside to vary it. That power was inconsistent with the proper exercise of

---

* Editor's note: Reported at [2010] 2 EGLR 93; [2010] 30 EG 64

the court's discretion to grant a declaration as to the construction of clause 3.5. Arnold J concluded:

> 58. … Milebush claims no private law right enforceable against either Tameside or Hillingdon. It seeks declaratory relief against Tameside on the footing that it is directly affected by the interpretation placed upon the Principal Agreement. But why should the court grant Milebush a declaration against Tameside in circumstances where, even if Milebush was right on the construction of clause 3.5 as it presently stands, Hillingdon would retain a discretion to decide not to enforce, or to vary, clause 3.5? It would be a pointless exercise. Counsel for Milebush had no real answer to this point.

More detail

[17] The provision of a service road at the rear of the authorised properties was prompted by Hillingdon's pedestrianisation of Uxbridge High Street and the consequent loss of vehicular access to them via the High Street.

[18] The material terms of clause 3.5 under which the right of way over the service road was to be granted were "in fee simple with or without vehicles to inure in perpetuity… over the Service Road for the benefit of the Authorised Properties for the purpose of obtaining access for servicing the rear of those properties only and in accordance with the terms of this clause…" and, under clause 3.5.2, "upon such reasonable terms as may be imposed by the Developer".

[19] The service road was completed around 2004-05. The development was first occupied around March 2008. A security gate was installed. It was kept locked. Access to the part of the service road beyond the gate is controlled by a security guard. Parts of the High Street were fully pedestrianised in stages over the years. Other parts, including those adjacent to the authorised properties, were subject to strict traffic controls. By May 2009, vehicular access to the authorised properties via the High Street was materially restricted.

[20] Following Hillingdon's service of a notice on them, Tameside wrote to Milebush on 21 May 2009 stating that they were prepared to grant to it a right of way pursuant to clause 3.5.2 of the principal agreement on terms and conditions stated in the letter. The terms and conditions were that the right could not be used as a pedestrian emergency exit or for any purpose other than allowing access and egress to service the rear of Milebush's properties (that is, loading and unloading of goods); that it could be used only between the hours of 7.30am and 6pm Monday to Friday; and that access would be controlled by the operation of the security gate on the service road.

[21] Those terms and conditions did not suit Milebush. It contended that Tameside were not entitled to impose those conditions on the grant of the right of way. Milebush principally seeks a right of way in order to provide a pedestrian emergency exit for a restaurant and wine bar and/or other of its High Street properties.

[22] The judge decided the question of construction of clause 3.5 first. Having decided it in favour of Tameside, he could simply have refused the relief claimed by Milebush on the ground that Milebush had lost the substantive issue of construction.

[23] However, the question of discretionary declaratory relief was fully argued and the judge ruled on it. The judge held that it was not the sort of case for the exercise of the court's discretion. Questions of relief are commonly decided after the court has ruled on the substance of the dispute. The judge was not invited to make a preliminary ruling on the availability of a declaration in this case. The course taken by the judge is not open to objection. He had heard full argument from each side on construction and relief. Both sides wanted a decision on each point. In those circumstances, and in view of the possibility that a different view might be taken on appeal, it was sensible for the judge to decide both issues.

[24] It is also sensible for this court to decide both issues. If the court agrees with the judge on the declaration issue, it would theoretically be possible to dismiss the appeal without deciding the construction issue. But that course would have the unfortunate consequence that Milebush would be left with rulings on construction, which it says are wrong, but which would give rise to an issue estoppel between it and Tameside in other proceedings.

[25] I will depart from the order followed by the judge and take G the declaration issue first. It raises a general point to which the legal argument has been principally directed.

A. Declaration issue

Court of Appeal's role

[26] On the appeal against the ruling that this is not an appropriate case for declaratory relief, I have a preliminary comment about the role of this court.

[27] This appeal is against a decision of the judge in the exercise of his discretion. He did not decline jurisdiction. Under the heading "The appropriateness of declaratory relief", he took account of various factors relating to the claim by Milebush for a discretionary declaration H on the footing that it was directly affected by the construction placed by Tameside on clause 3.5.

[28] The discretionary factors considered by the judge in that section of his judgment ([47] to [58]) were that, as was conceded, Milebush was not a party to the principal agreement and could not enforce clause 3.5 against Tameside as a private law right (for example, by specific performance of an obligation to make the grant of the easement). The judge also said that section 106 did not provide for enforcement of a planning obligation by a beneficiary of that obligation, such as Milebush, providing instead for the enforcement of the planning obligation by the planning authority themselves (that is, Hillingdon); J that Milebush was not claiming any public law right against Tameside; that Hillingdon had the power to further their planning objectives by deciding to release, or not enforce, or to vary clause 3.5; and that, in those circumstances, the grant of the declaration claimed would be a pointless exercise.

[29] This court, as is well known, does not interfere with the judge's discretion unless it is established that he acted on a wrong legal principle or if, for some other reason, his decision was plainly wrong.

Submissions of Milebush

General points K

[30] Mr Mark Warwick, appearing for Milebush, contended that the judge's refusal to grant a declaration was wrong in principle. Mr Warwick took the court chronologically through the authorities to support his contention that this action was a legitimate course to take. It was not, he said, an abuse of process, even if some other course could have been taken. Judicial review was available only on the basis that Hillingdon had made a decision and that there were grounds for challenging the lawfulness of the decision. No relevant decision about the service road had been taken by Hillingdon, which had taken a neutral stance.

[31] Mr Warwick criticised the judge's reasons for refusing a L declaration. He submitted that it would be proper for this court to grant a declaration in the terms of the draft order prepared by him at the request of the court. The draft includes a declaration about the use of the service road as a pedestrian emergency exit.

[32] Mr Warwick pointed out that Tameside had fully engaged in the construction issue before the judge. Tameside had not asked the judge to rule as a preliminary point on the unavailability of declaratory relief. The refusal of a declaration by this court would leave the parties in an uncertain state about what in Arnold J's judgment was binding on them, and would open up the uninviting prospect of further litigation. M

[33] The judge, said Mr Warwick, was wrong to describe the context in which a declaration was sought as "novel" and "inappropriate". Novelty was not in itself a good ground for refusing a declaration. It was not suggested by the judge that there was any other appropriate way of resolving a dispute that affected Milebush's property interests. The position was that Tameside were required by the principal agreement to grant to Milebush rights over the service road. Tameside accepted that they were obliged to grant rights of way. There was no suggestion from Hillingdon that they would make the grant of a declaration futile by deciding not to enforce the rights or to vary them. The dispute called for a decision, and a decision called for a declaration. Depriving Milebush

A of a declaration would leave a serious and unnecessary lacuna. Tameside would be placed above or beyond the law.

Legal submissions

[34] The present state of the law on declaratory relief under CPR 40.20 is that it is available in diverse circumstances in both public and private law proceedings. It is available whether or not any other remedy is claimed. To illustrate the width of the declaratory jurisdiction, Mr Warwick cited a number of authorities.

[35] In *Attorney-General* v *Gouriet* [1978] AC 435, at p501, Lord Diplock described the jurisdiction to grant declaratory relief as B discretionary, useful and more and more extensively used, but warned that it is subject to limits. A declaration could be only of legal rights.

[36] Lord Diplock explained, at p501E:

> So for the court to have jurisdiction to declare any legal right it must be one which is claimed by one of the parties as enforceable against an adverse party to the litigation, either as a subsisting right or as one which may come into existence in the future conditionally on the happening of an event.

[37] He added, at p501G, that, although it was not necessary for a person claiming declaration to have a subsisting cause of action:

> C the jurisdiction of the court is not to declare the law generally or to give advisory opinions; it is confined to declaring contested legal rights, subsisting or future, of the parties represented in the litigation before it and not those of anyone else.

[38] The general tenor of later cases is that this passage in Lord Diplock's speech should not have been treated literally as precluding the grant of a declaration on the construction of a contract to which the claimant was not a party: *cf Meadows Indemnity Co Ltd* v *Insurance Corporation of Ireland plc* [1989] 2 Lloyd's Rep 298, at pp304-305 and 309, where the court refused a declaration relating to the *validity* of a contract of insurance to which the claimant was D not a party. In *Mercury Communications Ltd* v *Director-General of Telecommunications* [1996] 1 WLR 48, at pp58-59 (the Court of Appeal judgments at [1994] CLC 1125 were also consulted during the hearing), the proceedings were by way of originating summons for declarations on the true construction of a licence granted in 1994 by the Secretary of State for Trade and Industry to BT. The statutory framework was for an agreement to be made between Mercury and BT, with the director general acting as arbiter. The director general made a determination of terms and conditions affecting Mercury, which contended that the director had misinterpreted phrases in condition 13 of the BT licence, which was incorporated in clause 29 of the Mercury/BT agreement. E Mercury sought a declaration on the true construction of condition 13. The House of Lords rejected the application by the director to strike out the proceedings on the grounds that the matters raised by the originating summons were matters of public law to be raised, if at all, under RSC Ord 53; that the relief sought would not settle the matters between the parties; and that the declaration could not be granted in the absence of other interested parties. The House rejected the submissions that the issues fell exclusively within the competence of the director and were academic or hypothetical. The determination of the director raised a question of construction and therefore of law. The director would not be doing what he was asked to do if he made a determination on the F basis of an incorrect interpretation. The House of Lords treated the dispute as one between Mercury and BT, on which the court could make a declaration on the construction of the condition incorporated in the relevant clause affecting the operation of the agreement between the parties.

[39] *Re S (Hospital Patient: Court's Jurisdiction)* (No 1) [1996] Fam 1 was a procedural case. The point was whether the plaintiff with a genuine and legitimate interest in obtaining a decision was able to bring proceedings concerning the wellbeing of a patient, which raised a serious justiciable issue. Millett LJ observed that declarations were limited to the resolution of justiciable issues about legal rights that were the subject of a real and present dispute between the parties, but it was not a requirement for a declaration that the legal right should be claimed to be vested in one of the parties. In that case, the fact that the legal rights were those of the patient and not of the claimant was not a bar to the claimant invoking the court's declaratory jurisdiction.

[40] Neuberger J, in *Financial Services Authority* v *Rourke (t/a JE Rourke & Co)* [2002] CP Rep 14, said that, when considering whether to grant a declaration, the court should take into account what was just for both parties, whether the declaration would serve a useful purpose and whether there were special reasons why the court should not grant a declaration.

[41] In *Feetum* v *Levy* [2005] EWCA Civ 1601; [2006] Ch 585, in [82], Jonathan Parker LJ observed that things had "moved on" since *Meadows* and that declaratory relief as to the effect of a contract should not be refused on the ground that the claimants were not parties to the contract: see [81] of that judgment.

[42] Mr Warwick submitted that the present position on discretion to grant a declaration was summarised in the principles listed by Aikens LJ in *Rolls-Royce* (see below). This case, Mr Warwick said, falls within those principles and within the broad tests laid down in the other cases cited.

[43] In applying the principles to this case, Mr Warwick submitted that it is appropriate to grant the declaration claimed. Tameside accept that they are obliged to grant the rights set out in clause 3.5 in favour of the authorised properties. Hillingdon want to ensure that Tameside, as successors in title, would comply with their obligations. There was a real and present dispute between Tameside and Milebush as to the construction of clause 3.5. Milebush's interests were affected by the way in which the clause was construed. The Chancery Division was the proper forum to decide questions of the construction of a deed or other document affecting real property. The academic feature of the case was to be found in Tameside's argument that Hillingdon might not enforce, or could release, or could decide to vary, clause 3.5. Hillingdon have never suggested that they would do other than accept the decision of the court on construction. To deprive Milebush of a declaration would be a serious omission in the law.

Discussion and conclusion

[44] I agree with Mr Warwick that the discretion to grant a declaration now covers a wide range of cases. The authorities show how it may be granted in private law proceedings about the disputed construction of a document affecting the claimant, even though the claimant was not a party to it.

[45] However, I have not been persuaded that there was any error of principle by the judge or that he was plainly wrong, or that there are any grounds for disturbing his decision to decline a declaration.

[46] First, the judge gave a proper self-direction in law on the principles governing declaratory relief, by reference to the judgment of Aikens LJ in *Rolls-Royce plc* v *Unite the Union* [2009] EWCA Civ 387; [2010] 1 WLR 318, in [120]. The judge considered the submission for Milebush that, although it was not a party to the principal agreement and had no cause of action against Tameside, it was directly affected by the issue of construction of clause 3.5.

[47] Second, all of the factors taken into account by the judge were relevant to the exercise of his discretion. In my judgment, the most important factor was that this is, in reality, not a private law dispute about the construction of a deed for the grant of a private right of way, or about its legal enforcement of an agreement, or even about the nature and extent of an easement: it is about the planning objectives of a planning authority and about the performance of planning obligations on which the decisions resting with the planning authority are important, even paramount.

[48] I agree that they are matters affecting Milebush, but they are public law planning matters that, if they are to be decided by a court at all, ought to be decided in judicial review proceedings to which Hillingdon are a party, being both the planning authority and the significant enforcing party to the deed with the original developers from which Tameside derives title. Hillingdon, as the planning authority, are the only party entitled to enforce the principal agreement. Their planning decisions and actions are challengeable by judicial review

to ensure that the proper procedures have been followed and that the relevant decisions have been lawfully made: see *Attorney-General ex rel Scotland* v *Barratt Manchester Ltd* (1992) 63 P&CR 179, at pp195-197, cited by the judge in [55] and [56]. In judicial review proceedings by Milebush, a declaration on the meaning and effect of clause 3.5 could be made; Hillingdon would be a party to the proceedings; they would be bound by the result, as would Tameside, which could be joined as an interested party.

[49] In my judgment, that is the correct characterisation of the dispute. The result is that judicial review is the proper procedure and the Administrative Court is the proper forum for this case and for future cases of this kind. In these private law proceedings, Hillingdon have, in general, adopted a neutral position. They declined to confirm formally the terms proposed by Tameside or the terms proposed by Milebush. They contended that it was not appropriate to be required to decide between Tameside and Milebush on the precise terms of the grant. They confirmed that "[they] would want to ensure that a successor on title of the Developer would comply with the obligations that are binding on it as set out in the s106 Agreements". They stated that they regard the precise terms of grant as ultimately a matter to be dealt with by "private parties taking a sensible and measured approach".

[50] However, without prejudice to their central contention that it was inappropriate for them to have been involved at all with these proceedings, Hillingdon stated their position in para 10 of their defence, agreeing with Tameside that servicing the rear of the authorised properties is separate from emergency access.

[51] Despite Hillingdon's declared formal neutrality in their action, I am of the view that the subject matter of this dispute is the enforcement of the planning obligation regarding the service road and that only Hillingdon are legally entitled to enforce it against Tameside. Hillingdon, as the planning authority, appear to be in broad agreement with the construction of Arnold J and this court on the main area of dispute. Even if they were not, they could vary clause 3.5 to give it a different construction.

[52] I agree with Ms Judith Jackson, appearing for Tameside, that this private action for a declaration is an inappropriate proceeding by Milebush. It is an attempt, to which the court should not lend its assistance, to secure via Tameside, against which it has no enforceable private rights, a right of way over the service road for the purposes of a pedestrian emergency exit without regard for Hillingdon's planning objectives.

[53] As explained earlier I would not, in the present circumstances of this case, rely on the unavailability of a declaration as a reason for denying an appeal against the judgment of Arnold J on the construction of clause 3.5. I therefore turn to the construction issues.

B. Construction issues

Judgment

[54] On the question of construction of clause 3.5, the judge held that: (a) the use of the service road did not extend to use for the purpose of a pedestrian emergency exit from the rear of the authorised properties; and (b) the issue of what are reasonable times, and operation of the security gate, and the provision of a key, were matters for expert determination in the event of a dispute between Tameside and Hillingdon.

[55] The judge approached the construction of the principal agreement by ascertaining the intention of the parties from the words used read in the light of the background circumstances known to the parties. The judge said that it was clear that the intention behind the grant of rights of way in clause 3.5 was to compensate the owners of the authorised properties for the loss of vehicular access via the High Street as the result of pedestrianisation.

[56] In holding that the purpose of obtaining access for servicing the rear of the authorised properties did not extend to use for the purpose of a pedestrian emergency exit, the judge said that pedestrianisation of the High Street had no adverse effect on emergency egress from the authorised properties; that emergency access was not within the natural meaning of use for obtaining access for "servicing" the rear of those properties; that this interpretation was supported by the limitation of its exercise to all reasonable times and by the permitted terms allowing the developer control of the access to the service road by a security barrier.

[57] As for the limitation of the exercise of the right of way to Monday to Friday 7.30am to 6pm, the judge said that the issue of whether those times were unreasonable could not be judged in the abstract: it was a question of fact for expert determination in the event of a dispute between Tameside and Hillingdon, but there was no such dispute at present.

[58] The judge took the same approach to the question of whether it was unreasonable to impose control of access by the operation of the security gate, in particular by obtaining an appointment in advance for a security guard to open the gate and Milebush's request for a copy of the key to the padlock.

Submissions of Milebush

[59] The case for Milebush advanced by Mr Warwick on the nature of the user of the service road is that the road is constructed to the standard of a public highway. There is no physical obstruction to any particular user. The court should start from the position that *all uses* of it are permitted. It is stated to be "with or without vehicles" and it involves providing a service at the rear of the authorised properties, which would include user to enable repair and maintenance of the existing buildings, user to enable new or replacement buildings to be constructed, user for loading and unloading, user for emergency access and egress, with or without vehicles, and user by a disabled person.

[60] It is said that the judge erred in rejecting use for the purposes of a pedestrian emergency exit. He overlooked or failed to give weight to the very wide scope of the right. In ordinary usage, a road for servicing properties involves emergency access, especially for residential occupants, and an emergency exit does not require unimpeded access.

[61] As for the provision permitting the developer to control access to the service road by the operation of a security barrier, the judge erred in holding that the system was reasonable. It is too restrictive. It derogates from the grant to Milebush. It involves Milebush having to contact Tameside, seeking use of the service road, waiting for a response, attending at an appointed time to meet the security guard and the collection of keys by the security guard and opening the gates by him. Milebush wants keys to the lock. The judge was wrong to regard the matter as one of fact for the expert. It is a question of mixed law and fact for which an expert was not the appropriate procedure now that Hillingdon had made it clear that they did not wish to become involved in the dispute between Milebush and Tameside. Milebush was not a party to the principal agreement and cannot utilise the contractual procedure for expert determination.

Discussion and conclusion

[62] In my judgment, the context of the principal agreement is important in the construction of clause 3.5. Pedestrianisation of the High Street created the need for alternative vehicular access for servicing the High Street properties at the rear. I agree with the judge that, in those circumstances, the provision of an emergency pedestrian exit at the rear of the authorised properties has nothing to do with servicing the rear of the properties. It is outside the contemplation of clause 3.5 of the principal agreement.

[63] As for the operation of the security gate to control the access to the service road, I also agree with the judge. The gate currently involves telephoning a security guard to come and unlock the gate. That is consistent with servicing use. It has not been shown to be unreasonable in all the circumstances and any dispute about the reasonableness of the provision is subject to the provision for expert determination. The complaint by Milebush that it was not a party to the principal agreement containing the expert determination procedure is yet another indication of how inappropriate it is for Milebush to be seeking relief against Tameside in a private law action for the construction of a planning obligation in a deed to which it is not a party.

Result

[64] I would dismiss the appeal: first, on the basis that, in these private law proceedings and in the circumstances of the case, the judge was entitled to conclude that it was not appropriate to exercise his discretion to grant Milebush a declaration in the form sought by it or in any other form; and, second, on the basis that the judge's construction of clause 3.5 of the principal agreement was clearly correct.

Giving the second judgment, **Moore-Bick LJ** said:

[65] I agree that the judge's conclusions on the questions of construction in this case were correct and that, to that extent, the appeal should be dismissed. However, I wish to say something about the form of procedure that was adopted in this case and, in particular, about whether this is a case in which it would have been appropriate for the court to grant declaratory relief of the kind sought by the appellant, Milebush Properties Ltd (Milebush).

[66] The agreement between the original developers and Hillingdon Borough Council (the council) under section 106 of the 1990 Act, which was subsequently varied to add the clauses that have given rise to these proceedings, is binding on the respondents, Tameside Metropolitan Borough Council (Tameside), as successors in title by virtue of the provisions of section 106(3), and is enforceable against them by the council. The agreement is a planning agreement and was required by the council in order to promote plans for the pedestrianisation of part of Uxbridge High Street that would remove or significantly restrict vehicle access to various properties facing onto that road. The construction of a new service road to the rear of the properties would provide alternative access.

[67] Under the terms of the agreement of 29 October 1997, the developer covenanted with the council that it would grant rights of way over the service road for the benefit of certain "authorised properties" for the purpose of obtaining access for servicing the rear of those properties. One of the authorised properties was 276-280 Uxbridge High Street, which is owned by Milebush. The agreement was, of course, entered into by the council partly, perhaps even primarily, for their own benefit, in as much as it was seen as a way of countering objections to their pedestrianisation plans from owners of the properties most likely to be affected. However, Milebush and the other owners and occupiers of the authorised properties, although not parties to the agreement, were parties for whose benefit it was made, in the sense that the grant of a right of way would provide a permanent means of access on foot and by vehicle for the purposes of providing necessary services. Indeed, it could be said that one of the main objects of the agreement was to provide them with that benefit.

[68] Tameside, as a party to the agreement, have expressed a willingness to honour the obligation contained in the covenant and have offered to grant a right of way on certain terms. Milebush says that Tameside's view of the scope of their obligation is too restrictive and that difference of opinion has given rise to a dispute. One question that arises on this appeal is how that dispute ought to be resolved.

[69] Since Milebush is not a party to the agreement, it cannot enforce it directly against Tameside and, since the council are not trustees of the obligation for the owners and occupiers of the authorised properties, Milebush cannot bring proceedings against them to require them to enforce the covenant for its benefit. Two avenues therefore appear to be open: one is for Milebush to ask the court to resolve the issue of construction by granting a declaration; the other is for Milebush to ask the council to require Tameside to grant a right of way in terms that it contends correctly reflect the agreement (no doubt bearing any associated costs) and, if they decline to do so, to seek judicial review of their decision. Those proceedings would give rise to two issues: whether the construction that Milebush places on the agreement is correct and whether the council's decision not to enforce the agreement was unlawful. Tameside would almost certainly be joined as an interested party, since they would be directly affected by the outcome.

[70] In a case where the dispute is confined to a legitimate difference of opinion about the scope of the covenant between Tameside, as prospective grantors, and Milebush, as prospective grantee, to engineer proceedings by way of judicial review seems to me to be unnecessarily cumbersome and not the most effective way of resolving a relatively simple dispute. It is said, however, that the court cannot grant declaratory relief in a case of this kind, or at any rate should decline to exercise its discretion in favour of doing so, either because the matter is one of public law rather than private rights, or because it would serve no useful purpose because it is possible at any time for Tameside and the council to vary the agreement. It was the latter consideration that ultimately led the judge to refuse to grant declaratory relief in this case, an aspect of the matter to which I shall have to return at a later stage.

[71] The jurisprudence relating to the court's jurisdiction to grant relief by way of declaration, and the circumstances in which it should be exercised, has developed significantly over the past 40 years. The leading authorities have already been reviewed by Mummery LJ and I can therefore refer to them quite shortly. Although there are earlier cases that emphasise that the court cannot be asked to declare whether a particular scheme or proposed course of action is lawful or complies with some particular legislation (a proposition that I have no difficulty accepting), a convenient starting point for this purpose is *Gouriet* v *Union of Post Office Workers* [1978] AC 435, in which Lord Diplock, at p501D-F, pointed out that the nature of the relief is a declaration of rights. He expressed the view that:

> a court of civil jurisdiction is concerned with legal rights only when the aid of the court is invoked by one party claiming a right against another party, to protect or enforce the right or to provide a remedy against that other party for infringement of it, or is invoked by either party to settle a dispute between them as to the existence or nature of the right claimed. So for the court to have jurisdiction to declare any legal right it must be one which is claimed by one of the parties as enforceable against an adverse party to the litigation, either as a subsisting right or as one which may come into existence in the future conditionally on the happening of an event.
>
> The jurisdiction of the court is not to declare the law generally or to give advisory opinions; it is confined to declaring contested legal rights, subsisting or future, of the parties represented in the litigation before it and not those of anyone else.

[72] It is, perhaps, worth noting in passing the reference there to rights that may come into existence in the future, since it might be said in this case that, if the council were to take steps to enforce the covenant, private law rights and obligations would come into existence in the future directly between Milebush and Tameside.

[73] The need for there to be (in Lord Diplock's words) a legal right, which is claimed by one of the parties as enforceable against an adverse party to the litigation, led this court in *Meadows Indemnity Co Ltd* v *Insurance Corporation of Ireland plc* [1989] 2 Lloyd's Rep 298 to refuse to grant a declaration to the claimant reinsurer as to the effect of the underlying contract of insurance, (despite the fact that its interests were affected by the outcome) on the grounds that it was not a party to the contract. Neill LJ, who gave the principal judgment, expressed some reluctance to reach that conclusion and clearly considered that it would have been beneficial to have been able to decide the issue before the court.

[74] Tameside placed some emphasis on *Attorney-General ex rel Scotland* v *Barratt Manchester Ltd* (1992) 63 P&CR 179. The issue in that case was whether a planning agreement, under which the planning authority and the owner of certain land agreed that it should be retained as a private open space, created a public right enforceable at the suit of the Attorney-General by way of relator action. The court held that the agreement did not give rise to a public right and the limited discussion of the jurisdiction to grant declaratory relief does not take the matter any further for present purposes. No one has suggested that the agreement in this case created a public right of the kind that can only be enforced by the Attorney-General.

[75] *Mercury Communications Ltd* v *Director-General of Telecommunications* [1996] 1 WLR 48 does, however, take the matter further. British Telecom (BT) held a licence for the provision of telecommunications services, clause 13 of which required it to enter into an agreement with any other regulated operators licensed to provide similar services that wished to connect into its systems, and

regulated, among other things, the charges that it could make for doing so. Mercury, another licence-holder, wished to enter into an agreement with BT for that purpose and negotiations began over the terms on which it should be allowed to do so. The parties agreed that the director general as regulator should determine certain terms of that agreement if they could not themselves agree. The director general had indicated his view of the effect of clause 13 of BT's licence, which was critical to any determination he might be called on to make. BT considered that view to be in accordance with clause 13, but Mercury did not. It therefore brought proceedings for a declaration as to the correct interpretation of that clause. The director general and BT sought to strike out the proceedings as an abuse of the process of the court.

[76] The judge at first instance (Longmore J, as he then was) refused to strike out the proceedings, but this court held by a majority (Dillon and Saville LJJ, Hoffmann LJ dissenting) that the proceedings were an abuse of the process. It is of interest in the context of the present case to note that Dillon LJ reached that conclusion on the ground that a party cannot obtain a declaration to get an abstract point decided in order to strengthen its hand in negotiations. He pointed out that there was no dispute between Mercury and BT and held that there was no ground on which Mercury could come to the court for merely declaratory relief. Hoffmann LJ considered that the question was primarily one of procedural convenience and that, since the director general had already made clear his interpretation of clause 13, it was valuable to establish by declaration the correct basis for the continuing negotiations between Mercury and BT. Saville LJ accepted that there was a real issue to be determined between the parties and that the question was whether it would be just to allow the proceedings to continue. He recognised that the alternative was to allow the director to make a determination that could be challenged by judicial review (which he regarded as a time-consuming and expensive exercise for no good reason, given that the proceedings were already before the court), but thought that to entertain the proceedings was tantamount to rewriting the agreement between the parties.

[77] The House of Lords reversed that decision. Lord Slynn, with whom the other members of the appellate committee agreed, rejected the suggestion that the proceedings raised a hypothetical or academic question because the director general had already made his views clear. He considered the meaning of clause 13 to be a key element in the negotiations between Mercury and BT and thought that it was good sense for the matter to be determined at that stage in that way. In my view, the decision strongly supports the conclusion that the court can, and in some circumstances will, determine a dispute relating to the construction of an instrument if both parties to the proceedings have a sufficient interest in the outcome, even if one or other is not a party to it or does not acquire any rights under it.

[78] *Re S (Hospital Patient: Court's Jurisdiction) (No 1)* [1996] Fam 1 raised the question of whether the court could exercise its jurisdiction to grant declaratory relief in order to determine who of two persons was entitled to have responsibility for the care of a patient who was incapable of expressing his own wishes in the matter. Although the question arose in a context far removed from that of the present case, some passages in the judgments are of importance, because, as Sir Thomas Bingham MR (as he then was) pointed out, at p8E, the only issue for determination by the time the appeal was heard was whether the plaintiff had sufficient standing to invoke the court's jurisdiction to grant declaratory relief. Having reviewed a number of cases in which declaratory relief had been granted to doctors, health authorities and others in relation to the care of patients he said, at p18G:

> It cannot of course be suggested that any stranger or officious busybody, however remotely connected with a patient or with the subject matter of proceedings, can properly seek or obtain declaratory or any other relief (in private law any more than public law proceedings). But it can be suggested that where a serious justiciable issue is brought before the court by a party with a genuine and legitimate interest in obtaining a decision against an adverse party the court will not impose nice tests to determine the precise legal standing of that claimant.

[79] Millett LJ, having referred to the passage in the speech of Lord Diplock in *Gouriet*, to which I referred earlier, said, at pp21H-22C:

> Since that decision the courts have developed the jurisdiction to grant declaratory relief in a number of cases which, though distinguishable from the present, are nevertheless not altogether dissimilar to it. We have now reached a position where the court is prepared in an appropriate case to fill much of the lacuna left by the disappearance of the parens patriae jurisdiction by granting something approaching an advisory declaration. In my judgment, the passage which I have cited from Lord Diplock's speech in the *Gouriet* case [1978] AC 435, 501, can no longer be taken to be an exhaustive description of the circumstances in which declaratory relief can be granted today. It is to be regarded rather as a reminder that the jurisdiction is limited to the resolution of justiciable issues; that the only kind of rights with which the court is concerned are legal rights; and that accordingly there must be a real and present dispute between the parties as to the existence or extent of a legal right. Provided that the legal right in question is contested by the parties, however, and that each of them would be affected by the determination of the issue, I do not consider that the court should be astute to impose the further requirement that the legal right in question should be claimed by either of the parties to be a right which is vested in itself.

[80] A little later, at pp22H-23A, he said:

> The dispute raises a justiciable issue; it concerns the legal rights of the patient; all proper parties, including the patient, are before the court; and the determination of the issue affects the rival claimants and their rights and obligations to the patient. In my judgment the court is entitled and bound to decide it.

[81] It is quite true that these statements of principle are found in a case that was concerned with the development of the jurisdiction to grant declaratory relief in a particular and somewhat specialised context in which it had already proved to be of great practical utility. None the less, as the references to *Gouriet* show, it is no more than a branch of the original tree and I see no reason why the jurisdiction should not be invoked in a principled way in other cases where it may prove equally useful. *Mercury* was not cited to the court in *Re S* (it had been decided only a week earlier), but the two decisions seem to me to be entirely compatible.

[82] Other authorities to which we were referred, including *Bank of Scotland* v *A Ltd* [2001] EWCA Civ 52; [2001] 1 WLR 751, serve to demonstrate that the jurisdiction is capable of being utilised in new and unusual situations when a proper case arises. Lord Woolf, giving the judgment of the court in that case, said, at pp767F-768A:

> In his first Hamlyn Lecture given in 1949, *Freedom Under the Law*, Sir Alfred Denning identified the challenge facing the court as being to develop "new and up-to-date machinery" (p116). The first element of the machinery identified in the lecture was the remedy of declaratory relief. The court's power to make a declaration (or "declaration of right") was derived from the Court of Chancery and was originally supposed to be restricted to declaratory judgments as to existing private rights (see *Guaranty Trust Co of New York* v *Hannay & Co* [1915] 2 KB 536, which sets out the early history). Sir Alfred Denning saw the need to develop its scope in order to control the abuse of executive power, and over the half-century which has elapsed since his lecture it has performed a crucial function in the emergence of the modern law of judicial review. The development of declaratory relief has not however been confined to judicial review. Doctors and hospitals have increasingly been assisted by the ability of the courts to grant advisory declarations. It was at one time thought that an interim declaration could have no practical purpose. The developments in other jurisdictions showed this was not the situation. Now the Civil Procedure Rules acknowledge that just as interim injunctions can be granted so can interim declarations. RSC Ord 15, r 16 still remains part of the Civil Procedure Rules. Its transitional life is about to come to an end. The Rule Committee has approved a new rule, CPR r 40.20, which omits any mention of "rights". It merely states: "The court may make binding declarations whether or not any other remedy is claimed."

[83] In *Financial Services Authority* v *Rourke* [2002] CP Rep 14, the present Master of the Rolls, then Neuberger J, held that the court had an unfettered power to grant declaratory relief whenever it is appropriate to do so, having regard to the need to do justice as between the parties and to whether to do so would serve a useful purpose.

[84] In *Feetum* v *Levy* [2005] EWCA Civ 1601; [2006] Ch 585, the claimants were members of a limited liability partnership that had given a counter-indemnity to a third party as part of the arrangements for securing a substantial loan to the partnership. Acting under a power given by a debenture, the third party had purported to appoint administrative receivers over the partnership who, under their statutory powers, had required the claimants to submit a statement of affairs and to comply with certain other requests. The claimants sought a declaration that the appointment of the receivers was invalid because no event had occurred to cause the partnership's insolvency. The defendants argued that the claimants were not entitled to obtain such relief because the declaration concerned the rights of the LLP alone. Having referred to *Gouriet*, *Meadows*, *Re S* and other authorities, Jonathan Parker LJ, with whom Sir Peter Gibson and Ward LJ agreed, rejected that argument, both because the claimants were directly affected by the existence of any duty to comply with the statutory provisions and because it would not be right to refuse declaratory relief simply because the claimants were not parties to the contract whose meaning they were seeking to have clarified. The court observed that things had moved on since *Meadows*.

[85] The last case to which I need to refer is *Rolls-Royce plc* v *Unite the Union* [2009] EWCA Civ 387; [2010] 1 WLR 318, in which the court was asked by Rolls-Royce to declare whether it would be illegal under the Employment Equality (Age) Regulations 2006 for it to include within a selection matrix for redundancy a length of service criterion that had been agreed with the union as part of two collective agreements. The court held by a majority that, although there was no immediate dispute between the parties, the point was not academic and that a decision would serve a useful purpose: see the judgments of Wall LJ (as he then was), at pp332E-333G, and Arden LJ, at pp358H-359C.

[86] Aikens LJ dissented, primarily because he considered that, since those who might be directly affected by the use of the length of service criterion were not before the court and would therefore not be bound by the decision, a declaration would serve no useful purpose. However, his judgment, in [120], contains the following summary of the principles that he considered were to be derived from the earlier cases:

> (1) The power of the court to grant declaratory relief is discretionary.
>
> (2) There must, in general, be a real and present dispute between the parties before the court as to the existence or extent of a legal right between them. However, the claimant does not need to have a present cause of action against the defendant.
>
> (3) Each party must, in general, be affected by the court's determination of the issues concerning the legal right in question.
>
> (4) The fact that the claimant is not a party to the relevant contract in respect of which a declaration is sought is not fatal to an application for a declaration, provided that it is directly affected by the issue; (in this respect the cases have undoubtedly "moved on" from *Meadows*.)
>
> (5) The court will be prepared to give declaratory relief in respect of a "friendly action" or where there is an "academic question" if all parties so wish, even on "private law" issues. This may particularly be so if it is a "test case", or it may affect a significant number of other cases, and it is in the public interest to decide the issue concerned.
>
> (6) However, the court must be satisfied that all sides of the argument will be fully and properly put. It must therefore ensure that all those affected are either before it or will have their arguments put before the court.
>
> (7) In all cases, assuming that the other tests are satisfied, the court must ask: is this the most effective way of resolving the issues raised? In answering that question it must consider the other options of resolving this issue.

[87] Although I am generally in agreement with that summary, I think, with respect, that para (2) is expressed somewhat too narrowly. In *Gouriet*, Lord Diplock himself recognised that the dispute could relate to rights that might come into existence in the future upon the happening of an event and in *Mercury* (which does not appear to have been cited to the court) the dispute did not directly concern the existence or scope of any private rights or obligations vested in the parties themselves. It concerned the correct interpretation of BT's licence, a document that stood in the realm of public rather than private law, but which was central to the negotiations between Mercury and BT.

[88] In my view, the authorities show that the jurisprudence has now developed to the point at which it is recognised that the court may in an appropriate case grant declaratory relief even though the rights or obligations that are the subject of the declaration are not vested in either party to the proceedings. That was certainly the view of the court in *Re S* and it is also the clear implication of the observations in *Feetum* and *Rolls-Royce* that things have moved on since *Meadows*. In *Mercury*, it was not considered relevant that BT had rights under the licence and it was no bar to the proceedings that Mercury did not. To that extent, the position is mirrored in this case, in which Tameside have obligations under the agreement but Milebush has no rights. I can see no reason in principle why the nature of the underlying obligation should be critical, although there may well be other reasons why, in the particular case, a declaration should not be granted. The most important considerations are likely to be whether the parties have a legitimate interest in obtaining the relief sought, whether to grant relief by way of declaration would serve any practical purpose and whether to do so would prejudice the interests of parties who are not before the court.

[89] In the present case, the fact that the agreement is a planning agreement made for public purposes is not, in my view, of itself sufficient to preclude the court from granting declaratory relief at the suit of Milebush. As one would expect, Tameside are willing to honour their obligations, whatever they are, although they reserve the right to seek to persuade the council to vary the agreement if, on a correct interpretation, its effect is wider than was supposed. They are entitled to do that, but no attempt has been made to clarify or vary the agreement as a means of disposing of the litigation, so I think one is entitled to assume that, at the moment, the council are not unduly perturbed by the prospect that Milebush might succeed. The fact that Tameside are willing to honour their obligation by making a grant in favour of Milebush in accordance with the terms of the agreement means that the only thing holding up the process is an authoritative decision on the scope of that obligation. Indeed, even if Tameside and the council are both dissatisfied with the outcome of the proceedings and wish to revise the agreement, they will be better placed to do so when they know what it means. On the face of it, therefore, it is in the interests of all concerned to know what the true scope of the obligation is. How that may most economically and effectively be determined then becomes an important question.

[90] In my view, the present proceedings for a declaration are a sensible and economical means of achieving that end. The principal objection to that course is that the obligation, being a planning obligation, exists in the public law sphere and should be enforced, if at all, only by means of proceedings for judicial review. In my view, however, that is not a good reason for the court to refuse to entertain a claim for a declaration in the circumstances of the present case. A similar objection was made in *Mercury* (see pp56B-E) but it was rejected both by the Court of Appeal and the House. The circumstances of that case were admittedly different in so far as the dispute could be viewed as one relating to the contract between the parties, but the principle underlying the decision is that the court should be guided by considerations of justice and convenience rather than being bound by a rigid adherence to certain forms of procedure. Lord Slynn expressed this as follows, at p58A-B:

> Moreover it cannot be said here, in my view, that the procedures under Order 53 are so peculiarly suited to this dispute (as they would be in a claim to set aside subordinate legislation or to prohibit a government department from acting) that it would be a misuse of the court's process to allow the originating summons to continue. On the contrary it seems to me that the procedure by way of originating summons in the Commercial Court is as least as well, and may be better, suited to the determination of these issues than the procedure by way of judicial review.

[91] In my view, the same can be said in this case and I think it would be wrong to hold that the court cannot or should not entertain an action for a declaratory relief simply because the rights in question arise in the sphere of public rather than private law. Moreover, I do not think that it is entirely correct to characterise these proceedings as an

attempt to enforce the agreement, in the sense that Milebush is seeking to compel the council as a public body to take certain action. The council have not hitherto shown any disinclination to allow Tameside to perform their obligations under the agreement and Milebush does not seek an order requiring them to take action of any kind. All it seeks is a declaration as to the meaning of the agreement. It recognises that it is dependent on Tameside's willingness to act honourably for obtaining the practical benefit it seeks, but has reasonable grounds for thinking that Tameside will not disappoint it.

[92] Milebush also accepts that it cannot prevent Tameside and the council from varying the agreement, and it was that possibility that led the judge to conclude that a declaration would serve no useful purpose. That led to the somewhat bizarre outcome below that the court heard argument on, and determined, the issues of construction in a manner capable of giving rise to an issue estoppel binding on Milebush, Tameside and the council but, having done so, declined to embody its decision in the grant of a formal declaration. In my view, that was a most unsatisfactory outcome. Moreover, if it were thought that no useful purpose would be served by granting a declaration as to the meaning of the agreement (a conclusion that could be drawn from the nature of the agreement), there was surely no point allowing the court's time to be wasted in hearing argument about its meaning. The claim should have been struck out as an abuse of the process.

[93] For the reasons that I have given, I think that there were good reasons to decide the issues of construction raised in this case and that the judge was right to entertain the proceedings. I also think that he decided the issues of construction correctly. However, in my view, he was clearly wrong to hold that no useful purpose would be served by granting declarations giving formal effect to his decisions and, in so far as his refusal to do so involved an exercise of discretion, I consider it to have been flawed as a result of a failure to recognise and apply the correct principles of law.

[94] I would therefore allow this appeal to the limited extent of varying the order below by granting declaratory relief in favour of Milebush that reflects the limited extent to which it succeeded before the judge on the issues of construction.

Jackson LJ said:

[95] I agree that this appeal should be dismissed for the reasons given by Mummery LJ.

Appeal dismissed.

Welwyn Hatfield Borough Council *v* Secretary of State for Communities and Local Government and another E

SUPREME COURT

[2011] UKSC 15

6 April 2011

LORD PHILLIPS, PRESIDENT, LORD RODGER, LORD WALKER, LADY HALE, F
LORD BROWN, LORD MANCE AND LORD CLARKE

Town and country planning — Residential use — Enforcement action — Limitation periods — Town and Country Planning Act 1990 — Planning permission granted for erection of hay barn for agricultural storage — Building fitted out and used as dwelling — Section 171B(2) of 1990 Act — Whether residential use immune from enforcement action after four years as change of use to single dwellinghouse — Whether landowner entitled to certificate of lawfulness of existing use

In December 2001, the appellant council granted planning permission G
to the second respondent to erect a hay barn to be used for agricultural storage only. Although the exterior of the building resembled a barn, the second respondent had fitted it out internally as a three-bedroom dwelling, with all utilities, and had always intended to use it as such. He moved into the building in August 2002 and lived there continuously for four years, without the appellants' knowledge.

In August 2006, the second respondent applied for a certificate of lawfulness of existing use, on the ground that the four-year time limit imposed by section 171B(2) of the Town and Country Planning Act 1990 for taking enforcement action against a change of use
to a single dwellinghouse had expired. The appellants refused the H
application but a certificate was subsequently granted by a planning inspector appointed by the first respondent. The appellants obtained an order of the High Court quashing the certificate on the ground that the barn had never been used other than as a dwelling such that a relevant change of use had not occurred: see [2009] EWHC 966 (Admin); [2010] JPL 352.

The Court of Appeal reversed that decision and reinstated the certificate. It held that the use of the building as a dwelling represented a change of use from either: (i) the agricultural storage use permitted under the relevant planning permission, even though there had been no
such prior use; or (ii) a period of "no use" between the completion of J
the building and the date on which the second respondent had taken up residence. It further considered that the four-year time limit under section 171B(1) applied to enforcement against the construction of the building, which had been built in breach of planning control because its physical features were those of a dwelling rather than the hay barn for which planning permission had been granted: see [2010] EWCA Civ 26; [2010] 1 EGLR 98; [2010] 13 EG 84.

The appellants appealed to the Supreme Court. They disputed the Court of Appeal's interpretation of section 171B(2) and further argued that even if the second respondent's development fell within
that provision, he should be barred from relying on it as a matter of K
public policy because he had deliberately deceived them in his planning application.

Held: **The appeal was allowed. (1) The second respondent's building had been designed and built as a house in every respect except its external appearance. It should not be treated as having a character that the person who erected it had not intended it to have. Therefore that building, as constructed and completed, was not the barn for which planning permission had been given and could not be regarded as having any permitted use. Accordingly, there had been no "change of use" from a permitted use to a use as a single dwellinghouse so as to fall within section 171B(2). The building fell within the provisions of section 171B(1) of the**

A **1990 Act dealing with unauthorised building operations. (2) Nor could the situation be analysed as involving a change of use from a period of "no use" prior to the second respondent's residency. Use as a dwellinghouse should not be judged day-to day but on a broader and longer-term basis. The correct approach in such cases is to look at the matter as a whole and ask what use the building has, or of what use it is. It would be artificial to say that a building has or is of no use, or that its use is other than as a dwellinghouse, when its owner has built it to live in and intends to move in shortly. In that context, the court should not place too much stress on "actual use", with its connotations of daily domestic activities. Accordingly,**
B **the commencement of residential use in such circumstances does not involve a change of use to use as a dwellinghouse and section 171B(2) does not apply. (3) That approach to section 171B(2) does not produce anomalous results. Although it means that the four-year enforcement period in section 171B(1) will apply to the unauthorised building operations, while the 10-year period in section 171B(3) will apply to the residential use of the resulting building, that is consistent with a proper understanding of the scheme of section 171B. Protection from enforcement in respect of a building and its use are potentially very different matters. In respect of use, section 171B(2) applies a four-year period only**
C **where this involves a change of use to a dwellinghouse; the focus on the concept of "change of use" rather than merely "use" is deliberate. Any resulting unfairness might, in an appropriate case, be covered by more general public law controls on administrative action by way of planning enforcement; once a planning authority has allowed the four-year period for enforcement against the building to pass, principles of fairness and good governance might, in appropriate cases, preclude it from subsequently taking enforcement stops to render the building useless.**

***Per curiam*: The principle that a person should not be permitted to profit from his own wrong is one of public policy and is not**
D **necessarily confined to cases involving a crime but can extend more widely. Whether conduct will disentitle a person from relying on an apparently unqualified statutory provision should be considered in its context, with regard to any nexus between the conduct and the statutory provision in question. The language of section 171B(2) cannot have been intended to cover a case involving positive deception consisting of the making of fraudulent planning applications designed to avoid enforcement action within any relevant four-year period.**

The following cases are referred to in this report.

E *Arun District Council* v *First Secretary of State; sub nom R (on the application of Arun District Council)* v *First Secretary of State*; *First Secretary of State* v *Arun District Council* [2006] EWCA Civ 1172; [2007] 1 WLR 523; [2007] 1 P&CR 10; [2007] JPL 237

Backer v *Secretary of State for Environment* (1984) 47 P&CR 149; (1982) 262 EG 549; [1983] JPL 167, CA

Booth v *Warrington* (1714) 2 ER 111, HL

Bulli Coal Mining Co v *Osborne* [1899] AC 351, PC

Cleaver v *Mutual Reserve Fund Life Association* [1892] 1 QB 147, CA

Gibbs v *Gould* (1882) LR 9 QBD 59

Hartley v *Minister of Housing and Local Government* [1970] 1 QB 413; [1970] 2 WLR 1; [1969] 3 All ER 1658; (1969) 68 LGR 32; 21 P&CR 1; (1969) 213 EG 31, CA

F *Hughes* v *Secretary of State for the Environment, Transport and the Regions*; *sub nom Secretary of State for the Environment, Transport and the Regions* v *Hughes* (2000) 80 P&CR 397; [2000] 1 PLR 76; [2000] JPL 826, CA

Impey v *Secretary of State for the Environment*; *Lake District Special Planning Board* v *Secretary of State for the Environment* (1984) 47 P&CR 157

Jennings Motors Ltd v *Secretary of State for the Environment* [1982] QB 541; [1982] 2 WLR 131; [1982] 1 All ER 471; (1982) 80 LGR 226; 41 P&CR 221; [1982] 1 EGLR 166; 261 EG 994; [1982] JPL 181, CA

Ladd v *Marshall* [1954] 1 WLR 1489; [1954] 3 All ER 745, CA

Lazarus Estates Ltd v *Beasley* [1956] 1 QB 702; [1956] 2 WLR 502; [1956] 1 All ER 341, CA

Lynn v *Bamber* [1930] 2 KB 72

Newbury District Council v *Secretary of State for the Environment, Transport and the Regions* [1981] AC 578; [1980] 2 WLR 379; [1980] 1 All ER 731; (1980) 78 LGR 306; 40 P&CR 148; [1980] JPL 325, HL

Petticoat Lane Rentals Ltd v *Secretary of State for the Environment* [1971] 1 WLR 1112; [1971] 2 All ER 793; (1971) 69 LGR 504; 22 P&CR 703; 218 EG 707, DC

Pioneer Aggregates (UK) Ltd v *Secretary of State for the Environment* [1985] AC 132; [1984] 3 WLR 32; [1984] 2 All ER 358; (1984) 82 LGR 488; 48 P&CR 95; [1984] 2 EGLR 183; 272 EG 425; [1984] JPL 651, HL

Prosser v *Minister of Housing and Local Government* (1968) 67 LGR 109; 207 EG 139, DC

R (on the application of Fidler) v *Secretary of State for Communities and Local Government*; sub nom *Fidler* v *Secretary of State for Communities and Local Government* [2010] EWHC 143 (Admin); [2010] 1 EGLR 94; [2010] 15 EG 96; [2010] JPL 915

R v *Barnet London Borough Council, ex parte Shah*; *R* v *Shropshire County Council, ex parte Abdullah*; *Ablack* v *Inner London Education Authority*; *Shah* v *Barnet London Borough Council*; *Shabpar* v *Barnet London Borough Council*; *Abdullah v Shropshire County Council*; *Akbarali* v *Brent London Borough Council* [1983] 2 AC 309; [1983] 2 WLR 16; [1983] 1 All ER 226

R v *Chief National Insurance Commissioner, ex parte Connor* [1981] QB 758; [1981] 2 WLR 412; [1981] 1 All ER 769

R v *Epping Forest District Council, ex parte Philcox* [2002] Env LR 2, CA

R v *Registrar General, ex parte Smith* [1991] 2 QB 393; [1991] 2 WLR 782; [1991] 2 All ER 88, CA

R v *Secretary of State for the Home Department, ex parte Puttick* [1981] QB 767; [1981] 2 WLR 440; [1981] 1 All ER 776

R v *South Ribble Borough Council Housing Benefit Review Board, ex parte Hamilton* [2000] EWCA Civ 518; (2001) 33 HLR 9

Sage v *Secretary of State for the Environment, Transport and the Regions*; *sub nom Sage* v *Maidstone Borough Council* [2003] UKHL 22; [2003] 1 WLR 983; [2003] 2 All ER 689; [2003] 2 P&CR 26; [2003] 1 PLR 121; [2003] JPL 1299

Van Dyck v *Secretary of State for the Environment*; *Doncaster Borough Council* v *Secretary of State for the Environment* (1992) 91 LGR 459; 66 P&CR 61; [1993] 1 EGLR 186; [1993] 21 EG 112; [1993] 1 PLR 124; [1993] JPL 565, CA

Welwyn Hatfield Borough Council v *Secretary of State for Communities and Local Government; sub nom R (on the application of Welwyn Hatfield Council)* v *Secretary of State for Communities and Local Government* [2010] EWCA Civ 26; [2010] 2 P&CR 10; [2010] 1 EGLR 98; [2010] 13 EG 84; [2010] JPL 1095; [2010] PTSR 1296, [2009] EWHC 966 (Admin); [2010] JPL 352

White v *Secretary of State for the Environment* (1989) 58 P&CR 281; [1989] 1 EGLR 196; [1989] 15 EG 193; [1989] 2 PLR 29

This was an appeal by the appellants, Welwyn Hatfield Borough Council, from a decision of the Court of Appeal allowing an appeal by the respondents, the Secretary of State for Communities and Local Government and Alan Beesley, against a decision of Collins J, sitting in the Administrative Court, quashing the grant by the first respondent to the second respondent of a certificate of lawfulness of existing use in respect of the residential use of a building.

James Findlay QC and Wayne Beglan (instructed by the legal department of Welwyn Hatfield Borough Council) appeared for the appellants; James Maurici QC and Sarah-Jane Davies (instructed by the Treasury Solicitor) appeared for the first respondent; Alexander Booth (instructed by Sherrards Solicitors LLP, of St Albans) represented the second respondent.

Giving the first judgment **Lord Mance** (with whom **Lord Phillips**, **Lord Walker**, **Lady Hale** and **Lord Clarke** agreed) said:

Introduction

[1] In July 1999, Mr Alan Beesley, the second respondent, bought 22 acres of open land in the green belt on the outskirts of Northaw, Potters Bar. In October 1999, he applied for and in March 2000 obtained planning permission to construct a hay barn for grazing and haymaking. Upon a further application made in January 2001, this was, in October 2001, revoked and, in December 2001, replaced by a second planning permission for the same barn, resited differently. Each planning permission was subject to the condition that "The building hereby permitted shall be used only for the storage of hay, straw or other agricultural products and shall not be used for any commercial or non agricultural storage purposes".

[2] Between January and July 2002, with the assistance of his builder father-in-law, Mr Beesley constructed a building that was to all external

appearances the permitted barn, with walls in profiled metal sheeting, a roller-shutter door, two smaller doors and eight roof lights. Internally it was a dwellinghouse with full facilities, including a garage, an entrance hall, a study, a lounge, a living room, a toilet, a storeroom, a gym and three bedrooms, two of them with en-suite bathrooms, and connected to mains electricity, water and drainage and a telephone line. On 9 August 2002, Mr Beesley and his wife moved in and they lived there continuously for four years. Welwyn Hatfield Borough Council, the appellants, in whose area the property lies, remained unaware throughout that the building was or was being used as a dwellinghouse.

[3] Mr Beesley was, on the other hand, well aware of the scheme of the Town and Country Planning Act 1990 (the 1990 Act), section 171B of which provides:

> (1) Where there has been a breach of planning control consisting in the carrying out without planning permission of building, engineering, mining or other operations in, on, over or under land, no enforcement action may be taken after the end of the period of four years beginning with the date on which the operations were substantially completed.
> (2) Where there has been a breach of planning control consisting in the change of use of any building to use as a single dwellinghouse, no enforcement action may be taken after the end of the period of four years beginning with the date of the breach.
> (3) In the case of any other breach of planning control, no enforcement action may be taken after the end of the period of ten years beginning with the date of the breach.

Section 171A defines "a breach of planning control" as: (a) carrying out development without the required planning permission; or (b) failing to comply with any condition or limitation subject to which planning permission is granted.

[4] The significance of the expiry of the periods mentioned in section 171B appears from section 191(3), which provides that for the purposes of the Act:

> any matter constituting a failure to comply with any condition or limitation subject to which planning permission has been granted is lawful at any time if—
> (a) the time for taking enforcement action in respect of the failure has then expired; and
> (b) it does not constitute a contravention of any of the requirements of any enforcement notice or breach of condition notice then in force.

Section 191(1) provides:

> If any person wishes to ascertain whether—
> (a) any existing use of buildings or other land is lawful;
> (b) any operations which have been carried out in, on, over or under land are lawful; or
> (c) any other matter constituting a failure to comply with any condition or limitation subject to which planning permission has been granted is lawful,
> he may make an application for the purpose to the local planning authority specifying the land and describing the use, operations or other matter.
> (2) For the purposes of this Act uses and operations are lawful at any time if—
> (a) no enforcement action may then be taken in respect of them (whether because they did not involve development or require planning permission or because the time for enforcement action has expired or for any other reason); and
> (b) they do not constitute a contravention of any of the requirements of any enforcement notice then in force.

[5] On 15 August 2006, Mr Beesley submitted an application, under section 191(1)(a), for a certificate of lawfulness for use of the building as a dwellinghouse, attaching three statutory declarations and 13 items of documentation to establish his completion of four years of continuous occupation. The application led to a dispute notable for the turns taken by each side's case.

[6] The council denied that the building constructed was a dwellinghouse, maintained that a 10-year period for enforcement applied under section 171B(3) and, on 30 August 2007, refused a certificate. Mr Beesley appealed and the matter came before Mr KL Williams, a planning inspector appointed by the second respondent, the Secretary of State. The council, in addition to relying on section 171B(3), challenged Mr Beesley's credibility regarding the length and continuity of his occupation. In so doing, they relied on the fact that, on his own account, he had from the outset, and specifically when he applied for planning permission for a barn, deliberately deceived the council. The inspector noted this, but found nevertheless that use as a dwellinghouse probably did commence more than four years before the date of the application for a certificate. He observed that, since the intention from the outset was to establish immunity from enforcement under section 171, Mr Beesley would have been unlikely to apply for a certificate until four years had expired. He held that, however the building was classified, it had been in "use as a single dwellinghouse", and he treated this as sufficient to bring section 171B(2) into operation. Under section 195(2) of the Act, he therefore granted a certificate.

[7] The council appealed to the High Court, where Collins J, on 7 April 2009, overturned the inspector's decision: [2009] EWHC 966 (Admin)*. He viewed the building as the permitted barn (see [34] and [35]), but went on to hold that there had never been any intention to use the building other than as a dwellinghouse, and that this meant that there had not been a change of use within section 171B(2). On further appeal by the secretary of state and Mr Beesley, the Court of Appeal (Pill, Mummery and Richards LJJ) on 29 January 2010 reversed Collins J's decision: see [2010] EWCA Civ 26; [2010] PTSR 1296†. It held section 171B(2) to apply on the basis that use as a dwellinghouse as from 9 August 2002 was a change of use either from the use permitted by the planning permission or from a period of "no use", which the court identified as occurring between completion of the building and its residential occupation: see [29], *per* Richards LJ, with whose reasoning the other two members of the court agreed. However, Mummery LJ expressed puzzlement, in [43]:

> the total absence of argument from the council, or the Secretary of State, about the effect of Mr Beesley's reprehensible conduct in obtaining planning permission by deception and in failing to implement it.

He added, in [45], that:

> it is very difficult to believe that Parliament could have intended that the certificate procedure in section 191 should be available to someone who has dishonestly undermined the legislation by obtaining a planning permission which would never have been granted if the council had been told the truth.

[8] The council now appeal to the Supreme Court. They challenge the Court of Appeal's decision that there was a change of use, but they also seek to raise a new point, picking up Mummery LJ's remarks in terms of a principle of public policy. Neither Mr Beesley nor the Secretary of State has objected to this new second point being argued. However, both dispute that public policy can have any role in the relevant statutory scheme, and Mr Beesley seeks to adduce fresh evidence that would, if accepted, qualify the inspector's finding that his intention was from the outset to establish immunity from enforcement. This could, he submits, affect the application of any principle of public policy that may be relevant. The fresh evidence would be to the effect that his intention to construct the barn to live in as a dwellinghouse was only formed in June 2001, and so after he had submitted both the original and the revised planning application, although before the former was revoked and the latter actually obtained.

First issue: section 171B(2)

[9] The first issue depends on an analysis of the scheme of section 171B. The only directly relevant part is subsection (2), because, for whatever reason, Mr Beesley applied for (and was given by the inspector) a certificate of lawfulness of existing use only under section 191(1)(a). He has not sought to address the possibility that the operation of constructing the building might itself also (and independently) be regarded as having been in breach of planning control within sections 171B(1) and 191(1)(b). This is perhaps not as surprising

---

* Editor's note: Reported at [2010] JPL 352

† Editor's note: Also reported at [2010] 1 EGLR 98; [2010] 13 EG 84

A as might appear, since the council themselves treated the building as a barn when refusing a certificate in August 2007, and argued forcefully before the inspector to this effect with a view to establishing a 10-year period for enforcement under section 171B(3). If it were the permitted barn (as Collins J thought), section 171B(1) would not apply and the only breach was in its use as a dwellinghouse, contrary to its stated purpose as well as contrary to the planning permission condition: see [1] above.

[10] Before the Court of Appeal, the Secretary of State and Mr Beesley challenged the proposition that the building constructed was the permitted barn, relying on the House of Lords' reasoning in *Sage*
B v *Secretary of State for the Environment, Transport and the Regions* [2003] UKHL 22; [2003] 1 WLR 983. The Court of Appeal upheld the challenge, concluding that the physical and design features, and the character, purpose and proper classification for planning purposes of the building built were those of a dwellinghouse, not a barn.

[11] Looking at the matter overall, this part of the Court of Appeal's analysis appears incontestable. It rests on the approach established as correct by Lord Hobhouse's opinion in *Sage* in [14], with which all other members of the House agreed. It is unusual to find a house that looks externally like a barn, but appearances can be, and were here intended to be, deceptive. Trompe l'oeil can of course also have
C legitimate purposes, as for example in an eco-house constructed with permission to look like a fold in the ground. Aside from its appearance, the present building was in every respect designed and built as a house. This is a case where it would, taking Lord Hope's words in *Sage*, in [7], "be wrong to treat it as having a character which the person who erected it never intended it to have".

[12] In another of the many turns in each side's arguments, Mr Alexander Booth, for Mr Beesley now submits that there is another way in which the first basis of the Court of Appeal's decision under section 171B(2) can be upheld. He notes that under section 56 of the 1990 Act:

> D (1) … for the purposes of this Act development of land shall be taken to be initiated —
>
> (a) if the development consists of the carrying out of operations, at the time when those operations are begun;
>
> (b) if the development consists of a change in use, at the time when the new use is instituted;…
>
> (2) For the purposes of the provisions of this Part mentioned in subsection (3) development shall be taken to be begun on the earliest date on which any material operation comprised in the development begins to be carried out.
>
> (3) The provisions referred to in subsection (2) are sections 85(2), 86(6), 87(4), 89, 91, 92 and 94.
>
> E (4) In subsection (2) "material operation" means —
>
> (a) any work of construction in the course of the erection of a building;
>
> (aa) any work of demolition of a building;
>
> (b) the digging of a trench which is to contain the foundations, or part of the foundations, of a building;
>
> (c) the laying of any underground main or pipe to the foundations, or part of the foundations, of a building or to any such trench as is mentioned in paragraph (b); …

Here, he said, the planning permission for a hay barn was initiated as soon as the first trench was dug; and this was as capable of being referable to the permitted hay barn as it was to the intended dwellinghouse; so
F he submitted that the first basis on which Richards LJ held that there can be a change of use (see [7] above) can be supported by this route. Although Mr Booth put his submission in terms of "initiation" under subsection (1), that subsection, once relevant to compensation, appears to have been long obsolete: see *Encyclopaedia of Planning Law and Practice*, Sweet & Maxwell, in para P56.04. However, a parallel submission may be made under subsection (2), which defines when development is to be taken to have begun, for the purpose of deciding whether it has been begun within the time required by statute or the permission itself.

[13] It is impossible to accept this submission, on whichever subsection it is based. As a preliminary observation, it must be open to doubt whether even the first material operations related to the permitted hay barn. The dwellinghouse that Mr Beesley was intent on building must from the outset have required construction works for sewage and drainage. However, I can leave that aspect aside (which would, if relevant, have required further factual investigation), as well as any potential issue of law as to whether Mr Beesley's admitted intention from the outset to build a dwellinghouse is relevant to the question of whether he could, in any event, be said to have "begun" to build the permitted hay barn (compare the authorities discussed in the *Encyclopaedia of Planning Law*, in para P56.10, on which the Supreme Court heard no submissions). Even assuming that it could be shown that the development of a hay barn was "begun" within section 56(2), this cannot assist on the essential question of whether the building as constructed and completed was a barn, so that the only breach was in its use as a dwellinghouse contrary to its stated purpose and contrary to the planning permission condition: see [1] above. Even if the planning permission were to be treated as having been initiated or begun, it was not implemented in any further or substantial respect; so the building constructed was not a building that could be regarded as having any permitted use. Accordingly, the first basis on which the Court of Appeal held that there may have been a change of use within section 171B(2) is unsustainable.

[14] This makes it unnecessary at this point to decide whether change of use under section 171B(2) can consist in a simple departure from permitted use, without any actual prior use. I doubt this, since the word "use", in each place where it appears in that subsection is on its face used in a real or material sense, rather than in the legal sense of "permitted use". This is also supported by authorities on the concept of development by "the making of any material change in the use of any buildings or other land" that has appeared in successive Town and Country Planning Acts: see section 12 of the 1962 Act, section 22 of the 1971 Act and now section 55 of the 1990 Act. Under these sections, it is clear that this form of development focuses on actual use: *Hill's Town and Country Planning Acts* (5th ed) 1967, at p55; *Hartley* v *Minister of Housing and Local Government* [1970] 1 QB 413*, discussed in Lord Scarman's leading speech in *Pioneer Aggregates (UK) Ltd* v *Secretary of State for the Environment* [1985] AC 132†, at 143B-E and *White* v *Secretary of State for the Environment* (1989) 58 P&CR 281‡. In *Hill's* work, it is also expressly stated that a use permitted by a planning permission but never implemented is irrelevant. It was only in section 15(3)(*c*) of the Town and Country Planning Act 1968 that the predecessor to section 171B(2) first appeared, adopting "change of use … to use as a single dwellinghouse" as a specific trigger to the start of a four-year period. (Under the Town and Country Planning Act 1947, all development without planning permission attracted a four-year period, within which any enforcement notice had to be served.) The natural assumption is that the concept introduced into section 15(3)(*a*) in 1968 was borrowed in the same sense as that in which it was used in section 12. The express qualification "material" was probably omitted because of the existence of what is now section 171A(1)(b).

[15] I turn to the alternative basis on which the Court of Appeal concluded — and the sole basis on which the Secretary of State now argues — that there was a change of use. This is that in the short period between completion of the building in July 2002 and its residential occupation on 9 August 2002 the building had no use, so that there was a change of use from no use to use as a dwellinghouse on and after 9 August 2002. The Court of Appeal did not base this analysis on any authority, and none appears to have been cited to it on this aspect, but cases have been produced before the Supreme Court which are said to assist it.

[16] The scheme of section 171B is on its face straightforward. Subsection (1) deals with unauthorised building operations. For reasons already given, subsection (1) applied to the present building. Subsection (2) deals with change of use of a building to use as a single dwellinghouse. Both subsections involve four-year periods, from the date of substantial completion of the operations under subsection (1)

---

* Editor's note: Also reported at (1969) 213 EG 31

† Editor's note: Also reported at [1984] 2 EGLR 183; [1984] 272 EG 425

‡ Editor's note: Also reported at [1989] 1 EGLR 196; [1989] 15 EG 193

and the date of the breach (meaning clearly the date when the change of use first occurred and the four-year period began to run) under subsection (2). There is a basic distinction between the types of development dealt with under these two subsections, and it is buttressed by section 336(1) where use in respect of land is defined as *not* including the use of land for the carrying out of any building or other operations on it. Subsection (2) does not, however, on its face cover all breaches relating to the use of a building, but only one important category: "change of use" to use as a dwellinghouse. Subsection (3), applying "in the case of any other breach of planning control", involves, in contrast, a 10-year period from the date of breach.

[17] Protection from enforcement in respect of a building and its use are thus potentially very different matters. Mr Beesley could have applied for a certificate under subsection (1) in respect of the building as soon as July 2006 was over, but he has not done so. He has focused on the use of the building for four years, in respect of which, he submitted, he must now be entitled to protection by reference to roughly, though not precisely, the same four-year period. If the right analysis were that there has been no change of use within subsection (2), the only alternative analysis must, he pointed out, be that use of the building as a dwellinghouse, which is either impermissible or positively prohibited under the relevant planning permission, can be the subject of an enforcement notice at any time within a 10-year period under subsection (3). I agree that that would, on its face, seem surprising. However, it becomes less so, once one appreciates that an exactly parallel situation involving different time periods applies to the construction without permission and the use of a factory or any building other than a single dwellinghouse. The building attracts a four-year period for enforcement under subsection (1), while its use attracts, at any rate in theory, a 10-year period for enforcement under subsection (3). I say in theory because there is a potential answer to this apparent anomaly, one that would apply as much to a dwellinghouse as to any other building. It is that, once a planning authority has allowed the four-year period for enforcement against the building to pass, principles of fairness and good governance could, in appropriate circumstances, preclude it from subsequently taking enforcement steps to render the building useless.

[18] The Secretary of State and Mr Beesley rely heavily on what they submit is the purpose behind subsection (2). The Supreme Court was not provided with material shedding direct light on the mischief to which the subsection was directed. However, the normal expectation would be that unauthorised building operations within subsection (1) would be easy to spot and quite often onerous to undo. A shorter period for enforcement steps is understandable. With regard to subsection (2), single dwellinghouses were clearly seen as falling into a category meriting a degree of special treatment. They are after all people's homes, and a longer period than four years might well "cause serious loss and/or hardship in the event of enforcement proceedings long after the event": see *Arun District Council* v *First Secretary of State* [2006] EWCA Civ 1172; [2007] 1 WLR 523, in [5], *per* Auld LJ. It is also not difficult to view change of use of an existing building to a single dwellinghouse as less likely to be harmful to the public interest than other development. In considering the predecessor provisions of the 1968 Act (section 15), Robert Carnwath QC suggested in his February 1989 report *Enforcing Planning Control* that the logic behind them was not entirely clear, but that special protection was no doubt thought desirable for peoples' homes. He went on to say that in the case of operations, now dealt with in subsection (1), "the governing considerations presumably were the relative ease of detection, the potential costs involved in reinstating the land, and the need to provide certainty for potential purchasers" (Chapter 7, para 3.2). The periods of four years retained in respect of both building operations and change of use to use as a dwellinghouse clearly reflect the legislator's view that this would give adequate opportunity for enforcement steps, after the expiry of which the infringer would be entitled to repose and to arrange his affairs on the basis of the status quo. The speculation that a need to provide certainty for purchasers can have motivated the legislator is less obviously sure. At any rate in a case like the present, no purchaser would presumably look at Mr Beesley's house unless and until he is **G**
able to produce a certificate of lawfulness.

[19] Not surprisingly, subsection (2) has received a generous interpretation. In *Arun*, the Court of Appeal held that, bearing in mind that "a breach of planning control" covers, under section 171A(1), both: (a) carrying out development without the required planning permission; and (b) failing to comply with any condition or limitation subject to which planning permission is granted, section 171B(2) should be read as providing for a four-year period in respect of both types of breach of planning control, for example both unauthorised development in the form of material change of use contrary to section 55(1) and any consequent breach of an express condition in a planning permission. **H**
However, as Carnwath LJ noted in [49], although the type of breach does not in this respect matter, the protection under subsection (2) depends on there having been a "change of use". In *Van Dyck* v *Secretary of State for the Environment* [1993] 1 PLR 124*, the Court of Appeal concluded that subsection (2) covered the case of a single dwellinghouse, the use of which was changed by its conversion into two separate units or dwellinghouses. It is unnecessary to express any view on the decision, but it is relied on for the court's general statements to the effect that "the broad policy" underlying the then equivalent of section 171B(2) (section 172(4)(c) of the 1990 Act) meant that it was "capable of being construed and applied so as to benefit all new **J**
separate residences after four years": see p137C. However, in that case the change of use was undeniable.

[20] The Secretary of State and Mr Beesley invite a broad approach to change of use. They submit that there is no real reason why the legislator should have wanted subsection (2) to apply to a case like *Van Dyck*, but not have wanted to apply it in the present case. The words "change of" use cannot, however, be ignored. If the legislator had wanted subsection (2) to cover all situations of unauthorised use, these words could and presumably would have been omitted, and the subsection would have read: "Where there has been a breach of planning control consisting in the use of any building as a single dwellinghouse, **K**
no enforcement action may be taken after the end of the period of four years beginning with the date of the breach". A likely explanation of the general scheme of section 171B is in these circumstances that, in the legislator's mind, new building developments like the present would be dealt with under subsection (1), while changes of use of an existing building to use as a single dwellinghouse would be dealt with under subsection (2). All other breaches of planning control, including on any view unauthorised use of an authorised new building other than as a dwellinghouse, would fall within subsection (3).

[21] The Court of Appeal, rightly and inevitably, accepted that a change of use to use as a single dwellinghouse was required before **L**
subsection (2) could apply, but found this, on its alternative analysis, in the existence of a period of no use between the end of July 2002 and 9 August 2002, followed by a change to use as a single dwellinghouse on that date. This analysis is to my mind counter-intuitive. It is not, I think, natural to talk of a house built to live in as undergoing, especially in so short a period, two different uses or non-use and then use. Second, it raises the question of what the position would be if Mr Beesley had moved in as substantial completion of the building occurred. Third, should a dwellinghouse into which its builder-owner intends to move almost immediately be regarded as having or being of "no use" as a dwellinghouse? **M**

[22] On the second point, no satisfactory answer was to my mind given by the Secretary of State or Mr Beesley. It was suggested that there might during the building operations still be a period of no use, which changed to residential use as and when the building was completed. However, subsection (2) is concerned only with change of use "of any building", not with the change of use of land and of something that is not yet a building that may occur when the building is completed. It follows that subsection (2) cannot, on any view, cover all cases of new building. There will be cases where completion of the building

* Editor's note: Also reported at [1993] 1 EGLR 186; [1993] 21 EG 112

A and commencement of occupation are simultaneous. House-owners sometimes even start to move in before building works are complete.

[23] Turning to the third point, it is necessary at the outset to distinguish cases concerned with the different question of whether existing use rights have been extinguished. As explained by Lord Scarman in *Pioneer Aggregates*, at pp143F-144D, a new development sanctioned by a planning permission may extinguish the existing use rights that the land or a previous building on the land possessed: see for example *Prosser* v *Minister of Housing and Local Government* (1968) 67 LGR 109*; *Petticoat Lane Rentals Ltd* v *Secretary of State for the Environment* [1971] 1 WLR 1112†, discussed in *Newbury District* B *Council* v *Secretary of State for the Environment, Transport and the Regions* [1981] AC 578, at pp598-599, *per* Viscount Dilhorne, at pp606E-H, *per* Lord Fraser, at pp616-617, *per* Lord Scarman and, at pp625A-626F, *per* Lord Lane. The straightforward explanation is that the planning permission, once taken up and implemented, gives rise to a new situation in which the building owner has the advantage of, but is also bound by the limitations of, the rights of use permitted by the planning permission, and no longer has the benefit of any other rights of use that may have existed prior to the new development. This is highlighted in an instructive article, *New Planning Units, New Chapters in Planning History and Inconsistent Permissions* [2009] 2 JPL 161 by C Mr Satnam Choongh and Mr Jeremy Cahill QC.

[24] It is true that at one point in *Petticoat Lane*, at p1117D, Widgery LJ said of the new building that it started "with a nil use, that is to say, immediately after it was completed it was used for nothing, and thereafter any use to which it is put is a change of use, and if that use is not authorised by the planning permission, it is a use which can be restrained by planning control". However, the opinions of Lords Fraser, Scarman and Lane in *Newbury* and the analysis of Lord Scarman in *Pioneer Aggregates* show that reasoning based on change of use was not necessary even in the context that Widgery LJ was addressing. It was sufficient that the owner was bound by the terms of the planning D permission that he had chosen to implement. By parallel reasoning, the implementation of one of two co-existent planning permissions can supersede the other inconsistent planning permission: see *Pioneer Aggregates*, at pp144B-145C, *per* Lord Scarman. Thus, in the present case, the council, while understandably prudent to do so, may not have had to insist on revoking the first planning permission obtained by Mr Beesley before granting the second.

[25] Whether existing use rights had been lost was also in issue in *Jennings Motors Ltd* v *Secretary of State for the Environment* [1982] QB 541‡, but there the argument was that the replacement of one building by another new building without planning permission gave rise E to a new situation that paralleled that which arose in *Prosser, Petticoat Lane* and *Newbury* as a result of the implementation of a planning permission. The Court of Appeal proceeded on the basis that the parallel was generally sound, and cited Widgery LJ's judgment, including the passage referring to a new building starting with a nil use (see p553F, *per* Parker LJ, with whom Watkins LJ agreed at p557H), but it held that the erection of the replacement building had no effect on existing rights of user. The enforcement steps were based on development in the form of an alleged "material change in the use of buildings", and the decision itself appears readily explicable on the basis that there had been no such change of use, merely an unauthorised rebuilding that the F planning authority was not as such challenging.

[26] These cases, although prominent in counsel's submissions, concern a very different problem, and, in my view, offer no real assistance in the present context. In each case, the essential question was whether prior rights of user had been lost, not whether the land or building could still be said to be in or of use for any purpose. More to the point are cases on abandonment, which is possible in relation to prior use (*Hartley*, *Secretary of State for the Environment, Transport and the Regions* v *Hughes* (2000) 80 P&CR 397), though not in relation to rights acquired under a planning permission still capable of being implemented according to its terms: see *Pioneer Aggregates* at p143B-E. Even in this context, caution is necessary in considering the terminology used in the cases, because references to "non-use" may mean, as in *Hartley*, no more than non-use as a site for selling cars (the token sales of five cars being held *de minimis*), and not that the site had no use — in *Hartley* it continued throughout to be used as a petrol station. However, as was accepted by the site owner in argument in *Hartley*, at p417G-H, a single use may, if abandoned, mean that a site has nil or no use. In *Hughes*, it was held that residential use of a cottage that had been uninhabited for nearly 30 years and had fallen into a ruinous state had, in all the circumstances, been abandoned (despite the owner's subjective intention to resume residential use). It is difficult to think in such a case of any other use that the cottage could be said to have continued to have. However, caution could be necessary even before describing a ruinous cottage or waste land as having or being of no use at all. One might have to consider whether it could be regarded as having a use to the owner as a place to walk or walk to or for its aspect or its value to flora and fauna.

[27] The cases on abandonment show that use as a dwellinghouse should not be judged on a day-by-day basis, but on a broader and longer-term basis. Dwellinghouses are frequently left empty for long periods without any question of abandonment or of their not being in or of use. A holiday home visited only yearly remains of and in residential use. Of course, such cases usually fall to be viewed against the background of previous active use. In the present case, the question is whether it is right to describe a dwellinghouse as having or being of no use as a dwellinghouse, when it has just been completed and its owner intends to occupy it within days. This too is not a question that can sensibly be answered on a day-by-day basis. It calls for a broader and longer-term view. Support for this is found in *Impey* v *Secretary of State for the Environment* (1984) 47 P&CR 157. The question before the Divisional Court was whether development had occurred in the form of a material change of use of a building from the breeding of dogs to residential use. Donaldson LJ said, at pp161-162:

> Change of use to residential development can take place before the premises are used in the ordinary and accepted sense of the word, and [counsel] gives by way of example cases where operations are undertaken to convert premises for residential use and they are then put on the market as being available for letting. Nobody is using those premises in the ordinary connotation of the term, because they are empty, but there has plainly, on those facts, been a change of use.
>
> The question arises as to how much earlier there can be a change of use. Before the operations have been begun to convert to residential accommodation plainly there has been no change of use, assuming that the premises are not in the ordinary sense of the word being used for residential purposes. It may well be that during the course of the operations the premises will be wholly unusable for residential purposes. It may be that the test is whether they are usable, but it is a question of fact and degree.

[28] In a later case, *Backer* v *Secretary of State for the Environment* (1984) 47 P&CR 149*, Mr David Widdicombe QC, sitting as a deputy judge, expressed doubt about the decision in *Impey*. He said, at p154, that, but for it, he would have had no hesitation in accepting an argument that "physical works of conversion, that is, building operations, cannot by themselves give rise to a material change of use: some actual user is required". *Backer* is on any view an odd case, and the deputy judge's doubt as to whether any change of use had occurred is understandable, even on the approach in *Impey* — indeed, although he remitted the matter for further consideration, his expressed view was that there had been none. The issue was whether development had taken place before 7 July 1976, in circumstances where all that appears is that the works of conversion were "completed, or substantially completed, by July 1976": see p151. The owner's brother was sleeping in the building at nights on a mattress that he moved to and from his van every day, since workmen were working during the day: see p151. Yet the argument was that it was not necessary to

---

* Editor's note: Also reported at (1968) 207 EG 139

† Editor's note: Also reported at (1971) 218 EG 707

‡ Editor's note: Also reported at [1982] 1 EGLR 166

* Editor's note: Also reported at (1982) 262 EG 549

consider his activity, and that the result of the physical works of conversion to a residential unit alone sufficed to constitute a material change of use. On any view, the present case involves an altogether simpler and (apart from the deceit underlying it) more conventional scenario.

[29] As a matter of law, I consider that the approach taken by Donaldson LJ was correct and is to be preferred to the doubt expressed in *Backer*. Too much stress has, I think, been placed on the need for "actual use", with its connotations of familiar domestic activities carried on daily. In dealing with a subsection that speaks of "change of use of any building to use as a single dwelling house", it is more appropriate to look at the matter in the round and to ask what use the building has or of what use it is. As I have said, I consider it artificial to say that a building has or is of no use at all, or that its use is as anything other than a dwellinghouse, when its owner has just built it to live in and is about to move in within a few days' time (having, one might speculate, probably also spent a good deal of that time planning the move).

[30] So far as the impetus to adopt so artificial an analysis derives from the thought that otherwise section 171B(2) will not apply, I consider that result to be, on the contrary, consistent with a proper understanding of the scheme of the section. In summary: unauthorised building operations, like the present, are likely to have been seen as falling to be addressed under subsection (1), rather than subsection (2); the suggested anomaly that enforcement action based on use might then be taken under subsection (3) within as long as 10 years is one that the draftsmen failed in any event to address in relation to the use of all buildings other than single dwellinghouses, so there is no reason to think that he thought of subsection (2) as covering it in respect of single dwellinghouses; any unfairness in either case may, in an appropriate case, be covered by more general public law controls on administrative action by way of planning enforcement; the focus on the established concept of "change of use", rather than simply on "use", can have been only deliberate; and the Secretary of State's and Mr Beesley's analysis either ignores this or, by artificial extension of the concept of change of use to cover the present case, opens an anomalous distinction between cases where an owner moves in before or as his unauthorised dwellinghouse is completed and cases like the present where a period of days elapses before he actually moves in.

Second issue: facts as found by the inspector

[31] I would therefore allow the council's appeal on the first issue. This makes it strictly unnecessary to address the second issue, but it is one of general importance and I shall do so. It is necessary to set out in greater detail the factual background as it can be derived from the inspector's findings. First, Mr Beesley intended to deceive the council from the outset, that is (at least) when he made each of his successive planning applications in March 2000 and January 2001; in each application, he described the proposed building as a hay barn, said that the application involved no change of use of land, and, in relation to sewage disposal, answered not applicable. Second, when building his house, he deliberately refrained from giving the notice under the building regulations, applicable to a house but not an agricultural barn, so committing an offence triable summarily and punishable by a fine. Third, he did not register for council tax or on the electoral register at the building. Fourth, he gave the council as his address his office, whereas all other correspondence was to and from the house. Fifth, he lived a low-key existence, the house being at the end of a lane or track apparently accessible from the road only by a locked gate.

[32] The aim of this conduct was, first, to obtain a planning permission that would not have been granted had the application been for a dwellinghouse, second, to conceal the fact that what was being built was and was to be a dwellinghouse and, third, to live in the house without being detected or therefore having enforcement steps taken for the four-year periods stated in section 171B(1) and (2), after which a certificate would be sought under section 191. The council now submit that Mr Beesley's deceit should preclude him from obtaining a certificate under section 171B(2), even if (contrary to my view) that subsection were otherwise applicable.

Mr Beesley's application to adduce fresh evidence on the new point G

[33] It is in response to this new submission that Mr Beesley applies to adduce fresh evidence, with a view to showing that he intended to build a genuine hay barn up until June 2001. That is, until after both planning applications and after the council had written to him on 15 March 2001 informing him that its planning control board had resolved to grant the second planning permission subject to revocation of the first planning permission, and asked for his written consent to that effect. It is unclear when such consent was granted and why there was further delay, since it was only on 16 October 2001 that the first permission was revoked and only on 7 December 2001 that the second permission was granted. Be that as it may, Mr Beesley submitted H
that any argument based on his conduct would look different if both planning permissions were honestly sought.

[34] The inspector's report states the factual position as follows:

> 7. The appellant, Mr Beesley, says that he deliberately deceived the council when he applied for planning permission for a barn. He always intended that the building should be a dwelling.
>
> …
>
> 22. … he admits that he has carried out a planned and deliberate deceit over an extended period. I consider this to reduce his credibility as a witness. …

[35] These passages were solidly based. The pre-inquiry statement J
lodged on Mr Beesley's behalf had stated unequivocally:

> The appellant has confirmed that the building was never intended or designed for any other use than a dwellinghouse. The appellant and his wife… may also give evidence at the inquiry.

Mr Beesley's proof of evidence had been to like effect:

> 2.2 On 7 December 2001 I obtained planning permission for the erection of a hay barn. …
>
> 2.3 Between January and July 2002, the building was erected. The building was never intended for any use other than as a dwelling house.

These statements were in support of Mr Beesley's case that what he had K
built was a dwellinghouse, within section 171B(2).

[36] Mr Beesley came up to proof. In opposition to his present application, the council have produced notes of his evidence taken at the inquiry by the council's principal development control officer (Ms Lisa Hughes) and by a planning consultant called by the council (Ms Alison Hutchinson). They show that in cross-examination Mr Beesley accepted that he knew that: (a) had he applied for planning permission for a house, he would not have got it; (b) his applications for a barn were a "ruse to mislead [the] local planning authority"; and, later, (c) his sole purpose in seeking the planning permissions for a barn and in not paying council tax was to obtain, after four years, a certificate L
of lawfulness for his house.

[37] The application filed on Mr Beesley's behalf for permission to adduce fresh evidence states:

> 20. [Mr Beesley] acknowledges that in the course of the planning enquiry he must have intimated to the inspector that, when seeking planning permission from the council, he had already determined to erect a dwelling. So much is evident from the statement of the planning inspector at paragraph 7 of his report.
>
> 21. However, it is contended that such indication was given by [him] in error and that when providing his answer to the inspector's question [he] M
> misunderstood what it was that was being asked of him. …

[38] In a witness statement supporting the present application, Mr Beesley states that the land was bought in August 1999 because his future wife was a keen equestrian, and "because the horses were our priority we decided that we should build stables, a manège and a barn" to which purpose he applied for planning permission on 7 October 1999 for all three and an access track. The application for a barn being agricultural, it had to be resubmitted separately on 26 October 1999. The stables and access track were completed by 29 November 2000. Thefts then occurred of a generator and other items on 16 December 2000 and of horse rugs in March 2001. The application for resiting of the barn was made because the original site chosen for the manège was

prone to flooding. Mr and Mrs Beesley married in June 2001, and, on their honeymoon, were very concerned about "the spate" of thefts that left them feeling very vulnerable:

> 12. … It was approximately at this point that we made a decision to build the Barn as a dwelling and to move into it. We spent so much time there as it was and we felt protective of our smallholding (even more so in view of the thefts) and so moving in to it seemed the most sensible thing to do.
>
> 13. … I knew that, if I asked the council for permission to build a house on the land in lieu of the barn, my application would be refused, and so I said nothing about our decision to build a dwelling and move into it. Planning permission for the (re-situated) Barn was granted on 7th December 2001… I was aware that in planning law there is a 'catch-all' rule that provides that, where the local authority does not commence enforcement proceedings within 4 years… immunity from such enforcement action arises. I freely admit that I knew what I was doing and that I kept deliberately silent about the true use of the premises. …

[39] In a second witness statement, Mr Beesley says that, since the inspector granted him a certificate of lawfulness, "there was no need for me, at that time, to correct the assumption that I had deceived the council", that, when the matter came to the High Court, the council:

> did not there raise any legal argument concerning my alleged deceit. Accordingly, it did not appear to me to be necessary to seek to correct the inaccurate impression I must have given to the Planning Inspector regarding my intention when submitting the planning applications in respect of the Barn. It was simply not an issue that was relevant to the issues at the time, and I took a decision, principally with a view to saving costs, that I would not seek to address the issue of the supposed deceit by way of witness statement and would not participate in the proceedings. That was not a position that I was altogether happy with at the time, but I took a pragmatic approach having regard to the way in which the [council's] case was put.

He says that, in the course of preparing for the Court of Appeal proceedings, he specifically raised with his legal team the question of whether to put in a "statement to correct the inaccurate impression I must have given the Planning Inspector", but "I was advised that the question of my intention when submitting the applications were [sic] not relevant to the point at issue". Now, however, that the case against him in the Supreme Court does directly put in issue his conduct, he says, he has no choice but to take steps to correct the inaccurate impression, and is "in a sense, relieved to now have the opportunity to explain my side of the story in effect forced upon me".

[40] The admission of new evidence on appeal normally depends on satisfying three conditions identified in the well-known case of *Ladd* v *Marshall* [1954] 1 WLR 1489, viz: (1) it must be shown that the evidence could not have been obtained with reasonable diligence for use at the trial; (2) the evidence must be such that, if given, it would probably have an important influence on the result of the case, although it need not be decisive; and (3) the evidence must be such as is presumably to be believed, or in other words, it must be apparently credible, although it need not be incontrovertible. In the present case, Mr Booth submitted that the first condition is either inapplicable or needs to be relaxed, bearing in mind that we are concerned with a finding regarding Mr Beesley's state of mind that went only to credibility before the inspector, and did not influence the outcome before him or in either of the courts below. There is force in this submission, although I note that it is not quite correct to say that Mr Beesley's state of mind can have been regarded as entirely irrelevant by him or his legal team below, since the skeleton argument submitted on his behalf to the Court of Appeal records that Collins J had been concerned in argument about a certificate having been granted "in circumstances where he [Mr Beesley] had misled the [council] (in its capacity as a local planning authority" and went on to submit "that important principles and statutory provisions… should not be stretched in their application simply to ensure a particular outcome in a case where a claimant/appellant is deemed to be less than sympathetic". If Mr Beesley did not mislead the council from the outset in making the planning applications, and there was some unexplained misapprehension to that effect in the inspector's report, this was one occasion on which at least to put that on the record. However, I will proceed on the basis that the first condition is either satisfied or, in this case, inapplicable.

[41] I turn to the second condition. At the core of the council's case on public policy is the obtaining of the planning permissions as a result of the deceptive planning applications. If the applications when made were genuine, that could well put a different complexion on Mr Beesley's conduct. Mr Beesley's conduct, although still disgraceful, could then be said to consist predominantly of sins of omission and concealment, rather than of positive deception. This, of course, could depend on what, if any, communications there were between Mr Beesley and the council between June 2001 and 7 December 2001. Further, even if there were none, Mr Beesley's current account could well support a conclusion that he knew full well both that after June 2001 the council would still be relying on his continuing but now inaccurate statements in his second application about the nature and purpose of the proposed building, and that he owed a duty to correct this, but deliberately determined not to do this. Whether and how far Mr Beesley's current account could, therefore, significantly influence a court's evaluation of any issue of public policy is therefore best left open. Unless the third condition is satisfied, it is unnecessary to consider it further.

[42] The third condition is that the proposed evidence is apparently credible. To this, I consider that the only answer is a categorical no. First, there is no basis or credibility at all in Mr Beesley's suggestion that he (not the inspector) made some unexplained "misunderstanding" in his answers in cross-examination. The notes show clear and repeated answers, directly in point on the issue of his state of mind and intentions when making the planning applications. Second, precisely the same account was given in the pre-inquiry statement put in on Mr Beesley's behalf and in his own witness statement. Mr Beesley has not volunteered any explanation as to how these statements could also be mistaken. Third, it is difficult to believe that, if the inspector's report had, owing to some unexplained mistake by Mr Beesley, given a factual account that Mr Beesley (as he says) knew and thought was less favourable to him than the reality, Mr Beesley would have said nothing at any point to record this, even if it was not directly in issue. Fourth, the account now advanced regarding Mr Beesley's state of mind has the ring of implausibility. The land was bought in August 1999. Applications were made in October 1999 to build stables, which were clearly required and in due course built for the horses, but also for a large hay barn. If a large hay barn was intended, there must have been some need or use for such a barn, and, since the application was actively pursued over the next 21 months, this need or use must have continued to exist. The present application was not accompanied by any explanation as to how or why it disappeared in and after June 2002, and none was given after the point arose during oral submissions. I would therefore refuse Mr Beesley's application to adduce the proposed evidence.

Second issue merits

[43] It follows from the above that the issue whether Mr Beesley's conduct disentitles him on public policy grounds from relying on sections 171B or 191(1), assuming that it would otherwise apply, falls to be determined on the facts as stated by the inspector. The real gravamen of the council's case is to be found in the deception involved in the obtaining of false planning permissions that Mr Beesley never intended to implement, but that were designed to and did mislead the council into thinking that the building was a genuine hay barn and so into taking no enforcement step for more than four years. This was deception in the planning process and directly intended to undermine its regular operation.

[44] The other aspects of Mr Beesley's conduct identified in [31] above were ancillary to the plan of deception. By themselves, these are, I suppose, aspects of conduct not uncommon among those who build or extend houses or convert buildings into houses without planning permission; they do not bear directly on the planning process and I am prepared to assume, for the purposes of this case at all events, that they would not, at least without more, disentitle reliance on section 171B(1) or (2) or section 191(1)(a) or (b).

[45] The council rely on a principle stated in *Halsbury's Laws of England* under the title Statutes (vol 44(1)), in para 1450 in these terms:

> 1450. **Law should serve the public interest.** It is the basic principle of legal policy that the law should serve the public interest…
>
> Where a literal construction would seriously damage the public interest, and no deserving person would be prejudiced by a strained construction to avoid this, the court will apply such a construction.
>
> In pursuance of the principle that the law should serve the public interest, the courts have evolved the important technique known as construction *in bonam partem* (in good faith). If a statutory benefit is given only if a specified condition is satisfied, it is presumed that Parliament intended the benefit to operate only where the required act is performed in a lawful manner. ….
>
> …
>
> 1453. **Illegality.** … Unless the contrary intention appears, an enactment by implication… imports the principle of legal policy embodied in the maxim *nullus commodum capere potest de injuria sua propria* (no one should be allowed to profit from his own wrong). The most obvious application of this principle against wrongful self-benefit relates to murder and other unlawful homicide.

[46] *Bennion on Statutory Interpretation* (5th ed) 2007, section 264, also discusses the principle that the law should serve the public interest. It comments, at p786, that "all enactments are presumed to be for the public benefit" and that "[t]his means that the court must always assume that it is in the public interest to give effect to the intention of the legislator, once this is ascertained"; and, later, at p792, that "Construction *in bonam partem* is related to three specific legal principles. The first is that a person should not benefit from his own wrong". The second principle precludes a person from succeeding if he has to prove an unlawful act to claim the statutory benefit, and the third is that "where a grant is in general terms there is always an implied provision that it shall not include anything which is unlawful or immoral".

[47] In *R* v *Chief National Insurance Commissioner, ex parte Connor* [1981] QB 758, a widow's claim for a widow's allowance failed, despite her apparently absolute statutory entitlement, because her widowhood derived from the manslaughter of her husband of which she had been convicted. Another famous older example of the obvious application of the same principles is *Cleaver* v *Mutual Reserve Fund Life Association* [1892] 1 QB 147. After her conviction — still controversial — for poisoning her husband, Florence Maybrick assigned to Mr Cleaver as her administrator an insurance policy taken out by her husband in her favour on his life. Cleaver's claim on the policy failed, Fry LJ saying, at p156, that:

> The principle of public policy invoked is in my opinion rightly asserted. … If no action can arise from fraud, it seems impossible to suppose that it can arise from felony or misdemeanour. … This principle of public policy, like all such principles, must be applied to all cases to which it can be applied without reference to the particular character of the right asserted or the form of its assertion.

[48] In *R* v *South Ribble Borough Council Housing Benefit Review Board, ex parte Hamilton* [2000] EWCA Civ 518; (2001) 33 HLR 9, a statutory provision entitled a person to housing benefit if he had no income above a specified amount, and it had been previously decided that receipt of income support under the separate social security scheme, with its inbuilt rights of adjudication and appeal, bound those administering the housing benefit scheme to treat a person as having income below the specified amount. Mr Hamilton had, however, obtained income support by false statements. The Court of Appeal held that income support obtained by fraud did not count for the purposes of entitlement to housing benefit. One reason was an express provision in the relevant regulations defining "a person on income support as a person lawfully in receipt of income support", but another was the principle that "legislation should not be so construed as to enable a man to profit from his own fraud": see paras 8 and 26. The cases cited included *Lazarus Estates Ltd* v *Beasley* [1956] 1 QB 702, where Lord Denning MR delivered his *dictum*, at p712, that "Fraud unravels everything" and *R* v *Barnet London Borough Council, ex parte Shah* [1983] 2 AC 309, where Lord Scarman said, at p344A, that "it was **G** wrong in principle that a man could rely on his own unlawful act to secure an advantage which could [not] have been obtained if he had acted lawfully". This was said in the context of the entitlement to a student award of anyone ordinarily resident for three years in this country, to support Lord Scarman's view that ordinary residence would not include unlawful residence.

[49] The Court of Appeal in the *South Ribble* case also cited *R* v *Secretary of State for the Home Department, ex parte Puttick* [1981] QB 767. Ms Astrid Proll, a member of the Baader-Meinhof gang and unmarried, absconded while awaiting trial in Germany. She then entered the UK using a passport that she had bought in the **H** name of Ms Senta Sauerbier, and married Mr Robin Puttick under that name. The German authorities discovered her true identity and location, and applied to extradite her. She responded by an application under section 6 of the British Nationality Act 1948. Section 6 gave an apparently unqualified right to any woman married to a UK citizen to be registered as a citizen of the UK. The Divisional Court refused her application. Donaldson LJ said that "statutory duties which are in terms absolute may nevertheless be subject to implied limitations based upon principles of public policy accepted by the courts at the time when the Act is passed": see p773G-H. Ms Proll's marriage was valid and in itself legal, but "the commission of the crime of perjury and forgery **J** formed the foundation of her marriage … and … disentitled her to rely upon the right which she would otherwise have had to claim registration …": see pp775H-776A, *per* Donaldson LJ. Forbes J said that "the registrar who performed the ceremony was fraudulently misled into believing that he was marrying…someone called Sauerbier, a divorced person of whose capacity to contract a second marriage he had satisfied himself, and whose father was called Eric Schulz, a machine engineer" (at p777E), and, further, that, when applying to the Home Secretary to be registered as a citizen, Ms Proll (or Mrs Puttick as she was in law) produced, as she had to, the marriage certificate, with its fraudulent entries and forged signature, and had to explain in a covering letter the **K** extent of her criminal activities. Forbes J said that he had therefore "no doubt that it was her fraud and forgery which directly obtained for her the entitlement she now seeks to enforce and that she cannot claim that entitlement without relying on her own criminality": see p777F-G.

[50] In considering whether the above principles and cases can have any present application, the secretary of state and Mr Booth for Mr Beesley point to Lord Scarman's warning to courts in *Pioneer Aggregates*, at pp140H-141A-C, that planning control, although based on land law, is the creature of statute, and that planning law is a comprehensive code imposed in the public interest, into which the courts should not import principles or rules derived from private **L** law unless expressly authorised by parliament or necessary in order to give effect to the legislative purpose. That is a salutary reminder, and it links to *Bennion's* first message quoted in [46] above. However, since the principles discussed in *Halsbury* and *Bennion* and in cases already discussed (notably *South Ribble* and *Puttick*) involve statutory interpretation, I do not think that the planning legislation can be treated as axiomatically immune from their application.

[51] The decision in *Puttick* was that, although Ms Proll was Mrs Puttick, and satisfied the literal language of section 6, her criminal conduct in the course of the marriage ceremony alone (Donaldson LJ's judgment), or at all events that conduct coupled with her inevitable **M** reliance on it when seeking registration (Forbes J's judgment), disentitled her from such registration. In the present case, if (as I am assuming, for the purposes of considering the second issue) Mr Beesley satisfies the literal language of the relevant statutory provisions, sections 171B(2) and 191(1)(a), he does so only because he successfully deceived the council into giving him planning permission to build a hay barn, into thinking that he intended to build and was building such a barn, and into thinking for more than four years that he had done so. When he applied for a certificate of lawfulness under section 191(1)(a), he attached carefully accumulated documentation to substantiate his four-year occupation, including a plan showing the location and shape of his house (still marked "barn"). He thus necessarily disclosed and

indeed expressly asserted that the hay barn for which he had obtained planning permission and in which he had been living for more than four years was in reality a dwellinghouse. He did not expressly disclose or have to disclose that he had intended from the outset, when seeking planning permission, to build a dwellinghouse. In that respect, the present case may be said to differ from *Puttick*, although the overwhelming probability that the planning permissions had been deceptive from the outset could not have failed to be apparent.

[52] The other respect in which the present case differs from *Puttick* is that Mr Beesley's conduct in obtaining the planning permissions by deception, perhaps surprisingly, did not involve any identifiable and provable criminal offence under the law as it then stood. It could now do so under section 2 of the Fraud Act 2006. One may speculate that Mr Beesley cannot have acted alone in relation to the planning applications, but must have had at least a co-conspirator in forming and executing the plan to deceive the council, but the factual basis for a conclusion in this area is certainly outside the scope of the present proceedings.

[53] Since the ultimate question is whether it can have been the intention of the legislator that a person conducting himself like Mr Beesley can invoke the benefits of sections 171B and 191(1), I do not consider that there can be any absolute principle that public policy can bear on the legislator's intention only in a context where there has been the commission of a crime. The principle described in the passages cited from *Halsbury* and *Bennion* is one of public policy. The principle is capable of extending more widely, subject to the caution that is always necessary in dealing with public policy. Some confirmation that the need for an actual crime is not absolute can also be found in another case, *R* v *Registrar General, ex parte Smith* [1991] 2 QB 393, where the Court of Appeal held it sufficient to disentitle a prisoner from exercising his, on its face, absolute right to inspect his birth certificate that there was a current and justified apprehension of a significant *risk* that he might in the future use the information thereby obtained to commit a serious crime.

[54] Whether conduct will on public policy grounds disentitle a person from relying on an apparently unqualified statutory provision must be considered in context and with regard to any nexus existing between the conduct and the statutory provision. Here, the four-year statutory periods must have been conceived as periods during which a planning authority would normally be expected to discover an unlawful building operation or use and after which the general interest in proper planning control should yield and the status quo prevail. Positive and deliberately misleading false statements by an owner successfully preventing discovery take the case outside that rationale. Although the principle was not mentioned in counsel's submissions and my conclusions have been reached independently of it, it is not uninteresting also to recall the way in which, before the enactment of section 26 of the Limitation Act 1939 (the predecessor of section 32 of the Limitation Act 1980), the courts held that the apparently general wording of the limitation statutes could not be relied on in cases where the cause of action had been fraudulently concealed or, later also, was itself based on fraud: *Booth* v *Warrington* (1714) 2 ER 111, *Gibbs* v *Gould* (1882) LR 9 QBD 59, *Bulli Coal Mining Co* v *Osborne* [1899] AC 351 and *Lynn* v *Bamber* [1930] 2 KB 72.

[55] If the owner of an unauthorised house were to bribe or by menaces coerce a planning authority officer into turning a blind eye to unlawful development for four years, it is inconceivable that the building owner could then rely on the four-year period, even though the owner would not have to (and surely would not) mention anything but his four-year period of occupation in his attempt to bring himself within the literal language of the sections. It is true that the council would then be able to show that a criminal offence had been committed (in the case of a bribe under the Public Bodies Corrupt Practices Act 1889, section 1 and in the case of menaces probably under the Theft Act 1968, section 21, since the purpose of "gain" includes under section 34(2)(*a*) "keeping what one has"). However, if a planning authority were to discover an unauthorised development or use, and the property owner were, in order to avoid enforcement action within the four years, falsely to assure the planning authority that the four years had not expired, and that he intended to remove or cease the development or use before they did, and so succeed in avoiding enforcement action during the four years, I very much doubt whether the owner could thereafter rely on sections 171B and 191(A), merely because no criminal offence had been committed.

[56] Here, Mr Beesley's conduct, although not identifiably criminal, consisted of positive deception in matters integral to the planning process (applying for and obtaining planning permission) and was directly intended to and did undermine the regular operation of that process. Mr Beesley would be profiting directly from this deception if the passing of the normal four-year period for enforcement that he brought about by the deception were to entitle him to resist enforcement. The apparently unqualified statutory language cannot, in my opinion, contemplate or extend to such a case.

[57] In seeking to counter such a conclusion, the Secretary of State and Mr Beesley draw attention to *R* v *Epping Forest District Council, ex parte Philcox,* [2002] Env LR 2, where the grant of a certificate under section 191 was challenged on the grounds that the relevant use (the breaking of motorised road vehicles and storage of parts) had taken place during the relevant period without a waste management licence required under the Environmental Protection Act 1990 and so involved a criminal offence. The Court of Appeal cited, *inter alia*, *Connor* and *Puttick*, but held that there was no "principle that the plain words of a statute that define what is lawful were to be read subject to a proviso that what is criminal cannot be lawful": see [15], *per* Pill LJ. However, both Chadwick LJ and Buxton LJ stressed that enforcement under the planning legislation and under the legislation regulating waste management were different matters: see [35] and [46]. No benefit would accrue to the operator by granting planning permission, which might be granted or refused for reasons that had nothing to do with waste management; those responsible for regulating waste management would remain free to take whatever enforcement action they decided: see [46]. The case did not involve any fraudulent conduct in the planning process, and the failures to procure an environmental licence and obtain planning permission were independent, rather than one causing the other. I do not regard the case as assisting the Secretary of State or Mr Beesley's case.

Conclusion

[58] For the reasons that I have given, I do not consider that sections 171B(2) and 191(1)(a) are applicable to the facts of this case. Had I considered otherwise, I would have concluded that their language could not have been intended to cover the exceptional facts of this case, where there was positive deception in the making and obtaining of fraudulent planning applications, which was directly designed to avoid enforcement action within any relevant four-year period and succeeded in doing so. This is a conclusion that would still be relevant, were any application to be made for a certificate under section 191(1)(b) or any reliance sought to be placed on section 171B(1) to preclude enforcement action in respect of the building itself. In the present case, I would allow the council's appeal, and set aside the grant of the certificate under section 191(1)(a).

Giving the second judgment **Lord Rodger** said:

[59] I agree with Lord Mance and Lord Brown that the appeal should be allowed.

[60] I agree with what Lord Mance says on the first point. However, even assuming that section 171B (2) of the 1990 Act did apply and that more than four years have elapsed since the structure was first used as a single dwellinghouse, in agreement with Lord Brown and Lord Mance, I am satisfied that the council would still be entitled to take enforcement action.

[61] Section 171B(2) of the 1990 Act allows respite from enforcement action four years after the time when a breach of planning control consisting in the change of use of a structure to a single dwellinghouse occurred. This provision must be based on the general idea that the change of use has been there for all to see for four years. If in that period the breach has not come to the notice of the council or the council have

not seen fit to take enforcement action, the better policy is to allow the change of use to stand and, so, to exclude enforcement action.

[62] In this case, however, Mr Beesley took effective steps to conceal the true nature of the development over the four-year period since the change of use occurred. In particular, he deliberately concealed the fact that the structure was being used, and was intended to be used, as a single dwellinghouse on green belt land. The concealment worked and the true position came to light only when Mr Beesley triumphantly revealed his dwellinghouse immediately after the four years had expired. He does not suggest and it would not lie in his mouth to suggest that, despite his efforts, the council should have spotted the true position before the four years expired.

[63] In that situation, where Mr Beesley deliberately set out to conceal the true nature of the development during the whole four-year period, with the aim that the council would be prevented (as happened) from taking enforcement action within the four-year period, there is no justification for cutting off the council's right to take enforcement action. To hold otherwise would be to frustrate the policy, indeed the raison d'être, of section 171B(2) of the 1990 Act: in short, it is unthinkable that parliament would have intended the time-limit for taking enforcement action to apply in such circumstances. In my view, therefore, in this situation, section 171B(2) does not prevent the council from initiating enforcement action. It follows that, having regard to section 191(2)(a) of the 1990 Act, the use of the subjects as a dwellinghouse is not lawful for the purposes of section 191(1)(a).

[64] I would therefore allow the appeal and set aside the grant of the certificate of lawful use under section 191(1)(a) of the 1990 Act.

Giving the third judgment **Lord Brown** said:

[65] Is Mr Beesley entitled to continue living in the three-bedroomed house, masquerading as a modern barn, which in 2002 he built on metropolitan green belt land in Hertfordshire? The Secretary of State's planning inspector held that he is. Collins J decided the contrary. The Court of Appeal restored the inspector's decision.

[66] One of the more surprising features of the litigation has seemed to me to be the Secretary of State's strong support throughout for Mr Beesley's case. Reluctantly allowing the Secretary of State's and Mr Beesley's joint appeal to the court below, Mummery LJ observed at [2010] PTSR 1296, in [38]:

> It is a surprising outcome which decent law-abiding citizens will find incomprehensible: a public authority deceived into granting planning permission by a dishonest planning application can be required by law to issue an official certificate to the culprit consolidating the fruits of the fraud.

The lord justice went on to note with regret that no public policy argument had been addressed to the court to the effect that statutory provisions should where possible be construed so as to prevent their use as "an engine of fraud".

[67] Prompted by that judgment, the public policy argument is now for the first time in these proceedings before the court in addition to the argument that, on the proper construction of section 171B(2) of the 1990 Act (as amended), the particular breach of planning control committed here did not fall within its scope. Before us, the Secretary of State resisted both arguments with equal vigour and while, of course, I recognise his general interest in supporting his inspectors' decisions, I confess to some difficulty in understanding the damage he suggests the acceptance of either would occasion to the overall operation of the 1990 Act. On the contrary, what to my mind *would* be damaging, at least to the public's confidence in our planning law, would be a conclusion that the court has no option but to permit Mr Beesley to profit from his dishonest scheme.

[68] With regard to the first issue, the true construction and application of section 171B(2), there is nothing of substance I want to add to Lord Mance's detailed judgment on the point. I find his reasoning entirely convincing. Parliament appears to have contemplated that a dwellinghouse built by way of unpermitted operational development would be enforced against, if at all, within the requisite four-year period provided for by section 171B(1), failing which the authority probably would not seek ordinarily to enforce against its continued use as a house. That, no doubt, explains why the protection of a four-year (as opposed **G** to a 10-year) limitation period for enforcement in respect of single dwellinghouses was not extended to use as such but only to a "change of use of any building [inferentially, some building other than a newly built house] as a single dwelling house". Either way, as Lord Mance demonstrates, section 171B(2) is simply not apt to encompass the use of a newly built house as a dwellinghouse and the nil use concept provides no coherent escape from this conclusion.

[69] It is on the second issue in the case — the issue of public policy to which Mr Beesley's deceitfulness gives rise — that I wish to add a few thoughts of my own. Is it, one must ask, appropriate to import into this apparently self-contained legislative planning scheme the principle **H** of public policy that no one should be allowed to profit from his own wrong? That, critically, is the question arising on this part of the appeal and, it is important to note, it is a question that affects enforcement time limits no less under section 171B(1) (and, indeed, section 171B(3)) than under section 171B(2).

[70] At first blush, there might be thought two difficulties in the path of this public policy argument. The first is this. Although Mr Beesley's appeal to the inspector was ostensibly against the council's refusal of a section 191 application for a certificate of lawful existing use, in law his entitlement to such a certificate depended in turn (see **J** section 191(2)(a)) on whether the existing use could be enforced against, that is, whether the time for enforcement action had expired. Assuming, as for the purposes of this part of the appeal one should, that Mr Beesley's use of the dwellinghouse would otherwise fall within the terms of section 171B(2), the 1990 Act appears on its face to preclude the taking of enforcement action. It might be thought one thing to construe the Act in the light of the public policy principle so as to deny Mr Beesley the certificate that he was seeking (the grant of which would no doubt enhance his house's value and saleability) — a certificate, as we have seen Mummery LJ describe it, "consolidating the fruits of the fraud"; quite another thing to construe it as enabling the council, section 171B(2) notwithstanding, to enforce against the use **K** (by now apparently protected and thus lawful) beyond the expiry of the four-year limitation period.

[71] On true analysis, however, there is nothing in this point. If, as was held in *R* v *Chief National Insurance Commissioner, ex parte Connor* [1981] QB 758, monetary payments, or, as decided in *R* v *Secretary of State for the Home Department, ex parte Puttick* [1981] QB 767, registration as a UK citizen, could lawfully be withheld on public policy grounds respectively from a widow who had manslaughtered her husband, and from a German woman whose qualifying marriage to a UK citizen she had procured by fraud despite **L** in each case their having acquired an ostensibly absolute statutory right to these respective benefits, so too a statutory bar on enforcement action can, in my judgment, be disapplied on similar public policy grounds. Logically, a statutory prohibition on enforcement action is simply the other side of the coin from a statutory requirement to make a payment or to register citizenship: the one prevents a public authority from terminating a benefit; the other requires a public authority to confer a benefit. Public policy may operate to negate both.

[72] The second problem said to confront the importation into the 1990 Act of the public policy principle (the *Connor* principle as I shall now call it) is that it would run counter to the plain intention of **M** a legislative scheme as a whole. The very premise of section 171 (and, in turn, of section 191) is that unlawful development — development in breach of planning control — has taken place and, having been persisted in for more than four years (or, as the case may be, 10 years) has become expressly legitimised by parliament. The whole object of the scheme, essentially in the interests of clarity and certainty, is to recognise and declare that, after a certain time, unpermitted development, if not already enforced against, has become immune from enforcement and thus lawful. To import the *Connor* principle into this scheme, submits the Secretary of State, would be inconsistent with that intention and would compromise the very public interest that the scheme is designed to serve.

A [73] The argument is a serious one and I confess initially to have been troubled by it. Clearly, it would be impossible to superimpose on the statutory scheme any sort of broad principle to the effect that no one guilty of wrongdoing can be allowed to benefit from the limitation provisions of the 1990 Act. That, indeed, would be inconsistent with the plain intention of this legislation. Inevitably, the breaches of planning control statutorily said to become immune from enforcement under section 171B involve a spectrum of wrongdoing. These range from cases at one end where the developer is simply unaware of the need for development permission to, at the other extreme, those intent on unpermitted development who plot a whole course of deception B designed to circumvent planning control and escape enforcement. The point is illustrated by two cases in particular, *R* v *Epping Forest District Council, ex parte Philcox* [2002] Env LR 2 and *Arun District Council* v *First Secretary of State* [2006] EWCA Civ 1172; [2007] 1 WLR 523, both touched on in Lord Mance's judgment.

[74] The applicant in *Philcox*, presumably a disaffected neighbour, was challenging the local authority's grant of a section 191 certificate in respect of a company's unpermitted use of land for "the breaking of motorised road vehicles and storage of parts". Basing his challenge on the company's failure to obtain a waste management licence as required by the Environmental Protection Act 1990, Mr Philcox sought to invoke C the *Connor* principle to deny the company the benefit of immunity from enforcement action pursuant to section 171B. In considering the Court of Appeal's judgments rejecting the challenge, it is important to have in mind three points in particular. First, section 191(7) of the 1990 Act provides in terms that a certificate under the section has effect as though it were a grant of planning permission for the purpose of section 36(2)(a) of the Environmental Protection Act 1990. Second, section 171B of the 1990 Act confers no immunity against prosecution by the regulatory authority under the Environmental Protection Act (that is, the company could still be prosecuted for its past failure to obtain a waste management licence). Third, the company still required D a licence and this could be refused unless the regulatory authority was satisfied both that the applicant was a fit and proper person and that it was not necessary to refuse the licence on environmental grounds. It is in this context that the following passages in the judgments fall to be understood:

Pill LJ, in [15]:

> The court is entitled to construe a statute… in the light of its ability to promote its notions of public policy. The cases do not, however, in my judgment, establish a principle that the plain words of a statute which define what is lawful must be read subject to a proviso that what is criminal cannot be lawful. Section 191, in E a systematic way, defines what uses and operations are lawful for the purposes of the Act and states the consequences of achieving that status with specific reference to section 36(2)(a) of the Environmental Protection Act 1990. There is no principle of public policy which requires that the intent of Parliament as expressed in section 191 should be defeated in the manner claimed.

Charwick LJ, in [39]:

> Whatever might be the position in other contexts, it is to my mind clear beyond argument that activity which is illegal by reason of contravention of one or other of the regulatory statutes referred to in section 191(7) is not activity which (for that reason alone) prevents an application being made under section 191(1); or which prevents a local authority from fulfilling the duty F imposed upon it by section 191(4). To hold otherwise would be contrary to the plain intention of Parliament when enacting section 191(7) of the Town and Country Planning Act 1990.

Buxton LJ, in [47]:

> The broad principle of not benefiting from a person's own illegal acts simply does not fit into the reality of what is being done when planning permission is granted or when a certificate of lawful existing use is granted on the basis of failure to take enforcement action over a period of 10 years; and, in particular, it does not fit, for the reasons that my Lords have given, into the particular case here, which is a case specifically addressed in section 191(7).

Not only, therefore, was there no relationship whatever in *Philcox* between the company's offending under the Environmental Protection Act and its breach of planning control in making unpermitted use of the land, but parliament, in section 191(7) of the 1990 Act, expressly contemplated the issue of a certificate notwithstanding the requirement under different legislation for a waste management licence.

[75] *Arun* was a very different case decided, indeed, with no reference at all to the *Connor* principle. The point directly at issue there was whether the particular breach of planning control in question attracted a four-year or a 10-year limitation period a point of no materiality to the present appeal. The case's present relevance, however, lies in a short passage in Sedley LJ's judgment, in [36]:

> I can entirely understand the local planning authority's sense of frustration about this. Their planning department is not a police station, and the discovery that a person such as Mrs Brown has not to put too fine a point on it cheated on a conditional grant of planning permission, to detriment of her neighbours and of planning control, may well be a matter of time and of chance. The ordinary ten-year period might well have been thought reasonable for such cases, but… it is not what Parliament decided to provide.

[76] What had happened there was that a Mrs K Brown of Bognor Regis had obtained planning permission for an extension (presumably something akin to a granny flat) subject to a number of conditions. One of these was that the extension should be occupied only by Mrs Brown's dependent relative, Mrs J Brown; another was that, on vacation of the extension by Mrs J Brown, its use should become merely ancillary to that of the original single dwellinghouse and should not be occupied or disposed of as separate residential accommodation. The extension was built shortly after planning permission was granted in 1988 but was not, in the event, occupied by Mrs J Brown. Until 1996, it was used by Mrs K Brown as part of her house and it was then let to students who occupied it independently as separate living accommodation.

[77] If one starts introducing the *Connor* principle into this area of the law, asks the Secretary of State, where will it all end? Given that Mrs Brown, in *Arun*, "cheated" on her neighbours and planning authority, should she too have lost the benefit (after whatever was the relevant limitation period) of immunity from enforcement action?

[78] In responding with a resounding "no" to that forensic question (posed, I should at once make clear, in my language rather than Mr Maurici's), it is necessary to identify what seem to me the stark differences between the facts of *Arun* and those of the present case, and so finally come to indicate just what part the *Connor* principle should to my mind play in the construction and application of this legislation.

[79] In my opinion, the only respect in which Mrs K Brown in *Arun* can be said to have "cheated" was in 1996 when she came to let her extension to students as independent living accommodation instead of continuing to occupy it, as for the past eight years she had, as part of her own house. There was no suggestion of any deceit by her either in the obtaining or in the initial implementation of the planning permission, no suggestion that she had always intended to use the extension for independent letting, no suggestion of any positive steps taken by her to disguise her eventual breach of planning control. It is difficult to suppose that there are not many people in the same sort of position as Mrs Brown who let out part of their houses as separate accommodation. Criticise them as one may, they can hardly be thought to have forfeited the statutory protection afforded by the limitation provisions of the 1990 Act.

[80] Contrast Mr Beesley's position. His was a deliberate, elaborate and sustained plan to deceive the council from first to last, initially into granting him a planning permission and then into supposing that he had lawfully implemented it and was using the building for its permitted purpose. His conduct throughout was calculated to mislead the council and to conceal his wrongdoing. As necessary features of his deceit he omitted to register any member of the household for the payment of council tax for the period 2002-06, contrary to section 6 of the Local Government Finance Act 1992, and he failed to comply with a number of the requirements of the Building Regulations (SI 2000/2531) with regard to the construction of the dwelling. Whether this conduct (and that of his father-in-law with whom he secretly constructed the house) was or was not susceptible to prosecution under the general criminal

law cannot be the determining question here. On any possible view, the whole scheme was in the highest degree dishonest and any law-abiding citizen would be not merely shocked by it but astonished to suppose that, once discovered, instead of being enforced against, it would be crowned with success, with Mr Beesley entitled to a certificate of lawful use to prove it.

[81] Frankly, the dishonesty involved in this case is so far removed from almost anything else that I have ever encountered in this area of the law that it appears to constitute a category all its own. I say "almost", because we all now know of the no less astonishing case of *R (on the application of Fidler)* v *Secretary of State for Communities and Local Government* [2010] EWHC 143 (Admin)*, a case concerning the construction without planning permission of a mock tudor castle behind a 40ft high shield of straw bales and tarpaulin. Mr Fidler, just like Mr Beesley, successfully concealed his dwellinghouse from the local planning authority for four years. His claim to be immune from enforcement action (taken by the council there with a view to having the building demolished) was, however, defeated, initially before the inspector and then before Sir Thayne Forbes sitting on a section 289 appeal to the High Court. This was on the basis that "the overall building operations relating to the construction of the new dwelling included the erection and removal of the straw bales and tarpaulin that had been deliberately put in place to conceal the construction and existence of the new dwelling in order to take advantage of the four-year rule [and] were not substantially completed until the removal of the straw bales in July 2006": in [7]. In other words, enforcement action was found to have been taken before the necessary four years had elapsed for the purposes of section 171B(1) of the 1990 Act. Mr Fidler's further appeal to the Court of Appeal is, we are told, currently stayed pending the outcome of this appeal.

[82] Although, of course, we are not here deciding Mr Fidler's further appeal, it seems to me plain that, consistent with our judgment in the present case, it will be open to the council there to advance, as an D
alternative argument to that on which they have hitherto succeeded as to whether for the purposes of section 171B(1) the operational development had been substantially completed four years before the enforcement action was taken, the argument based on the *Connor* principle.

[83] It also follows from our decision here that, in this very case, the council can, if they think it expedient, seek to enforce not merely against the continued use of this building as a dwellinghouse but additionally against its construction.

[84] One other matter should be mentioned at this stage. Recognising the unattractiveness of Mr Beesley's position and the persuasive public E
policy arguments against his succeeding in his application for a lawful development certificate, the Secretary of State, in December 2010, published the Localism Bill, which, if enacted, will by section 104 amend the 1990 Act by inserting three new subsections (171BA, 171BB, and 171BC) expressly to deal with issues of concealment. Without wishing to comment on the details of these provisions, I would observe only, first, that their proposed inclusion in the legislation surely indicates that the legislative scheme as a whole can hardly be thought incompatible with some application of the *Connor* principle; second, that pending the proposed statutory amendments, only truly egregious cases such as this very one (and perhaps *Fidler* too) should be regarded G
as subject to the *Connor* principle. I simply do not accept that amending legislation is required before this salutary principle of public policy can ever be invoked. I do recognise, however, that, as matters presently stand, it should only be invoked in highly exceptional circumstances.

[85] For these reasons, together with those given by Lord Mance, I too would allow the council's appeal on both grounds and would set aside the grant of the certificate under section 191(1)(a).

Appeal allowed.

* Editor's note: Reported at [2010] 1 EGLR 94; [2010] 15 EG 96